Naval Warfare in the Age of Sail

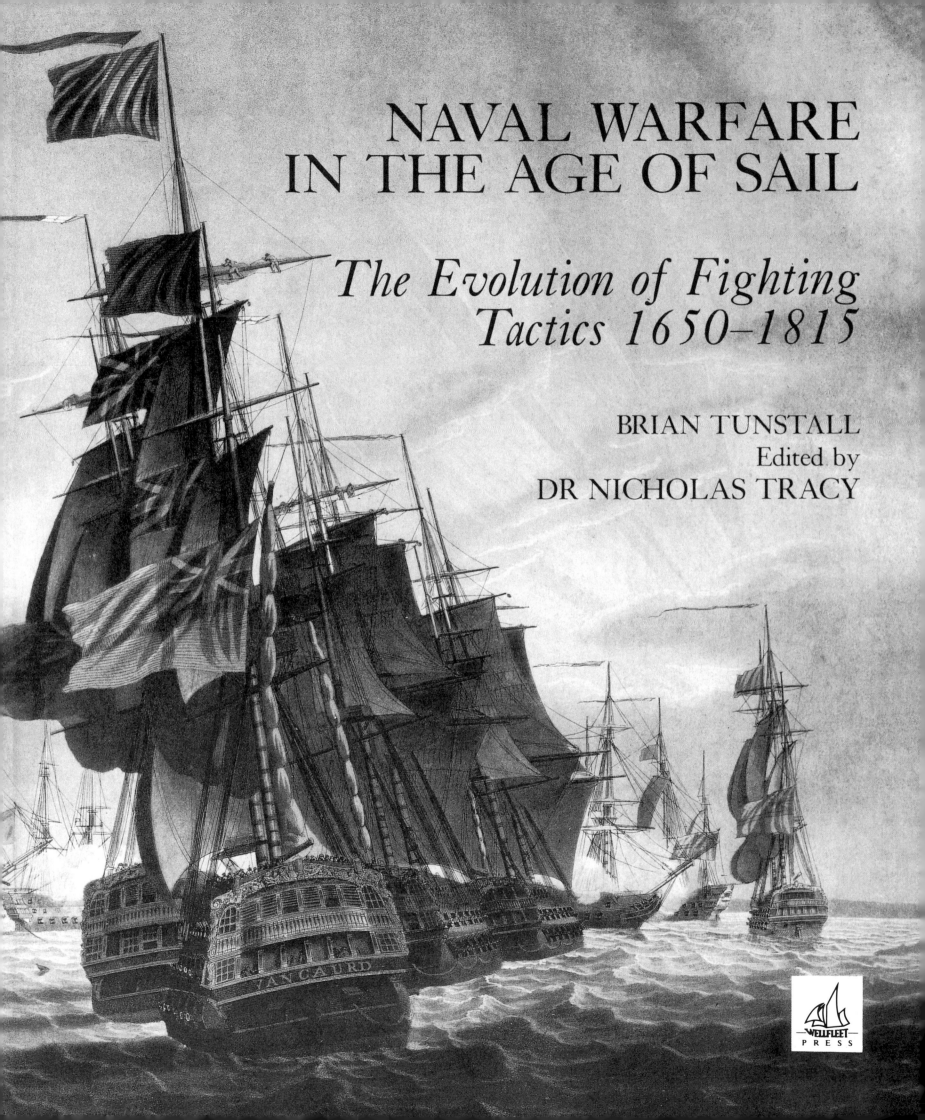

NAVAL WARFARE
IN THE AGE OF SAIL

*The Evolution of Fighting
Tactics 1650–1815*

BRIAN TUNSTALL
Edited by
DR NICHOLAS TRACY

WELLFLEET
PRESS

Frontispiece: *Nelson's fleet sailing into action at the battle of the Nile.*
NMM

This edition published in 2001 by

WELLFLEET PRESS
A division of Book Sales, Inc.
114 Northfield Avenue
Edison, New Jersey 08837

© BRIAN TUNSTALL.
With © NICHOLAS TRACY 1990

First published in Great Britain in 1990 by
Conway Maritime Press
A division of Chrysalis Books Plc, London.

ISBN 0-7858-1426-4

Printed and bound in Spain by Bookprint, S.L., Barcelona

Contents

William Cuthbert Brian Tunstall FSA

This book represents the most sustained contribution to Naval History of a considerable scholar. After coming down from St John's College, Cambridge, Tunstall taught at Oundle School, and at the Royal Naval College, Greenwich. During the war he served as a naval expert in the BBC. Ultimately he became Senior Lecturer in International Relations at the London School of Economics.

He made his scholarly reputation as a naval historian. Not only was he Sir Julian Corbett's son-in-law, but he also gathered, preserved and catalogued Corbett's papers. He wrote a strong, realistic account of Admiral George Byng while he was editing the Byng Papers (three volumes) for the Navy Records Society. Of the commercial edition of the biography, a reviewer observed, 'Mr Tunstall moves easily under the burden of heavy research' – happy sentence. Tunstall also wrote a life of the Great Commoner, the Elder Pitt, which represented one of the first attempts to use psychiatric insights in a responsible way in biography. For much of the 1930s he was Honorary Secretary of the Navy Records Society, and he blasted Sir Geoffrey Callendar's bland patriotic pap with that little gem *The Realities of Naval History*. The book startled the naval establishment, along with Callendar, and doubtless contributed to Tunstall's departure from the College in the late 1930s.

After the war he wrote three celebrated chapters in the *Cambridge History of the British Empire*. He taught not only international relations, but also national policy in relation to sea power, to a generation of graduate students. He was a superb teacher, invariably advancing new knowledge or fresh approaches under the guise of tentative suggestions. His mind was as powerful as his deployment of its products was unorthodox.

Tunstall's work on signals derived directly from the work of Corbett. He carefully and systematically organised and added to his father-in-law's collection of signal books. He traced the whereabouts of complimentary collections, and in his latter years had come to know where all were to be found. But the signal books, Corbett's work, and Tunstall's general knowledge of naval history came together as he began to examine the changes in tactics that had taken place since 1654, a reassessment based on a detailed appraisal of the signal books themselves. He did not neglect the framework that Corbett had pioneered. Nevertheless, he looked at every change in usage pragmatically, and observed the struggle of the practitioners of the signaller's art as a part of fleet policy with ultimate national purposes. It is not too much to say that the book he has produced has irrevocably changed the world. His view was that the relationships of naval strategy (including tactics) at sea to state policy cannot be appreciated without a thorough knowledge of the significance for fleet tactics of fleet signalling.

This is, therefore, an important work and Dr Tracy's service to naval historical scholarship is, consequently, hard to overestimate.

DON SCHURMAN

Author's Preface

THE FIRST move towards a systematic study of British naval tactics was made by Sir Julian Corbett in 1905. In that year, the centenary of Trafalgar, he edited a collection of documents for the Navy Records Society, entitled *Fighting Instructions, 1530-1816*. These documents showed for the first time something of the origins, development and continuity of tactical methods as pursued by a long line of naval commanders from the early Tudor ship captains to the authors of the Signal Book of 1816. Until that moment neither historians nor naval officers had appreciated that naval tactics had a history and lineage which could be traced and studied in serious terms.

As a result of Corbett's *Fighting Instructions*, more documents were brought to light, and in 1908 he edited a second volume, entitled *Signals and Instructions, 1776-1794*. This volume included selections from a valuable collection of manuscript and printed sailing and fighting instructions, orders of battle, tactical memoranda, signal books and projects for new systems of signalling, made by Admiral Sir Thomas Graves during the War of American Independence. Graves, who was a member of the famous naval family, was a comparatively young man at the time of the American War. His collection helps to trace the course of this period of important change in British tactics. He was afterwards second-in-command to Nelson at the Battle of Copenhagen (1801); he died in 1814. 'His representatives patriotically presented his collection to the [Royal] United Service Institution. A few of the documents found their way to the shelves of the Library, but the bulk remained together and uncatalogued.' They were discovered again by Sir John Laughton, whose original researches in British and French naval history, begun many years previously, had inspired Captain A T Mahan, USN, to begin writing *The Influence of Sea Power upon History*.

Corbett's two volumes of documents, together with his substantial editorial comments, provided a corpus of source material linking Blake, James Duke of York, Monck, Prince Rupert, Rooke, Vernon, Anson, Hawke and Rodney with Hood, Howe, St Vincent and Nelson. He drew distinctions between the ideas embodied in critical treatises, the instructions issued by admirals and the way tactics were employed in battle. Relationships were shown between traditional ideas and innovations and the reader glimpses something of the intellectual ferment which inspired the work of Howe, Hood and Kempenfelt. Nevertheless, these volumes were not intended to do more than provide a corpus of source material. They do not, and were never intended to constitute a history of British naval tactics. Attempts to isolate the editorial matter, and use it in the manner of a textbook, while at the same time ignoring the huge accretions of new source material, can only produce misleading and unscholarly results.

Before his death in 1922, Corbett had accumulated enough new material for a completely revised edition of both volumes. Since then, the flow of newly discovered documents has been very considerable and, as a result Corbett's work, can be seen in better perspective. Many of the distinctions and conclusions he drew have been confirmed. Several of his more speculative conclusions, based on brilliant guesswork, have been triumphantly vindicated. Documents have now appeared which he had assumed to exist but had never seen. In other cases, his views can be shown to be in need of serious revision, especially where his assumptions about the authorship and dating of documents were in error. It is a tribute to his great work as a naval historian that so many writers can still permit themselves to make completely uncritical use of conclusions based on a corpus of original source material amounting to less than a quarter of what is available today.

Apart from some important items acquired by the Admiralty Library and the Royal United Service Institution since 1908, the bulk of the new material now available for study is in the National Maritime Museum, Greenwich.[1] This great collection of documents is rich in French items, including the instructions and signals issued by the Comte de Tourville in 1690-3. It is not only the most important collection, in both quality and quantity, but it is also the best administered. The combined Corbett and Tunstall collection is now considerable and includes French, Spanish, Dutch and Italian items.[2] The collection formed by the late Commander Hilary Mead is specially important for its Howe, Kempenfelt and St Vincent items. Altogether it may be said that the amount of documentary material available today is more than three times the amount available to Corbett in 1908.

After 1908 the most important contributor to the history of British tactics was Admiral Sir Herbert Richmond. Though primarily concerned with the broader strategic, political and economic aspects of sea warfare, he dealt very fully with tactics in *The Navy in the War of 1739-48* (1920) and *The Navy in India 1763-1783* (1931). A few years later Lt-Commander J H Owen made a number of invaluable studies of British and French tactics in the American War using the original logs and journals of British ships. Rear-Admiral A H Taylor has recently made important and scholarly reconstructions of a number of fleet actions, including Trafalgar.[3] All students of the Dutch Wars are indebted to R C Anderson for his masterly elucidations of Dutch and English texts of fighting instruction, order of battle, signals and tactical narratives. *A History of Naval Tactics 1530 to 1930* by Rear-Admiral S S Robinson, USN and Mary L Robinson (US Naval Institute, 1942) is the only work in existence of its kind. Unfortunately, it is not an original history at all but a series of summaries and excerpts taken from a wide range of printed secondary sources.

In all his writings on naval history, of which the volumes of tactical documents formed only a very small part, Sir

---

1 The Royal United Services Institute collection has now been broken up and divided between the Ministry of Defence Library, Whitehall (Naval Library), and the National Maritime Museum, Greenwich.

2 The Corbett, Tunstall and Mead collections are now all in the National Maritime Museum.

3 *The Mariner's Mirror*, 36 (1950), no 4

Julian Corbett emphasised the importance of signalling at sea. He was indeed the first historian to draw attention to the fact that in addition to showing how battle orders were actually transmitted, signals often provide evidence of tactical ideas and plans otherwise undiscoverable. In dealing with signalling techniques, he relied very considerably on the resources of William Gordon Perrin, the Admiralty librarian, whose immense erudition was always at the disposal of even the most unsophisticated enquirer. Perrin's *British Flags* (1922) includes a valuable section on naval signalling systems and signal books but he did not live to write the full-dress work he intended on this subject. Vice-Admiral Lancelot Holland accumulated a valuable collection of instructions and signals, including those of the Comte de Tourville now in the National Maritime Museum. He was killed in HMS *Hood* in action with the *Bismarck* in 1941, and left only a very short study for publication.[4] Commander Hilary Mead was amongst the leading world authorities on the subject of flags generally, including all methods of visual signalling. His large collection of books and manuscripts includes many naval items but he did not make naval signalling his sole specialisation.

Of the French naval historians, G Lacour-Gayet and Charles de la Roncière have each made considerable use of original source material in their analyses of tactics. The first detailed study of particular tactical experience was made by Colonel Edouard Desbrière, whose *La Campagne Maritime de 1805 – Trafalgar,* published in 1907, is a monument of able and devoted scholarship. Nevertheless, it covers only the battles of Ferrol, Trafalgar and Cape Ortegal. In 1933 Constance Eastwick not only translated the whole work but re-examined the original French, Spanish and British documents and was thus able to add some useful corrections to Desbrière's work. The most original, stimulating and irritating book on naval tactics by a French author is *Les Idées Militaires de la Marine du XVIIIme Siècle: de Ruyter à Suffren* (1911) by Captain R Castex. It is based on research and written in brilliantly critical terms. Almost all that he says about the difference between tactical theory and tactical practice, and the doctrine or ethos underlying official pronouncements, is excellent. Unfortunately, much of his work is marred by violent exaggerations and a number of sweeping generalisations unsupported by evidence.

Such are the writers who, in company with Corbett, have contributed towards a better understanding of the tactical theories, and tactical and signalling techniques during the sea wars of the seventeenth, eighteenth and early nineteenth centuries. The present work has a comparative basis and is the first in which any attempt has been made to evaluate the new evidence, as well as to re-examine the old.

4 'The Development of Signalling in the Royal Navy', *The Mariner's Mirror,* 39 (1953), no 1

Editor's Preface

BRIAN Tunstall's great work, completing the analysis of naval tactics under sail which had been begun by his father-in-law, Sir Julian Corbett, was not published during his lifetime. The reasons will be apparent to anyone who cares to read his manuscript, which is part of the collection at the National Maritime Museum. Perhaps the convoluted style of Admiral Lord Howe, who did more than anyone to reform British naval tactics, laid a curse on the subject. In the posthumous preparation of this book for publication, editorial intervention in those sections describing tactical ideas and signalling systems has had to be so extensive that it virtually amounts to co-authorship. The objective has been to bring out as clearly as possible Tunstall's research and perceptive analysis. Since his death important books have appeared on various aspects of the naval history of the period covered by Tunstall's book, but the value of his work has not been materially affected. Because it is inevitable that scholarly attitudes will vary over the years, it has been the policy of this editor to present Tunstall's case without significant alteration, so that the text is a true statement of Tunstall's ideas, expressed as far as possible in his own words.

In contrast to the sections describing the efforts of tactical reformers, the sketches Tunstall has written of naval battles have required little if any substantive intervention by the editor. Their value is all the greater because they are tightly focused on the important tactical developments which occurred during these actions.

Over half the manuscripts to which Tunstall referred were in private collections, including his own and that of the Royal United Services Institute. These collections are no longer intact. It has been necessary to track down the manuscripts in their new depositories. In this task I have been helped by Roger Morriss of the National Maritime Museum, and David Brown of the Admiralty Library. Unfortunately, not all the manuscripts became part of public collections, and in those cases the references given in the following pages are to the original collections, which are no longer in existence.

Notes on the drawings

The drawings which accompany the text of this book have been drawn mainly from the information given by the author, but also with reference to other relevant sources, such as battle plans in standard works, or the original books in the case of theoretical tactics. A standard set of symbols has been used throughout. All battle plans are oriented to the north, except where a compass rose is used to indicate north. A version of the tricolour has been used to indicate French flagships, though of course that flag was not in use until the revolution of 1789.

Key

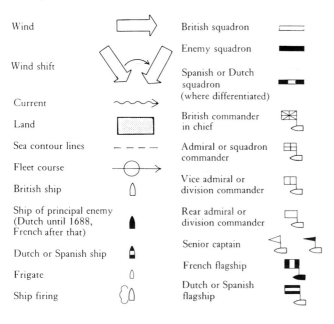

Wind	British squadron
Wind shift	Enemy squadron
	Spanish or Dutch squadron (where differentiated)
Current	British commander in chief
Land	
Sea contour lines	Admiral or squadron commander
Fleet course	
British ship	Vice admiral or division commander
Ship of principal enemy (Dutch until 1688, French after that)	Rear admiral or division commander
Dutch or Spanish ship	Senior captain
Frigate	French flagship
Ship firing	Dutch or Spanish flagship

Abbreviations

BM   British Museum
MM   Musée de la Marine, Paris
MOD Ministry of Defence Library, Whitehall (Naval Library)
NMM National Maritime Museum, Greenwich
PRO Public Record Office, Kew Gardens
RUSI Royal United Services Institute

INTRODUCTION:

The Art of the Admiral

NAVAL tactics are an admiral's art, and in a fleet propelled by wind and sails he exercised his art with difficulty. At no time could he make use of more than five-eighths of the sea around him, and when a calm prevailed he could not move his fleet at all. As he stood on the poop or quarterdeck of his flagship with the wind blowing, the remaining three-eighths of the sea were barred to him because his fleet could not sail directly into the wind. When close-hauled, that is to say when sailing as close as possible to the wind, his ship could not keep closer than six points of the compass on either the starboard or the larboard (port) tack. Hence progress directly into the eye of the wind could only be made by tacking across from one side to the other, and never closer than six points of the compass (67.5°). Even when running before the wind, it was generally best to go quartering to and fro rather than have the wind directly astern, since, in square-rigged ships, the sternmost sails would blanket the sails ahead and prevent them catching a stern wind.

The enemy were of course subject to exactly the same limitations. When two fleets sighted each other, one of them was usually to windward, and so held what was termed the weather gage. Since the weather gage automatically conferred some measure of tactical advantage, especially in the choice of whether, when and how to attack, the manoeuvres preliminary to a battle were often concerned with keeping, gaining or disputing the windward position.

In battle each admiral had to make a double set of calculations relating his fleet to that of the enemy in terms of distance, speed and course. Not only were there the wind and sea to consider, as they were at the moment, but also possible change in the force and direction of the wind and the influence, when near land, of tides and currents. If the wind dropped altogether the rival fleets might be left immobilised, half engaged, fully engaged or with the engagement over and still in sight of one another. Everything depended on the wind and the weather and, unless these were favourable, no battle could take place, nor could either fleet be formed in battle order.

The warships of the sailing period 1650–1815 were not built primarily for speed or manoeuvrability. They were wooden fortresses, propelled by wind and sails, furnished with two large batteries of guns, one on each broadside, and manned by a huge garrison, that is compared with their size and tonnage. They were heavily timbered, both for defence and to withstand the strain caused by the discharge of their own guns. They had stowage space for large quantities of food, stores, equipment and ammunition. They could stay at sea for many weeks, and they could withstand heavy weather. Each ship was a self-contained fighting unit of considerable power, capable of developing a heavy 'fire face' on each broadside, though with only a very few guns mounted to fire directly ahead and astern.

Success in war depended on an admiral's ability to organise a body of ships into a disciplined fleet, capable of obeying his instructions and signals. Only when his fleet was properly organised was he in a position to execute such tactics against the enemy as he deemed both possible and remunerative. Rapid changes of formation in the face of the enemy were useless and even dangerous if the captains were not sufficiently practised to carry them out. Nor could the admiral expect wholehearted support for a form of attack of which the captains did not approve or which they did not thoroughly understand. Much depended on the time available for exercising the whole fleet together at sea. If its composition was continually changing or if detachments had to be made for convoy work, watching enemy ports or minor operations, efficient organisation was not possible. This was especially the case at the beginning of a war. By the end, a common experience had generally been built up, thus allowing interchanges to take place between ships of different fleets and squadrons without any severe diminution of tactical efficiency. An anonymous admiral wrote of the 'Glorious First of June' (1794) that it 'was the *first* general action fought in the course of the war, and led to many glorious results; had it been the *last*, not one of the French ships would have been allowed to return to port'.

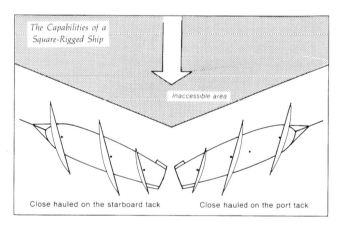

The Capabilities of a Square-Rigged Ship

Inaccessible area

Close hauled on the starboard tack   Close hauled on the port tack

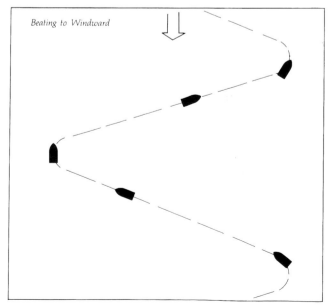

Beating to Windward

At sea each captain had to be capable of keeping his ship on the course set by the admiral and at the required distance – and, if necessary, on the required bearing from the next ship ahead or astern, or to starboard or larboard. Changes in sail carried by the admiral had to be promptly copied, with such adjustments as were necessary for correlating the known sailing powers of the individual ships of the fleet with that of the flagship. The larger the fleet the more difficult it was to control, and the longer it took to carry out evolutions. Tacking through the wind, and wearing or gybing (changing course with the wind astern), whether in succession or together, were movements which tested the seamanship and discipline of every single ship. If the fleet was in close order there was always danger of collision if the instructions were not properly observed. Station keeping was all-important, and correct alignment had always to be regained after tacking and wearing. Losing station by falling to leeward was the commonest kind of lapse, and British fleets always steered seven points from the wind when close-hauled, even though six points was possible, so as to enable ships which had fallen to leeward to work up again to their station. Well-trained French fleets – and before the Revolution most of them were – could manage six points off. A very high degree of corporate skill was needed to keep this alignment. Great responsibility rested on captains leading the fleet on either tack in line of battle ahead, that is, with the ships in file on ahead of another. This also applied to the leading captains if the ships were disposed in two or more parallel columns in line-ahead. Leading the line or lines and station keeping in general were particularly important and especially difficult when the ships were in line of bearing and in echelon.

All this skill in controlling the individual ships composing a fleet was required for one purpose only, the defeat of the enemy fleet. Evolutions could never be an end in themselves, though this is how they sometimes seemed to have appeared in the French service. Yet, without a properly maintained order of sailing, a fleet could be taken unawares, especially if it became dispersed during the night. Admiral Thomas Mathews failed to beat Admiral La Bruyere de Court off Toulon in 1744 largely because his fleet failed to get into station on the previous evening. Vice Admiral John Byron was beaten off Granada in 1779 by the Comte d'Estaing largely because his fleet had become dispersed in the night. Admiral Augustus Keppel was much praised after the Battle of Ushant (1788) for engaging the Comte d'Orvilliers without wasting time in making the signal for 'form line of battle.' This was certainly creditable but if he had had his fleet better in hand at day-break he would not have had to commit himself to a ragged attack. Lord Nelson did not have the whole of his fleet as close to him as he might have wished on the morning of Trafalgar (21 October 1805), especially when he decided to attack immediately. This disability was never put right and it can be argued that he might have done better to wait until his fleet was more collected before launching his attack.

Night action itself was shunned until Nelson risked it at the Nile (1798). There were signals for it in both navies, since it was recognised that admirals might have night action forced on them through chance encounters. Strong fleets shunned night fighting, however, because they could not use their strength effectively. Weak fleets equally shunned it because they feared that they might become heavily engaged and be unable to withdraw.

Failure to maintain a proper system of reconnaissance seems to have been one of the worst faults of the sailing fleet admirals. The Comte de Grasse risked complete defeat because he was surprised at the Chesapeake in 1781. Admiral François Brueys d'Aiguïlliers failed to use his frigates to keep a look-out for Nelson before the Battle of the Nile. The inadequacy of reconnaissance was to some considerable extent a result of the lowly status of private ships, which were unable to signal back to the admiral in the same terms in which he could signal to them. No reconnaissance system was of any great use unless the ships conducting it could make full and detailed reports as to the composition, strength and disposition of the enemy, their present course and seeming intentions and the amount of sail they carried. In the French service more advanced ideas prevailed. Nelson at Trafalgar had an excellent system which was able to employ Sir Home Popham's 'telegraphic' system of signalling.

The most perfect signalling system for look-out frigates was useless unless the frigates were sufficiently bold and cunning to get up with the enemy without being winged or captured. This meant being able to avoid or beat off the enemy's frigates. Hence the importance attached to the qualities of frigate captains, one or more of whom might even be senior to some of the captains of the ships-of-the-line.

Discussions often arose as to whether the admiral ought not to be in a frigate instead of in one of the most powerful ships in his fleet, which must by its very character become heavily involved in the pell-mell. Lord Howe actually tried it and so did Admiral George Rodney. Audibert Ramatuelle discussed it at length in his *Cours Elementaire de Tactique Navale*, 1802. Another advantage was that the admiral could carry out advance reconnaissance in person instead of having to rely on signals sent back to him, as Napoleon pointed out to the Comte de Ganteaume. In battle, however, his position would be equivocal, since he could hardly be judge of tactical movement in which he himself was not participating. If his fleet was engaging from windward he would have to be to windward of his own line and if from leeward, then to leeward, with the additional disadvantage of his view being blocked by smoke. In a light breeze the smoke from the guns hung over the battle area and prevented the admiral from seeing anything beyond the ships immediately around him. His fleet meanwhile might be strung out over a wide sea area, under conditions which made signalling difficult even if there was no smoke at all.

When attacking a fleet already formed in line of battle, it was customary to make a slow and stately approach. This enabled the attacking ships to maintain their alignment and not mask each other's fire by running too far ahead or accidentally steering across the line of approach of another ship. Even with a stiff breeze in their favour, they might well be advancing at only about four knots. It is difficult for modern readers to visualise the agonising slowness of most sea battles under sail. Hours and even days might be spent in trying to get into the desired order of battle before launching the attack. It is easy to see that under these circumstances the strain on the admiral was terrific and why it was that both Rodney and Howe were incapable of pursuit after their respective victories of the Saints (1782) and the First of June.

Diagrams printed in the tactical text-books and official instructions of the sailing-ship period naturally show ships carrying out evolutions in perfect alignment, though this was not always attainable in practice. For this reason diagrams of sea battles under sail are misleading; the reader is tempted to project the modern capacity for station keeping into the past, simply on the analogy of what modern machine-driven warships can do with the aid of radio and radar.

Even under sail, relative speed was always of great tactical importance, and this was often determined less by the ship's design than by the amount of weed and barnacles fouling her bottom. Prior to the introduction of copper sheathing, a fleet of newly cleaned ships, fresh out from dock, could outsail by a significant margin a fleet of ships exactly similar in build, but in need of scraping after many weeks at sea. The introduction of built-in speed differentials as part of steam-ship design was one of the most important influences on the development of naval tactics in the iron-clad period.

No one ever really knew the exact time at which things occurred in a sea battle under sail. Despite the efforts of the clockmaker John Harrison, chronometers did not appear until towards the end of the period, and even then they were not supplied by the Admiralty. Some officers bought their own at a cost of £100, a very large sum in comparable terms today; assuming that he was at sea continuously, the pay of a captain commanding a line of battle ship would vary between about £246 and £365 a year, depending on the rating of the ship. Although ships' logs may record the times of battle movements and of signals sent and received, these times are only of value in so far as they relate to the progression of time in a particular ship. Instruments were used for estimating the distance between one ship and another when formed in line of battle but there were no range finders for estimating the distance of the enemy's ships.

The sailing ship-of-the-line experienced the problem of bringing its armament into action far more than did the warships of the machine age. Even when weather conditions were ideal no ship-of-the-line could advance directly on its enemy without disabling its own batteries, which could fire only in a narrow arc on either beam. Systems of naval tactics were thus inevitably designed to bring fleets into action as rapidly as possible while exposing them to as little hostile fire as possible. This could be done by approaching the enemy's line end on, so that the enemy was prevented from firing his batteries; by an oblique approach which allowed the ships to begin firing on part of the enemy line even if they had to shift targets as they advanced; or by a bold head-on charge at the enemy line, as at Trafalgar, which reduced the time of exposure to long-range enemy fire.

Once fleets were engaged, there was more to gunnery than firing and reloading rapidly, although that was indispensably important. Effective gunnery depends on morale and training, either of which alone is useless. This was the lesson of the famous single-ship action between the *Chesapeake* and the *Shannon* in 1813. The latter frigate's crew not only had excellent morale but were the best trained body of gunners in the whole Royal Navy at that time. Her captain, Philip Broke, had been preparing them for a fight such as this for over six years.

It is frequently asserted that, whereas the British fired at the French hulls, the French fired at the masts, rigging and sails of the British ships. There is some truth in this, though Admirals Edward Vernon and Richard Kempenfelt were all for imitating the French. The hull was more difficult to hit but was the more decisive target. The piercing of the ship's sides for gunports made them comparatively vulnerable and a series of well-aimed shots would quickly damage and dis-mount guns, and kill and wound the gunners. Besides this, hull shots cut deep into the ship and might well crack or damage the base of the masts in addition to disorganising the control of the ship and making it difficult to bring up fresh powder and cartridges from the magazine. Waterline and below waterline shots started leaks which the carpenter and his mates might find difficult to plug. British ships were often so crank that their seams leaked through the strain of their own guns being fired, so a few well placed enemy shots would soon have them hard at work at the pumps. One of the most damaging methods of attack at close quarters was alternative depressing and elevating the guns. Under such conditions the guns could be fired with reduced charges so as to cause the maximum amount of splintering in the enemy ship.

Seen at a distance, the whole ship, rather than the hull alone, presented an easier target; in this case, damage aloft could stop a ship's advance or send her reeling out of the line. Although there were standard methods of strengthening the rigging and chaining the yards to prevent their fall, every-thing aloft was vulnerable. Sails could be slashed to ribbons, important rigging such as the stays could be cut, yards brought down, together with topmasts. Sometimes the whole mast itself splintered, cracked and finally broke at deck level. This generally meant a huge cluster of sail, woodwork and rigging over the side of the ship, entirely masking the guns in front of which it fell.

With anything like a sea running, it was difficult to aim at the enemy's hull at any distance as it kept partially disappear-ing from view. There was also the difficulty of timing the fire; the French generally fired on the upward roll of the firing ship.

Battle damage and battle casualties were often dispro-portionate. In the Battle of St Vincent (1797) there were only 73 men killed in the whole British fleet, whereas the *San Nicolas* alone had 144 killed. In the Battle of the First of June

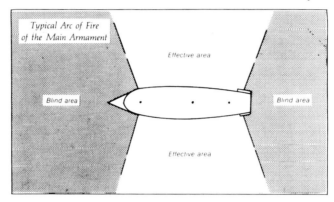

Typical Arc of Fire of the Main Armament

Effective area

Blind area

Blind area

Effective area

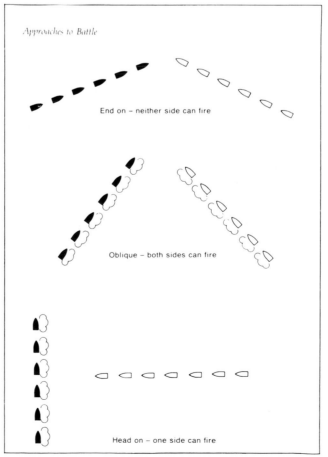

Approaches to Battle

End on – neither side can fire

Oblique – both sides can fire

Head on – one side can fire

the number of French killed in the six ships captured was more than three times that of the total killed in the whole British fleet. Once a ship with good gunners had the upper hand over another, the result was decisive, as is shown by an account of the sinking of the *Vengeur* at the First of June written by a lieutenant of the *Brunswick:*

As the *Vengeur* pass[ed] we split her rudder, stove her stern post into shivers; we saw the water pouring into her counter by tons. The *Ramillies* was yet waiting for her to settle more from us. Looking out of the stern ports, I could well perceive every precaution taken by them to train their muzzles clear of our stern, and point them to the Frenchman's quarter. At length went their broadside, and such a tremendous fire, in my life, I never saw or could have believed; I verily think every shot struck her, and as she was within pistol distance, the effect must have been terribly severe. . . .

We do not remember the mizen mase of *Le Vengeur* falling, when near the *Ramillies*. All her masts went soon after, but from the weight of shot thrown into her, viz. 3 rounds shot of 32 lb., i.e. 96 lb. of ball from each gun, driving home our coins at one time, watching their rising, to fire beneath the water line, next withdrawing the coins to elevate the muzzles and rip up her decks, this alternative mode of firing for two hours and a half alongside, we rather think it may probably happen her masts were cut by our shot from within board. The silence of their fire for the last two hours, shews the people were employed at their pumps, but all their efforts could not preserve their ship from sinking.[1]

Although British admirals paid lip-service to the idea that battles were decided by individual ships and squadrons pressing home their attacks – the notion of the 'pell-mell' – only a few attempted to fight a major battle without exercising close tactical control. Still fewer were the successes resulting from such pell-mell actions. Many British admirals honestly believed that a good cannonade in line ahead, parallel with the enemy's line at fairly close range, was a satisfactory enough form of engagement. Advantage would then rest with the fleet having the most effective gunnery and the best-formed line from which to apply it. Ships disabled aloft or heavily leaking from wind and water shots would try to withdraw from their respective lines; those in the line to leeward would manage this with greater ease. In the end the less effective fleet would withdraw altogether.

British admirals seem to have preferred these linear battles because they could see no other way of dealing with a well-drilled French fleet in a well-formed line. Splitting up the fleet, and attacking with separate squadrons, seemed less effective because it led to dispersion, risking piecemeal defeat; in any case it was not the best way of developing the maximum fire face against the enemy. The main reason, however, for not adopting a more elaborate form of attack was the difficulty of organising and executing anything requiring sophistication and finesse, especially in view of the limited British signalling system. The problem became greater once ships were damaged aloft; when any of the leading ships were thus disabled, and unable to keep station, the whole attack became disorganised. But, as John Clerk of Eldin so cogently argued in his *Essay on Naval Tactics* (1790-97), it was the slanting attack on the enemy's van, as so often practised, which proved unsatisfactory, rather than the linear engagement itself.[2]

When two fleets of unequal strength, or with sharp differences in gunpower, met in battle, there were various

devices by which the weaker fleet might space out its ships so as to leave some of the enemy ships without any target to fire at; since broadside guns had a very restricted arc of fire, this happened frequently. Nor was it easy for the ships left unengaged to swing themselves into position from which to open fire. By doing so they risked upsetting the forward movement of their own line of battle, which necessarily carried sufficient sail to maintain steerage way.

The least likely form of engagement to produce any substantial results was when the fleets passed each other on opposite tacks, as at the Battle of Ushant. This kind of fighting seems often to have taken place in the Anglo-Dutch Wars, but by the end of the eighteenth century it had become discredited, at any rate in the Royal Navy. A more promising scheme was for the stronger fleet to use its excess of ships to harry the van or rear of the weaker fleet by raking their bows or sterns and, if feasible, doubling on them. This involved getting right round onto the disengaged side of the enemy's van or rear, so putting them between two fires and forcing them to engage with both broadsides at once, a difficult task requiring high efficiency and morale. On these occasions the doubling ships had to be careful not to hit each other when firing at the enemy ships sandwiched between them.

A more sophisticated form of attack was by concentrating the bulk of the fleet against part of the enemy's line, and leaving the remainder practically unengaged. In this way the attackers might hope to bring a crushing superiority of fire against a smaller number of enemy ships and thus achieve decisive results before the unengaged ships could rally to the support of that part of their fleet already under attack. In such cases concentration on the enemy's rear seemed to offer the best results. Much depended, however, on the extent to which the attackers could get into sufficiently close order to bring their superior weight of fire to bear and yet avoid obstructing each other.

Still more sophisticated and ambitious was the manoeuvre of breaking the enemy's line. In theory this form of attack had many possible variations. It could be executed from windward or leeward, by all or part of the fleet, and at any point in the enemy's line. Lord Howe's special variation, which he actually attempted at the First of June with some measure of success, was for his whole fleet to cut through the gaps between all the enemy ships. In doing so, his ships could rake the bows and sterns of the ships respectively to starboard and larboard of them as they passed through and then indulge in a general pell-mell. It is, of course, obvious that if part of an attacking fleet cut through the enemy's line, both were liable to become separated into two parts; in other words, there were risks for the cutters as well as for the cut. The assumption was, however, that the cutters had superior fighting efficiency and morale, and that in the general confusion which ensued they would benefit either by retaining their tactical cohesion or in the pell-mell fighting. The assumption was also made that by forcing the ships in the line being cut to give way, the cutters would pass through intact as far as the rearmost ship, even though the cutting movement may have been initiated by a ship only part of the way along the line.

Despite the frequent issue of orders not to open fire too soon, or until signalled to do so, battles were often begun and sustained at a range as great as 1000 yards. This might happen when a ship in an attacking fleet was damaged aloft, and, swinging broadside on, opened fire more or less in self defence against the ship engaging her. This in turn might set off other ships and so prevent the rest of the attacking fleet getting much nearer, as otherwise they would be in danger of masking each other's fire. This emphasises the overriding importance of fire discipline. Long range cannonading also

---

1 Source unknown
2 *An Essay on Naval Tactics*, London, 1782 (a rare copy is in the Royal Naval College library, Greenwich).

*Two Methods of Breaking
the Line*

Breaking through
at a single point

Howe's method

Detail of ship passing through

fire along all the gundecks, if necessary on both sides of the ship. As the action continued, guns were put out of action by enemy shots and gun crews killed or wounded. Further casualties resulted when overheated guns recoiled so hard that they broke the double breeching ropes specially fitted for battle but possibly cut by shot. Marines and seamen exchanged musket fire from the forecastle, quarterdeck and poop, and also from the tops. Swivel guns, and later carronades mounted on the upper works and loaded with canister, could be used to sweep an enemy's deck clear of boarders forming up for an attack. Boarding was only possible when two ships were locked together; such a position was dangerous if either ship caught fire. Mutual bombardment might sometimes go on for several hours until casualties amongst gun crews in one ship reduced their rate of fire and finally left them unable to resist further.

Seamanship in a pell-mell meant, first, trying to rake an enemy ship by bringing the broadside of the attacking ship against the enemy's bow or quarter. It also meant using such movement as was possible with spars and sails shot away to get into a more favourable position or to engage a new opponent; this was also what was meant by mutual support, one ship coming up to relieve another, either by drawing the enemy's fire or by interposing the ship between the enemy and the hard-pressed ship.

Ships-of-the-line were seldom sunk in battle as a result of gunfire alone. The fate of the *Vengeur* at the First of June was exceptional. Ships which caught fire might blow up if the fire reached the magazine. Badly damaged ships might sink eventually as a combined result of shots below the waterline and heavy weather. Really crank ships were in danger of falling to pieces simply as a result of firing their own guns.

When, in 1689, the navies of England and France began that long series of contests which ended in 1815, their respective systems of tactics were very similar. Both navies used the 'line-ahead' formation, with each ship directly astern of the one ahead. In battle the opposing fleets were ranged roughly parallel so as to develop the maximum degree of broadside fire. French tactical doctrine, like the English, favoured the close-hauled line-ahead ('au plus près du vent') as a preliminary tactical formation. But, whereas the English system had been gradually developed through the experience of the three Dutch Wars, the French system had come into being as part of the naval expansion directed by Jean Baptiste Colbert (1619–83), statesman and virtual founder of the navy of Louis xiv. When the new French navy began to emerge, with a strong corps of well-built and well-gunned broadside battleships, it became possible, and indeed necessary for the fleet to adopt the line-ahead formation in battle though the formation's origins were acknowledged: 'Cet arrangement a été inventé par les Hollandois et les Anglois.'[3] So quick were the French to adopt what seemed the best tactics for their new broadside battleships that, by the Third Anglo-Dutch War of 1672–3, they were able to contribute a well-drilled and powerful squadron to the allied Anglo-French fleet.

As Colbert's reforms took root, French tactics came under the influence of the mathematicians and scientists, who were playing such an important part in French naval development as a whole. The navy of Louis XIV was a direct expression of the royal will. The new warships were works of art and science; art in the form of carvings and decorations designed by the leading sculptors, architects and painters of the day, science through the advice of the newly founded Académie Royale des Sciences. No longer could the construction of

occurred when one fleet was trying to keep another from carrying out some manoeuvre or was trying to preserve its own freedom of manoeuvre. 'Close range' is an elastic term, and seems to have meant anything under a quarter of a mile. 'Musket shot' and 'pistol shot', used in both the French and British services, are also inexact terms. Musket shot presumably meant the range at which the marines would usually open fire at an enemy ship, but this might differ greatly according to circumstances. It might well have been between 150 yards or more, even if the musket was not regarded as accurate beyond 80 yards. Pistol shot again might mean anything between duelling range, say 25 yards or less, and long-distance pistol shot up to 50 yards. On the whole it would seem that in the majority of battles the ranges may possibly have been slightly greater than had hitherto been supposed. We know of course that at the First of June, St Vincent and Trafalgar, some ships were actually touching their opponents, but this seems to have been exceptional. Though preferring longer ranges than the British, the French, once in a pell-mell, favoured boarding, believing that this was something at which they themselves excelled and the British feared.

In a real pell-mell, with the ships lying close together in no particular order, everything depended on efficient gunnery and seamanship and on what the British called 'mutual support'. In terms of gunnery this meant keeping up a steady

3 G Daniel (de la Compagnie de Jésus) *Histoire de la Milice Françoise*.
Paris, 1721, II, p735

ships be left to the traditional empiricism of untutored shipwrights. Much theoretical work was being done on the resistance offered by the sea to various shapes of hull, to problems of stability and buoyancy, and to the effect of wind pressure on sails of various shapes and sizes, set at varying heights from the deck and at varying angles to the centreline of the ship. From this study of the movement of the individual ship, as governed by wind and water, it was an easy step to the study of ships moving together in squadrons or fleets involving angular measurements of the ships from one another and the mathematical problems raised by changes of formation in relation to the wind.

In navigation, as in shipbuilding, the best scientific advice was obtained through the new Service Hydrographique and the new Paris Observatoire. The later editions of Georges Fournier's *Hydrographie*, really a work on pilotage, first published in 1643, became a fine example of contemporary science applied directly to maritime needs.[4] When it came to setting out in textbook form the various orders of sailing and orders of battle which the fleet should assume or the methods of changing from one to another, the mathematicians were heard with respect. Indeed it was only in France that the need for such a textbook was as yet recognised. The need produced the book, *L'Art des Armées Navales*, by Père Paul Hoste, a mathematician.[5] It proved however to be prejudicial to the French service in that it tended to substitute elaborate evolutions for actual battle tactics and, in so far as it dealt with battle at all, placed undue emphasis on defensive fighting and defensive needs. Hoste was a protégé of the great Comte de Tourville, and he wrote in the shadow of French defeats. No doubt his mathematically based love of complex patterns and his overwhelmingly defensive attitude were suited to the mood of the age in which he wrote.

Amongst the new discoveries of exceptional interest which have, however, been available for study for many years are the printed sailing and fighting instructions, signals, and orders of sailing, issued by Tourville during 1690–1 and 1693. They show an immense superiority over contemporary English methods of fleet organisation and signalling, though the actual battle instructions and battle signals show no advance on the ideas of James Duke of York or of Prince Rupert.

Eventually a doctrine emerged, this time strategic in intent, but fitting in only too well with defensive tactics: the extension and control of the world-wide French empire was to take precedence over winning battles. Battle tactics and battle opportunities, therefore, because subordinated to more far-reaching strategic enterprises. It is no accident that naval commanders such as the Marquis de La Jonquière, the Marquis de La Galissonière, the Comte de Bougainville, the Marquis de Vaudreuil and the Comte d'Estaing played a leading part in French imperial development.

These are some of the elements leading towards the development of a French view of naval tactics, a view which seems to have hardened as time went on and did not change until the Revolution. In England the situation was different. The hard fighting experience of the Anglo-Dutch Wars had led both to the consolidation of the line-ahead formation in battle and to various special methods of attack. As in the case of the Dutch, the greatest obstacle to tactical efficiency was the size of the fleets employed. With seventy, eighty, ninety and even more ships included in the line of battle, supported by twenty or thirty fireships and a host of auxiliaries, it was impossible to obtain tactical cohesion, especially as more than

10 per cent of the line ships were hired merchantmen. Nor was it possible to do much in the way of tactical training, as the fleets were based in ports comparatively close to the Dutch ports and anchorages. Broadly speaking, they set out with the intention of fighting and immediately returning to base for repairs. Although this did not always happen, the situation was very different from that on a distant station, where there was time and opportunity to develop a small fleet into a well-drilled tactical combination. In the Anglo-Dutch wars proper tactical control was impossible: ships followed their own squadron commanders or even the commanders of their squadron sub-divisions, who often acted quite independently, so that the fleet tended to become dispersed in a series of desultory piecemeal engagements. Tactical sophistications were unattainable because the preliminaries of tactical control had not yet been achieved. Nevertheless, it is perfectly clear that the aim of English commanders was to defeat the enemy, and that tactics were no more than a means to that end.

The so-called 'fighting instructions' (a term abused by many writers) never existed as a separate work. They were merely one chapter in what became the permanent 'Sailing and Fighting Instructions for Her [later His] Majesty's Fleet'. This book includes signals by day for when at anchor, weighing anchor, sailing and anchoring; a similar series for the night; distinguishing lights for the flags; instructions for sailing in a fog; instructions about ship precedence between 'younger' and 'elder' captains; signals for calling officers; instructions for masters, pilots, ketches, hoys and smacks, attending the fleet; the 'encouragement' for fireships and instructions for fireships.

The sailing instructions were initially very rudimentary, though they provide the only evidence we have of the signals used for controlling the fleet when not in the presence of the enemy. Gradually the sailing instructions came to include more details and, eventually, to provide a whole set of fleet cruising formations, with the methods of changing from one to another or into order of battle. From these we gain a picture of what a great fleet must have looked like when searching for the enemy and ready to deploy into line of battle according to the various methods available. They thus form a parallel set of documents to the fighting instructions.

It has become fashionable to attribute British tactical failures between 1704 and 1781 to a pedantic, formalistic and docile adherence to the order of battle in line-ahead formation. The chief reason for the line's disrepute was its association with the notions of defensive fighting arising from the experience of the Battle of Malaga in 1704. In the hands of General at Sea Robert Blake, Prince Rupert and James Duke of York, the line-ahead formation became the best means of bringing the maximum number of guns to bear on the enemy so as to defeat them. When Admiral Sir George Rooke and his contemporaries and successors began to emphasise the line's defensive virtues, those responsible for the earlier development of the line could be dubbed formalists. By contrast, those who put less stress on warlike order and still favoured the independent squadron movements of an earlier age came to be regarded as the truly offensive fighters. It is important to appreciate the changes which took place in the structure of the line of battle between 1672 and 1744 and the difference made by the increased proportion of three-deckers. Much of the criticism levelled against docile British admirals who failed to force a pell-mell, in contrast with dashing men such as Lord Anson, Admiral Edward Hawke and Admiral Edward Boscawen, takes no account of the fact that it needs two to make a battle. As Admiral John Jerius, Lord St Vincent, so aptly put it, 'I have often told you that two fleets of equal strength can never produce decisive events, unless

4 Georges Fournier, *Hydrographie*, Paris, 1643
5 See below, p59

they are equally determined to fight it out or the Commander-in-Chief of one of them so bitches it as to misconduct his line.'

General chase, the action so popular with writers of stirring tales, was not in itself a battle formation and, unless carefully controlled, could only invite defeat in detail. Nor was it quite such a rapid business as these robust writers would have us suppose. In a moderate breeze an attacking fleet might approach the enemy at a speed of between 2 and 3 knots, so as to make sure of keeping their alignment and not masking each other's fire. How many knots would general chase produce with a similar breeze? Hardly more than 6 or 7 – not a very breath-taking speed at which to 'hurl' a fleet at the enemy.

The American War of Independence was the highest competitive point in French and British tactics. Except when led by Admiral Pierre André de Suffren, French fleets never attacked, as indeed they had never attempted to do since 1704. Their tactics when opposed by fleets of equal strength were mainly defensive. Nevertheless, they outpointed the British again and again, Rodney's great victory at the Saints coming too late to alter the strategical issues of the war. The fleets depended on the new, unwise and unnecessary elaboration of French signalling and tactical formations which accompanied the naval revival following the defeats of the Seven Years War. In 1763 the Vicomte de Morogues produced in *Tactique Navale ou Traité des Évolutions et des Signaux*, a voluminous work whose sheer extent and complexity have defeated many critics. His influence on French tactics was disastrous. As late as 1806, the French navy rejected the proposed adoption of a true numerical system of signalling, by a vote taken at each of the naval bases. Instead, they voted to continue using the highly complicated system introduced by Captain du Pavillon in 1776. They retained this system until after the end of the Napoleonic Wars.

During the American War, a differentiation between British tactical traditionalists and reformers can be seen in both the form and the content of the additional fighting instructions and signals issued by the Admiralty (though not in title) and by particular admirals for particular fleets. It can also be seen in the examples provided by actual battle. In many respects the form governed the content. Signals as mere words coming at the end of what were often long-winded instructions could never be as effective as signals directing this or that immediate and direct manoeuvre divorced from any instructional element. What the British most needed was a pure signal book with signals for attacking and fighting the enemy in various ways. Instructions were absolutely necessary for fixing how certain battle manoeuvres should be executed, but there was no need to keep new battle signals chained to the instructions, especially when an instructional element was involved.

Rodney never issued a signal book in the whole of his career. His signals still lay embedded in the printed text of his instructions. The term 'Rodney's signals' refers to private signal books made by officers of his fleet for their personal use. By contrast, Lord Howe was of pre-eminent importance in the reform of British tactics and signalling. He was the first to separate signals and tactics, and he gave his fleet an official, printed signal book, the accompanying book of instructions being 'explanatory of and relative to the Signal Book herewith delivered'. This was as early as 1776, when he commanded on the North American station. In 1782 he applied his new system to the Channel Fleet. In 1790 he gave the Royal Navy a true numerical system of signalling. A lucky discovery of a set of printed additional sailing and fighting instructions, unsigned, undated and unissued, though clearly composed by Howe, provides an important link between his

activities in the Seven Years War and the American War.[6]

Howe also has the unique distinction of being the executor of his own themes. In the Battle of the First of June he attempted with some success his long planned cutting of the enemy's line at all points. He was then sixty-eight, the greatest age at which any British admiral has ever won a major battle. Howe, unlike Kempenfelt, refused to be dazzled by the elaborate complexities of the French tactical and signalling methods or the French theoretical treatises. Nevertheless, he could never have won the battle if the French had been unwilling to fight it out and it is to the credit of his opponent, the Comte de Villaret de Joyeuse that, with a comparatively untrained fleet, he unflinchingly accepted the full hazards of a real pell-mell.

Howe's triple success as tactician, signaller and commander-in-chief in battle is all the more remarkable in view of the appallingly long-winded obscurantism of several of his sailing and fighting instructions. When he gave oral instructions the result was sometimes as bad or even worse.[7] Despite this handicap, he succeeded mainly because he had only one great object in view: the defeat of the enemy.

It is sad to record that Kempenfelt fails to emerge from detailed investigation as anything more than a vigorous exponent of Howe's earlier ideas. The evidence is prolific and its implications unmistakable. Vice-Admiral Sir Roger Curtis, Howe's chief of staff, has been revealed, by a lucky discovery of manuscript working papers, as a complete traditionalist, so far as signalling is concerned. Their outlooks were entirely different.[8]

Admiral Sir Charles Henry Knowles was a great experimenter in tactics and signalling. His Set of Signals, printed in 1777, was an astonishing achievement for an officer aged only twenty-three.[9]

After the Revolution the tactical efficiency of French fleets declined, though unlike the fleets of the *ancien régime* they did not invariably try to avoid close action. By contrast, the need for a more offensive system of tactics was now proclaimed by the writers of theoretical treatises. Beginning with the Vicomte de Grenier in 1777, the demand for offensive tactics increased with the Comte d'Amblimont in 1788, Ramatuelle in 1802 and Le Chevalier Delarouvraye in 1815, who had *Delenda est Britannia* printed on the title page of his book. Inside the service itself, however, no changes took place. There is no evidence that the work of these men influenced service thinking – and, after 1802, it was in any case too late.

The final developments of British tactics and signalling, after the issue of the first official Admiralty books of 1799, were gathered together in the Admiralty Signal Book of 1816. In this book, Howe's system is abandoned in favour of that invented by Popham, whose *Telegraphic Signals or Marine Vocabulary* had already proved so useful to Nelson. This work was now adopted by the Admiralty and issued as a companion volume to the Signal Book.

Naval tactics are normally the expression of the tactical doctrine prevailing at the time in the particular service concerned. The doctrine is given official and systematic form in the manuals and books of instruction under which the officers are trained, and in the light of which they fight. We know, however, from experience in other fields that the actual working of a system cannot be understood solely by reference to a corpus of official regulations. The picture they provide is never complete, their interpretation depending on

6 Tunstall Collection, *NMM*, TUN/111 (formerly FI/5)
7 Sir John Barrow, *Life of Richard, Earl Howe*, London, 1838, pp118 9
8 Corbett Collection, *NMM*, TUN/7 (formerly S/MS/Am 7)
9 Sir Charles Henry Knowles, *NMM* S(H)25. There is also a mutilated copy in the Mead Collection, *NMM*, TUN/136 (formerly Mead/S/1).

professional practice, the 'custom of the service'. Beyond this again, there must always be a philosophy, a sentiment, a tradition, a prejudice or an ethos which both inspires the users of the system and governs their interpretation of it. Broadly speaking, the correlation between official thought and fighting practice is closest when morale and fighting efficiency are high. It is when defeat or internal disintegration has brought the spirit low that the official regulations take on the character of theoretical propositions divorced from reality. It is only too easy for a navy in decline to be issued or re-issued with fighting instructions and signals, detailing methods of attack which experienced officers recognise as unlikely even to be attempted.

In practice, no admiral could afford to ignore the strategic implications of his battle tactics. The instructions he carried with him and under title of which he held his command, specified, often in detailed terms, the strategic objects he was bound to pursue, and to some extent the methods by which he must attain them. Princes, ministers, admirals and generals of allied states must be persuaded and placated, and governors of neutral ports must be cajoled; while over all loomed the constant need to protect trade with frigates and to supply escorts for convoys. To fight or not to fight, and if to fight, when and in what manner, were not matters which could invariably be settled simply by noting the wind and watching the movements of the enemy.

British and French signals

Standard methods of signalling had become established in the English, Dutch and French navies well before the end of the seventeenth century and, apart from developments in codes and techniques, they remained unchanged until after 1815. Signals were originally conceived as a means by which the admiral communicated his immediate orders and future intentions to his subordinate flag officers and captains. They themselves could acknowledge or repeat his signals but not much more. Private ships had a limited number of signals which they could make direct to the admiral or repeat to him, mainly in terms of 'land discovered', 'strange sail sighted', signals for distress or for springing a leak. They could, of course, make such signals to each other and were indeed expected to do so. The full range of signals, however, belonged to the admiral alone. In the Royal Navy, in early days, he alone possessed a full set of flags. French private ships-of-the line possessed all the flags and repeated the admiral's signals. The admiral might appoint one or more repeating frigates, with flags, to stay on his disengaged side in battle, to repeat his signals so that the fleet could see them free from ships ahead or astern and the smoke overhanging the line of battle.

The admiral also had other means of communication. Where tactics were concerned and a special situation envisaged, he might call a council of war, often as a means of obtaining sanction for not obeying his instructions to the full. He might also call his flag officers and captains on board to explain his intentions, though not all admirals were prepared to risk this kind of conference, with the embarrassment of being outfaced in argument by able and aggressive subordinates. Alternatively, he might distribute special tactical instructions by boat; this seems generally to have been done by signalling each ship to send a boat to receive a copy.

By day, in port, he could signal by sail movements and guns, since flags might not blow out enough to be seen. At

sea he had various choices. For communicating with the fleet as a whole or with squadrons or even divisions, flag signalling was best, the signal generally being accompanied by the firing of one or possibly two guns to draw attention to it. Signals to individual ships were made by first hoisting the recognition pendant for that particular ship. Orders could also be sent to individual ships by boat, though this was an extremely wasteful method as it generally required an officer of the rank of lieutenant to take it either in writing or by word of mouth. Nevertheless, orders were often sent by boat in battle, especially if the fleets had only steerage way on, or it was dead calm so that a single boat might visit several ships in succession. Ships-of-the-line, frigates, fireships, fleet auxiliaries and boats, if near enough, could be given direct orders by speaking trumpet. Orders could also be sent by hailing a frigate, attendant cutter or ketch and ordering her captain or commander to deliver a spoken message to a flag officer or captain of a private ship.

In the battles of the Dutch Wars, when large fleets were spread out too far for signals to be easily recognised or understood, especially with smoke from the guns obscuring the view, the dispatch of orders by boat or some larger vessel seems to have been the most effective method. In fact it was only during the eighteenth century, when fleets on the whole were smaller and the technique of repeating frigates better developed, that the admiral could do much with flag signals once the battle was joined. Throughout the Dutch Wars, it seems doubtful if the respective commanders-in-chief ever exercised or expected to exercise much control over the fleet as a whole during the battle. This, of course, had an important bearing on the extent to which fighting instructions with signals attached, purporting to deal in different ways with various tactical situations, ever gained practical recognition.

Night signalling was always difficult and, in any case, fewer fleet movements could be attempted. In the early days signals for running into danger, strange sail sighted and recognition signals for 'losing company and meeting again', cutting and slipping, weighing anchor or anchoring, were the main ones. The basis of night signalling was with lights (lanterns), generally up to four, displayed in various places and patterns, together with guns (always fired from the same side of the position). 'False fires' were also used in addition to or as substitutes for lights. These were manufactured from gunpowder and wheat flour, in the proportion of 1lb to 6oz (8 to 3), contained in a tube made by rolling paper thickly round a stick, which was then withdrawn. The composition was then hammered home. A quick-match was used for firing. Later in the eighteenth century, blue lights were used, composed of a mixture of 7lb of saltpetre, 1lb 12oz of sulphur, and 8oz of blue orpiment.[10]

Sky rockets were also in use at least as early as the Third Dutch War. In the French navy they became divided into *fusées en étoiles*, *fusées en pluie* and *fusées en sermentaux*. They were especially useful towards the end of an action when fresh gunfire would attract little attention.

One of the chief difficulties of night signalling was that of not being able to see the signals made by the admiral when blanketed by his own sails or the sails of intervening ships. Much also depended on the lights displayed in each ship, by his orders, to guide the fleet on its immediate course, regardless of any attempt to change formation. The tactical disadvantage of night signalling was that it gave warning to the enemy of the presence of another fleet. Vernon realised this, and cut down the number of guns used for night signalling in the West Indies.

Fog signalling was even more limited in scope, guns being the chief method, with later sophistications embracing quick

---

10 The formulae are from the Admiralty Night & Fog Signals of the Victorian period.

and slow firing with measured intervals between individual guns, and with sequences of guns. For exchanging recognition at close quarters it was possible to give the respective pass-words by hailing with a speaking trumpet, just as in day time or at night. In the French service, auxiliary fog signalling methods included bugle calls, drums, bells, trumpets, fifes, muskets and swivel-guns, alone or in combination with signal guns.

In both the English and French navies, as well as in the Dutch, the early system of signalling was that each flag was known by its description and had no number or letter attached to it. A signal flag, therefore, only meant something when it was hoisted in a particular position. The English worked with very few flags. The French were more enterprising, largely because they had already developed a more elaborate series of instructions which, in turn, required more signals. In the English navy the positions for hoisting signal flags were the fore, main and mizzen topmast heads; the fore yardarm, the fore topsail yardarm and the fore shrouds; the main backstay, the main yardarm, the main topsail yardarm, the main shrouds and the main topmast shrouds; the mizzen peak, mizzen shrouds, mizzen topmast shrouds, the crossjack yardarm and the mizzen backstay; and the ensign staff. The French used only about eleven of these positions.

The English and French used different flags for signalling. The English flags were: Royal Standard, Union Flag, Red Ensign, red, white, blue, yellow, red and white stripes horizontal, red, white and blue stripes horizontal, yellow and white stripes horizontal, yellow and white stripes oblique, blue and white stripes horizontal, red and white stripes horizontal, white with a red cross, blue with a red cross, and red, white, blue and yellow pendants. The French flags were: red, white, blue, white with a red cross, red-white-red horizontal stripes, white and blue horizontal stripes, white-red-white horizontal stripes, white and red horizontal stripes, blue-white-blue horizontal stripes, white-blue-white horizontal stripes, white and blue horizontal stripes, white and red horizontal stripes, white with a blue border, Dutch Jack, white with a red border, white with a blue cross, blue with a white cross, white and red chequered, white and blue chequered, White, Red and Blue Ensigns, and red, white, blue, white and red, white and blue pendants.

Apart from flags, there was a sharp difference in the way in which signals were treated by the two navies. Tourville had already established the principle of issuing printed *signaux généraux* in addition to instructions. Thumb-indexed signal books with painted flags could be made from these for personal use. In the British service a somewhat similar system had prevailed until Rooke removed the signal indices from the Sailing & Fighting Instructions, leaving the details of the signals embedded in the words of the instructions. As a result, personal signal books made by extracting the actual signals from the instructions together with a few words of 'signification' became practically a necessity, whereas in the French service they were more of a luxury. Luckily for the British, there were few flags and far fewer signals, so that the construction of the books from the thumb-indexing standpoint was not difficult.

Captain of the Fleet

The chief of staff system developed less easily at sea than on land. From quite early times, however, it was recognised that a flag-captain to a commander-in-chief held a very responsible and difficult position. He had to assist the admiral as well as assume responsibility for the ship, which was generally the largest and strongest in the fleet and therefore needed the most supervision. During the Restoration period it

Positions for Signal Flags

Main topmast
Fore topmast
Main topmast shrouds
Main backstay
Mizzen topmast
Fore topsail yardarm
Main topsail yardarm
Mizzen topmast shrouds
Fore yardarm
Main yardarm
Mizzen backstay
Mizzen peak
Fore shrouds
Main shrouds
Mizzen shrouds
Crossjack yardarm
Ensign staff

became customary to appoint an additional captain to the commander-in-chief's ship, to act solely as his chief of staff and without any responsibility for the ship itself. He would usually be a senior man, with the same pay as a rear-admiral while he held the appointment.[11]

By 1731 it was established that this so-called First Captain of the Fleet should 'be esteemed as a Rear-Admiral, and take place at all Councils of War'.

Not all commanders-in-chief took advantage of this facility. Some, perhaps, felt no need for it; others did not qualify because of the size of their fleets. Nevertheless, there must have been many fruitful collaborations between commanders-in-chief and their respective chiefs of staff, of which the Curtis-Howe and Douglas-Rodney combinations are amongst the best known. Kempenfelt's letters to Admiral Sir Charles Middleton give a vivid description of the duties and trials of being a First Captain to Admiral Sir Charles Hardy and Admiral Sir Francis Geary, good officers in their time but worn out by age and unaccustomed responsibility.

In 1747 it was laid down that a command of fifteen British ships-of-the-line, or twenty partly British, was the minimum qualification for the appointment of a First Captain.[12] In 1795 this figure of fifteen British ships was repeated, nothing being said about combined fleets with allies.[13]

The main tactical duty of a First Captain seems to have been to help the commander-in-chief draw up orders of sailing and battle, and draft operational instructions. Sir Robert Calder seems to have supplied Sir John Jervis with important tactical memoranda but the full extent of their collaboration is not clear. The First Captain was also concerned with signals, though in special cases, such as that of

11 This arrangement dates from 1668 at the latest. It was confirmed by Order in Council, 26 June 1674. Sir Clowdisley Shovell was given power to appoint a second captain (to take charge of the ship) in case he had a prospect of engaging the enemy. Papers of Thomas Corbett, Admiralty; notes by Admiral Sir Edmond Slade for Sir Julian Corbett.
12 Order in Council, 22 March 1747
13 Order in Council, 6 November 1795

John McArthur, the organisation of the signal book and the drafting of the new signals might be done by the admiral's secretary. Nelson never had a First Captain; until Trafalgar, he did not qualify for one as he was not a commander-in-chief. At Copenhagen (1801) he used Captain Foley, to whose ship he had just shifted his flag, and also Captain Edward Riou, chosen, apparently, on the spur of the moment, as he and Nelson had not met before. On the morning of Trafalgar, Nelson gave certain staff duties to Captain Blackwood of the frigate *Euryalus*, including the power to signal to ships in the rearmost line of battle in the admiral's name.

With the growing complexity of naval operations, the position of First Captain became increasingly important, and his title was changed to that of Captain of the Fleet. By a regulation of 1806, the minimum size a fleet required to qualify was reduced to ten of the line, the officer appointed to be a rear-admiral or senior captain. He was empowered to give orders in the name of the commander-in-chief, though under his direction, to all officers, including flag-officers senior to himself. But he was not 'to alter, without his permission, the temporary position of the fleet, by signal or otherwise unless some evident necessity require it'.

The development of the chief of staff system in the French navy seems to have been on similar lines, and possibly started sooner. By the early eighteenth century it seems to have been customary to appoint a major to all fleet commanders, whether or not they had the titular style of a commander-in-chief. The major, and such aide-majors as might be appointed, were concerned both with guards and with signals, no doubt because of the security link. An aide-major's duties at sea were similar to those of a junior staff officer in the army.

At the beginning of the American War, staff officers were particularly important because of the increased size of the French fleets and the new developments in signalling. Pavillon's appointment as *major d'escadre* to d'Orvilliers in 1778 seems to have been made chiefly for signal purposes. One would like to know exactly what part de Vaugirauld played as *major d'escadre* to be Grasse in influencing French tactics during 1781–2, and what part Buer de La Charulière played as *major d'escadre* to de Guichen in 1780.

CHAPTER ONE:

The Development of Line Tactics in England

LITTLE is known about the tactical systems of the sailing fleets of the leading maritime states between the years 1500 and 1650, though there is useful evidence for the galleys. On the whole, broadside tactics lagged behind contemporary developments in shipbuilding and gun mounting. Despite the obvious implications of broadside fire, fleets still seem to have been organised to fight in line abreast, as if their main offensive power lay in the bows of their ships, and their main object to promote single-ship encounters in which grappling and boarding could still be decisive. The English, however, developed a practice of avoiding close quarters where Spanish infantry could get at them. To do so, they had inevitably to develop tactics to make best use of their artillery.

Philip II of Spain seems to have been well informed about English tactics. He wrote to Medina Sidonia, commander in chief of the Armada of 1588:

You should take special note, however, that the enemy's aim

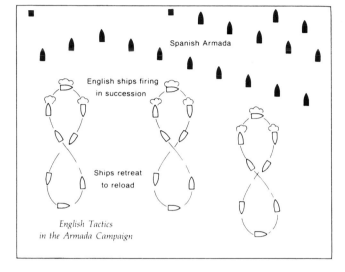

English Tactics in the Armada Campaign

1 Christopher Lloyd, editor, *The Naval Miscellany*, vol IV, London, 1952, p16

2 Julian S Corbett, editor, *Fighting Instructions*, London 1905 (hereafter cited: *Fighting Instructions*) p42

Sir Edward Cecil, Viscount Wimbledon (1572–1638); engraving by Simon Pasaeus. BM

will be to fight from a distance, since he has the advantage of superior artillery and of the large number of fireworks with which he will come provided; while ours must be to attack, and to come to grips with the enemy at close quarters, and to succeed in doing this you will need to exert every effort. That you might be forewarned, you will receive a detailed report of the way in which the enemy arranges his artillery so as to be able to aim the broadsides low in the hull and so sink his opponents' ships.[1]

According to two sets of orders issued, respectively, by or under the names of Sir Walter Raleigh and his cousin, Sir William Gorges, and most probably reflecting Armada or even earlier practice, English ships were to attempt to get to windward of the enemy and to follow one of the flagships 'or other leading ship within musket shot of the enemy'.[2] The leading ship, having fired her broadside into the 'windermost ship or ships of the enemy', was then to tack, followed in succession by the rest of the squadron or *ad hoc* group. As the sternmost English ship tacked, the leading ship would appear from the windward on the opposite tack to batter the enemy once more, forcing them 'to bear up and so entangle them and drive them foul one of another to their utter confusion'. This evolution would then be repeated. Accounts of the Armada operations, vague and piecemeal as they are, suggest that this is what happened on several occasions, the English squadron attacks being in rudimentary line-ahead formation, moving in a succession of figures of eight.

The only other tactical lesson learnt from the Armada operation was the use of eight fireships by the English. True, they were completely unsuccessful as a means of directly damaging the Spanish ships, but they certainly provoked confusion and even panic in the Spanish fleet and led to a precipitate flight from Calais and a loss of cohesion and control which was never properly recovered.

In 1625 Sir Edward Cecil, Viscount Wimbledon, commander-in-chief of the English fleet sent to capture Cadiz, issued sets of instructions with a strongly professional flavour. He and all his flag and squadron commanders were land officers, mostly drawn from the regiments serving in the Low Countries against Spain. He repeated the instructions of Gorges and Raleigh for the attack from windward, but combined these with orders for the whole fleet to 'follow in their several places, the admirals with the head of the enemy, the vice-admirals with the body, and the rear-admirals with the sternmost ships of the chase' (Article 35). This put the admiral in the van and not the centre. Another set of instructions prescribed detailed tactical arrangements. The

fleet was to be organised in three squadrons each of nine ships, each squadron being again sub-divided into three divisions, each of three ships. Three extra ships were to be in reserve. The allied Dutch squadron also taking part in the expedition was to be posted 'on the starboard side of the admiral [the post of honour] and observe their own order and method of fighting' (Article 2). The vice-admiral's squadron was to be to larboard of the admiral, the post of lesser honour, while the rear-admiral's squadron was to act as a general fleet reserve. Nothing was specifically stated about

English ships, mostly hired merchantmen, on the expedition to the Ile de Ré in 1627. NMM

Wimbledon's Fleet Disposition. 1625

Rear-admiral	Commander-in-chief	Vice-admiral

Reserve

the method of fighting except that the nine ships of each squadron 'should discharge and fall off three and three, as they are filed in this list' (Article 1). Should the enemy fall into confusion 'our rear-admiral and his squadron with all his divisions should lay hold thereof and prosecute it to effect' (Article 8).

More interesting for the future is Article 10:

That if any ship or ships of the enemy should break out or fly, the admiral of any squadron which should happen to be in the next and most convenient place for that purpose should send out a competent number of the fittest ships of his squadron to chase, assult, or take such ship or ships breaking out; but no ship should undertake such a chase without the command of the admiral, or at leastwise the admiral of his squadron.

This article was clearly intended to take advantage of any weakness shown by the enemy while, at the same time, curbing irregular pursuit by individual captains seeking plunder.[3]

When these instructions were discussed by Lord Wimbledon's council of war it was contended that the squadron organisation was too difficult to comply with, and it was only accepted on the understanding that it was to be regarded as perfectionist:

It was observed that it intended to enjoin our fleet to advance and fight at sea, much after the manner of an army at land, assigning every ship to a particular division, rank, file, and station: which order and regularity was not only improbable but almost impossible to be observed by so great a fleet in so uncertain a place as the sea. Hereupon some little doubt arose whether or not . . .[4]

In the autumn of 1628, after the murder of the Duke of Buckingham had removed the expedition's most obvious

commander, an attempt was made under the command of Robert Bertie, first Earl of Lindsey, to relieve the blockaded Huguenots in La Rochelle. His instructions repeat Article 10 of Lord Wimbledon's third set, about pursuit of the enemy, and add another: 'If the enemy be entangled among themselves, or forced to bear up, that an advantage may be had, then shall the rear-admiral with his squadron lay hold thereof and prosecute it with effect.'[5] The rear-admiral's squadron, being anyhow regarded as a reserve, is thus given the more positive task of exploiting weakness in the enemy's fleet (as Vernon provided for later) rather than that of assisting disabled or hard-pressed ships in his own.[6]

A further set of instructions issued at about this time and certainly after the Ré and La Rochelle expeditions which it mentions, provided for the vice-admiral's squadron to attack the enemy from windward, followed by the admiral's squadron[7] The rear-admiral's squadron was to act in support of any ships which were hard pressed or disabled. Ships were to maintain sufficient distance from each other as the situation required. If the rear-admiral happened to be nearest to the enemy and to windward of them he might begin the fight, unless ordered not to by the admiral.

If any ship or ships of the enemy shall break out and stand away then the Admiral of any squadron which shall be next them, is to give chase with all his force if it be needful and send out a sufficient number of his fittest ships to take them, but never to undertake such chase without command of the Admiral or at the least the Admiral of his squadron.

Lord Lindsey's instructions for the Ship Money Fleet in 1635 lay down an important though not very progressive principle, namely that the admiral, vice-admiral and rear-admiral should, if possible, engage the enemy flagships opposed to them, private ships being forbidden to do so. Instead they were 'to match themselves accordingly as they can, and to secure one another as cause shall require.'[8] This is interesting, first, because it seems to drop the whole idea of a

Algernon Percy (1602–68), tenth Earl of Northumberland and Lord High Admiral, artist unknown, in the style of Van Dyck. NMM

3 *Fighting Instructions*, pp52–64

4 *Fighting Instructions*, pp63–4

5 Source unknown

6 The great series of engravings by Jacques Callot illustrating the operations give a chaotic impression of contemporary tactics.

7 Unpublished notes from the Leconfield MSS by Sir Julian Corbett

8 Article 18; M Oppenheim, *The Naval Tactics of Sir William Monson in Six Books*, London, 1902–14 (hereafter cited *Monson*), vol iv, p8

The Battle of the Gabbard, 2 June 1653, by Heerman Witmont. NMM

*The English to windward at the
beginning of the St James's Day
battle, 25 July 1666, from the
Instructions for the better
Ordering of the Fleet, c1688.*
MOD

*Admiral Sir Thomas Allin
(1612-85), by Sir Godfrey
Kneller.* NMM

It is extremely difficult to gain any coherent picture of the
tactics of the Battle of St James's Day, 25 July 1666. Quite
early, the van and centre divisions of the English van (White)
squadron began to gain the advantage in this exchange of fire
over the Dutch van, and this advantage was never lost. The
English kept on pressing the Dutch and by evening had won
a complete victory, the two fleets being roughly equal at the
start. Admiral Sir Thomas Allin, commanding the English
van squadron, recalled the action in classical style:

We having the Van our people were very awk [reluctant] to
get into line and some never did, as Day, Sackler [captains] and
some others, but those shot through several of our ships
contrary to a strict order. We fell to fighting between 9 and
ten. There were two [Dutch] Vice Admirals with about 5
ships more, then an Admiral, then 6 ships and a Rear Admiral,
then another Admiral. Sir Tho Teddeman [Vice-Admiral of
the White and van squadron] fought bravely upon his party,
although the *St George* and *Anne* did him no service and the
Old James did us a little. The *Rich[ard] and Martha* went away

Calendar of State Papers Domestic, vol CLXIV, no 72, with the same date
and endorsed in a covering letter to Charles II from Rupert and Albemarle
dated 23 July 1666 (letter of 1 April and another undated to Sir John
Laughton in the Corbett Papers). V Vale discovered yet another copy, also
dated 16 July 1666 in Sir William Coventry's papers at Longleat (*The
Mariner's Mirror,* 38 (1952) no 3).

from us. The Rear-Admiral's division [of the van] did us little help.[51]

After the campaigning season was over and the main fleet laid up for the winter, James issued Sailing Instructions, dated Portsmouth, 16 November 1666.[52] They repeat, almost word for word (substituting 'Admiral' for 'General') the twenty-one Articles of the Commonwealth Instructions, but add four new articles dealing with manoeuvre, and with slipping cables. These instructions were reissued with Mathew Wren's countersignature at some point between 1667 and the spring of 1672, most likely during the preparation for the Third Dutch War early in 1672. They were accompanied by eight 'Additional Sailing Instructions', together with 'Instructions for Masters, Pilots, Ketches, Hoys and Smacks', signed James and countersigned by Wren. There is a signal for tacking in fog, five signals detailing squadrons or individual ships to chase a strange sail,

---

51 R C Anderson, editor, *Journals of Sir Thomas Allin 1660-1768*, London, 1939, vol i, pp277-8. 'Naval Operations in the Latter part of the Year 1666,' W G Perrin, editor, *Naval Miscellany*, vol III, London, 1928, contains an extremely spirited and Homeric account of the battle, pp8-13.

52 *NMM*, DAR/2; no copy of a countersignature

53 *NMM*, DAR/13 pp5-10 (letter book copies). A further reissue appears in the National Maritime Museum, DAR/9, where the original twenty-five sailing instructions and the eight additionals are consolidated in thirty-three articles. They are headed as issued by James, but probably belong to the period of the Third Dutch War.

and a general instruction that when the fleet brings to in the open sea, minor war vessels are to anchor to windward of the admiral.[53] At some stage immediately after the war a slightly less advanced issue of sailing instructions was made in which the Commonwealth issue of 1653 was reduced to fourteen articles. It refers to 'Instructions of Articles of Sailing given out by the Generals of the Fleet in '66'.

The Third Dutch War, 1672-3

Superficially the battles of the Third Dutch War, unlike the sea *Iliad* battles of the Second War, have an appearance of greater tactical coherence. This appearance, however, may be an illusion, arising from two quite different causes. First, official reports and admirals' well-kept journals not only exist but have been discovered and printed. Second, the tactical setting of all four of the battles happened to be along or between coasts and shoals, so each occurred within a navigational framework favourable for tactical analysis.

Throughout the war, England was in alliance with France and a substantial French squadron, amounting to practically a fleet, was present at each battle as part of the bargain. Tactical cohesion, as understood in Nelson's time, was precluded by the fact that the flagships outgunned the weakest ships included in the line of battle by more than three to one, and in weight of metal by a still greater proportion. On the Dutch side it was much the same, though the Dutch fleet was

The burning of HMS Royal James *at the Battle of Sole Bay, 28 May 1672, by Willem Van de Velde the Younger.* NMM

*Battle of Sole Bay –
The Dutch Approach*

*Battle of Sole Bay –
Engagement*

outnumbered by about eighty-two ships to sixty-two. Control of tactics depended, as usual, largely on divisional leadership. Direct orders could be, and indeed were, sent by boat or by smaller warships, but these again were mainly of local application and often miscarried or otherwise failed to obtain the desired conformity. A commander-in-chief in the midst of a smoke-obscured battle could not exercise any overall tactical initiative once the fleets were fully engaged; hence the importance of the instructions issued before contact with the enemy.

James's 'Instructions for the better ordering of His Majesty's fleet in Fighting' of 1672 contained twenty-six Articles.[54] These were little more than a reissue of his earlier fighting instructions of 10 April 1665, consolidated with the additional instructions of 18 April, though nothing is said about how to deal with disabled enemy ships (Article 10 of 1665). Article 6 of the new instructions enjoins the need to close gaps caused by the falling out of disabled ships from the line, and incorporated Rupert's Article 3 (for a flag officer to shift his flag to another ship).[55] The first thirteen Articles, as in the case of the whole issue of 10 April 1665, are largely a reissue of the Commonwealth instructions of 1653.[56] They are immediately followed by the three 'Further Instructions for Fighting' (signed James and countersigned by Wren) which as noted above, were first issued by Rupert and Albemarle on 18 July 1666.

In the first battle, fought on 28 May 1672 in and off Southwold (or Sole) Bay on the coast of Suffolk, the allied fleet was anchored along the coast in the following order, reading from south to north: the White squadron (30 ships) Admiral Abraham Du Quesne, Admiral Jean d'Estrées (in command), Admiral Treillebois des Rabesnières; the Red (28 ships) Spragge, James (now restored as C-in-C), Sir John Harman; the Blue (24 ships) Sir John Kempthorne, Sandwich (in command), Sir Joseph Jordan. When the Dutch were sighted approaching from the north-east there was a very light breeze from ESE or east by south. The allied fleet was anchored not in a straight line but in a slightly concave formation following the coast, with the flagships somewhat to windward of the rest.

At the start of the battle the allies, though not entirely taken by surprise, were not yet prepared for action. Nevertheless, the Blue and the Red squadrons were soon under sail to the north, that is in reverse order with the rear of the fleet leading. D'Estrées, seeing the English getting under sail, at once sent his chief of staff to ask for the Duke's orders for the White. The French were lying a little too close to sand banks to gain the wake of the Red without falling to leeward of the Dutch. D'Estrées, therefore, felt that he could best keep the wind by heading south, that is in the opposite direction to the two English squadrons. He was not disobeying the Duke in doing so, since no tack had been given him. His ships were in some disorder and, before he was properly lined up, the Dutch van of twenty-one ships under Vice-Admiral Adriaen Bankert was down on him from windward. A stiff action between the two squadrons followed and lasted all day, carrying them well to the south of the main battle area. No blame attaches to d'Estrées for separating himself from the Red and the Blue, nor for any subsequent lack of aggression. The French fought well and des Rabesnières, commanding the rear division, was killed.

54 *NMM*, DAR/13

55 *NMM*, DAR/2; printed in *Fighting Instructions*, p130

56 Another copy in the National Maritime Museum, DAR/9, actually incorporates the three Further Instructions of 18 July 1666 at the end, as Articles 27–29 respectively.

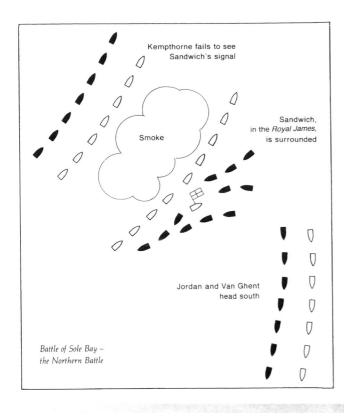

Kempthorne fails to see
Sandwich's signal

Smoke

Sandwich,
in the *Royal James*,
is surrounded

Jordan and Van Ghent
head south

*Battle of Sole Bay –
the Northern Battle*

To the north, things on the whole went ill for the English. De Ruyter brought the Dutch fleet down in line-abreast with a sprinkling of detached ships and fireships ahead. Bankert, seeing the French tacking to the south, turned his squadron into line-ahead on the larboard tack and became detached from his main body. Meanwhile Vice-Admiral Willem Joseph van Ghent with the rear came to the wind on the starboard tack to engage the English Blue squadron. Soon after, de Ruyter with the centre did the same, and engaged the Red. Owing partly to the oblique approach of the Dutch and partly to the slight concavity of the allied line, the English Blue squadron was engaged first. Partly, moreover, because of his more windward position, Lord Sandwich in the *Royal James* met the first severity of the Dutch attack. These facts largely determined the subsequent form of the battle. The Dutch pressed the *Royal James* with all the local strength they could muster. Sandwich sent a message to Jordan with the leading division to tack and weather the Dutch and so return to help him. Jordan had already tacked, his division so far being only lightly engaged; but having successfully weathered the Dutch, he stood on to the south, engaging van Ghent's squadron which was also tacking but keeping away to windward of Sandwich as he passed. Kempthorne came up astern with the Blue rear, passed to leeward of Sandwich and continued to the north, still engaged with the Dutch squadron opposite him at the start of the action. Being unable

The Battle of Sole Bay, 28 May – 7 June 1672, by Willem Van de Velde the Younger. NMM

to see Sandwich's plight because of the smoke, Kempthorne merely sent a boat 'to see what the matter was'.

Meanwhile, James in the *Prince* was hard pressed by the Dutch centre, having been caught somewhat to windward of the rest of the centre division of the Red. Because his flag captain, who was also his chief of staff, was killed, he was forced to shift his flag to the *St Michael*. He was soon heavily engaged once more. Having reached a point only just short of the shoals off Lowestoft, he tacked southward with the ships following him, and soon de Ruyter did the same. As a result both northern squadrons came onto the larboard tack heading south in divisional lines and groups. Jordan still led the English Blue. Van Ghent had been killed and his squadron opposing the Blue was now in some confusion. Astern of them came the English Red, still fighting de Ruyter. The wind meanwhile had veered to south-east, and so carried the fleets straight along the track they had pursued in the morning, the *St Michael* actually passing to leeward of the burning *Royal James*. By the late afternoon, James had been forced to shift his flag again to Spragge's *London*, the fleets continuing to head south in a fierce but tactically confused action. James kept up some kind of communication with the flag officers nearest him, but the general pattern appears still to have been divisional. By now both sides were utterly exhausted and had suffered heavy casualties, so that only a marginal change of fortune was required to bring victory or defeat. Bankert and d'Estrées were now sighted again to the south. De Ruyter judged the moment right to signal a general

withdrawal. Van Ghent was killed, two ships were lost and nearly 2000 killed or wounded. Yet, with an inferior force, he had dealt the allies a very shrewd blow. Their total casualties were also about 2000; the *Royal James* of 100 guns was burnt to the water's edge; and two flag officers, Sandwich and des Rabesnières, were killed.

Sir John Narborough recorded:

His Royal Highness went fore and aft in the ship and cheered up the men to fight, which did encourage them very much. The Duke thought himself never near enough to the enemy, for he was ever calling to the quarter-master which cunded [conned] the ship to luff her nearer, giving me commands to forbear firing till we got up close to them. Between 9 and 10 o'clock Sir John Cox was slain with a great shot, being close by the Duke on the poop. Several gentlemen and others were slain and wounded on the poop and quarterdeck on both sides of the Duke.

Presently when Sir John Cox was slain, I commanded as Captain, observing his Royal Highness's commands in working the ship, striving to get the wind of the enemy. I do absolutely believe no prince upon the whole earth can compare with his Royal Highness in gallant resolution in fighting his enemy, and with so great conduct and knowledge in navigation as never any General understood before him. He is better acquainted in these seas than many Masters which are now in the fleet; he is General, soldier, Pilot, Master, seaman; to say all, he is everything that man can be, and most pleasant when the great shot are thundering about his ears. [57]

Almost exactly the same words were spoken of de Ruyter.

The first Battle of Schooneveldt, 28 May 1673, by Willem Van de Velde the Elder (dated 1684). NMM

57 Narborough's Journal, 28 May 1672; *The Third Dutch War*, pp96–7

The Battles of the Schooneveldt and the Texel

Anglo-French strategy for the year 1673 was simple enough: to defeat the Dutch fleet and land troops behind the Dutch land forces resisting the French. This done, the allies would be able to blockade the southern Dutch ports and capture their returning East India merchant fleet. The allied command had been completely reorganised. Prince Rupert had superseded James as commander-in-chief but was to lead the fleet in the Red squadron; d'Estrées with the White was placed in the centre, apparently because his masters and pilots knew even less about Dutch coastal waters than did the English. Spragge had the Blue in the rear. The Dutch under de Ruyter had Cornelis Tromp in the van and Bankert in the rear.

The Dutch fleet lay to the north of Schooneveldt bank, in an area of open water lying north of Ostend and known as the Schooneveldt anchorage. To describe it as funnel-shaped or of any other particular shape is a little misleading, for whereas it has thin lines of sandbanks running roughly north east and south-west on its north side, its southern edges touch the main system of inshore sandbanks stretching from Flushing to Welcheren. Neither fleet possessed charts based on any kind of trigonometrical surveying. The Dutch charts were the better, but even they were totally unreliable, since they only recorded the traditional knowledge of pilots, shipmasters and fishermen. The French were inexperienced in Channel fighting, and had never fought in Dutch waters at all.

In the first battle, 28 May (OS), each side hoped to profit by the lack of sea room. De Ruyter, in what practically amounted to a defended anchorage, could use his local knowledge to entangle the allies in the shoals and at the same time offset his own disadvantage in numbers. Rupert, though lacking sea room in which to deploy his more numerous fleet to advantage, nevertheless hoped to drive the Dutch onto the Raan shoals by sending forward an advance force drawn from each of his three squadrons and supported by fireships. The wind which had been WSW in the early morning was WNW when the battle began, and was still veering. The allied fleet was in open water in a line abreast ranged roughly north and south, Red, White and Blue and with the Red and Blue somewhat advanced. The fleets were then nine miles apart, the Dutch ranged roughly parallel to the allies, with Tromp, de Ruyter and Bankert in order from north to south. No sooner had firing begun when the allied plan broke down; the advanced squadron proved a failure and the Dutch, instead of retreating towards Flushing, formed a line on the larboard tack. Tromp, to the north, beat off the advanced ships opposed to him and became engaged with Rupert's Red squadron. The French White, now in the centre, engaged a large part of Bankert's Dutch rear, which left the Blue with only the tail end of the Dutch line opposed to it. Meanwhile, de Ruyter, only weakly opposed, tried to push through the allied centre. He was held off by some of the French, who kept to windward. All this time Rupert and Tromp were fighting a detached action to the north. They eventually put about, however, and joined the general action to the south, in which both pairs of opposed squadrons and their respective divisions were becoming mixed. No ships-of-the-line were lost on either side. Fourteen fireships were expended without effect.

The beginning of the second Battle of Schooneveldt, 4 June 1673, from the Instructions for the better Ordering of the Fleet, c1688. MOD

The Second Battle of Schoonveldt was fought on 4 June (OS). The allied fleet was in reversed order, Blue, White and Red from north to south, and the Dutch in the same order as before, Tromp, de Ruyter and Bankert. This time the Dutch had the wind, which was north east. Rupert decided to push ahead but failed to inform d'Estrées, and presently the three Red divisions were ploughing through the middle of the Whites. Meanwhile, Tromp engaged Spragge, though he avoided close action; indeed, it was clearly in the interest of the Dutch not to become so heavily engaged that the allies could use their extra numbers to advantage. At the end of yet another confused and indecisive action in which no ships were lost on either side, the allies retired to the Nore to make repairs and replenish their ammunition. De Ruyter had again scored a tactical success and, in addition, had for the moment broken the allied blockade.

Spragge observed in his journal, on 1st June 1673, that had the Dutch attacked first, the allies would have been caught in disorder for lack of good reconnaissance. 'For which reason and to prevent such surprise you ought always to have your scouts out and your fleet always, if room, to anchor in the line you must sail, and always to sail in the line of battle, if by the wind.' In the case of a large fleet with enough sea room, the squadrons should sail in line abrest one astern of the other, with the divisional vice-admiral and rear-admirals on the starboard and larboard sides respectively.

In the rear or van [division of a particular squadron] I would never have above one ship in the van or the rear of either of the

The second part of the second Battle of the Schoonveldt, 4 June 1673, from the Instructions for the better Ordering of the Fleet, c1688. MOD

58 Spragge's Jurnal; *The Third Dutch War,* p321
59 *The Third Dutch War,* p327

Flags. My reason is, they being usually the most experienced men will better observe the times of tacking the fleet, and also their ships being commonly the best and strongest ships [of the] force will better endure being in the rear [which] is always the easiest cut off.[58]

This statement is of great interest – first, for its emphasis on reconnaissance; second, because it advocates tactical deployment when not necessarily in sight of the enemy; and third, for its reaffirmation that divisional vice-admirals and rear-admirals should be near the extremities of the squadron and not in the centre of their respective divisions. Only when fleets became sufficiently experienced did it seem wise to station the divisional flag officers in the centre of their respective divisions, as in the case of the squadron commander himself. Spragge also recorded in his journal on 21 July 1673,

The Prince [Rupert] placing himself in the van, the French in the middle, the line-of-battle, being of 89 men-of-war and small frigates, fireships and tenders, is so very long that I cannot see any sign the General Admiral makes, being quite contrary to any custom ever used at sea before, and may prove of ill consequence to us. I know not any reason he has for it except being singular and positive.[59]

This was an extremely apt criticism. Besides, if the fleet had to move in reverse order, the commander-in-chief, being then in the rear, would have less opportunity to gauge the tactical situation than if he were in the centre.

The tactics pursued at the Battle of the Texel, 11 August (OS) 1673, the last battle of the war, are even more difficult to elucidate than those of the first two. Both fleets were larger than before, ninety to sixty, but in respective gunpower much the same. Though at dawn the fleets were close in to

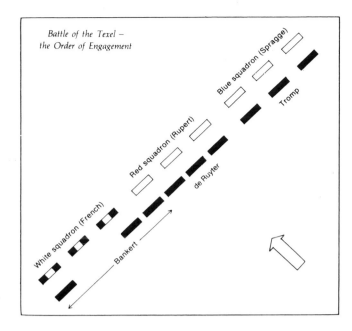

Battle of the Texel – the Order of Engagement

French with Bankert's roughly equal force (the Dutch being in reverse order), and ignoring d'Estrées's in the centre of the White. Thus Bankert's centre and rear, plus the whole of de Ruyter's centre squadron, were able to concentrate on the rear division of the White and the English Red. Roughly speaking, this enabled de Ruyter to bring five weak divisions of Dutch ships against four strong ones. In the rear, Spragge engaged Tromp squadron to squadron. D'Estrées laid great store by getting to windward of Bankert, and was at last successful when the wind gradually veered to south. The French by now were in some confusion, and Bankert was able to drop to leeward to assist de Ruyter against Rupert, these two squadrons having continued fighting together on a south west course regardless of their respective vans. There is no doubt that, with the appearance of a large part of Bankert's squadron, the English Red were hard pressed. Eventually both sides stopped firing and put about, steering north-east towards Spragge and Tromp, de Ruyter being to windward. The French followed well to windward of all the rest. The wind was now south-west, and Spragge and Tromp were still hard at it, disregarding the remainder of their fleets; indeed Spragge hove to quite early on, partly to challenge Tromp's squadron. By the time the remainder of the two fleets reached sufficiently far north to join in the Spragge-Tromp pell-mell, each admiral had had to shift his flag, and Spragge had been killed in the process. Rupert signalled for the whole fleet to form a line on the flagship, but obtained little response. D'Estrées did not obey and possibly misunderstood the signal. Rupert wrote, 'The enemy, when dark came, stood off to their own coast which I had reason to be glad of.' The Dutch had undoubtedly won the battle. They had forced the allies to retire to their base, and so completely broke the blockade of their own ports.

the Dutch coast immediately opposite den Helder and the Texel Channel, the battle was fought in open water. The Dutch squadron commands were as before. In the allied fleet, d'Estrées's White squadron took the van and Rupert with the Red returned to his orthodox station in the centre of the fleet, possibly as a result of Spragge's protest.

When the battle began, the wind was veering from southeast and the fleets were heading south-west by south, with the allied line in good order, the White leading. The Dutch at once showed great skill by engaging the van division of the

The last stage of the second Battle of the Schooneveldt, 4 June 1673, from the Instructions for the better Ordering of the Fleet, c1688. MOD

The first book of instructions

The name of James Duke of York, acting in his capacity as Lord High Admiral, appears as the issuing authority of what

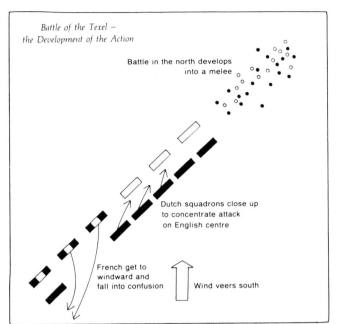

*Battle of the Texel –
the Development of the Action*

Battle in the north develops
into a melee

Dutch squadrons close up
to concentrate attack
on English centre

French get to
windward and
fall into confusion

Wind veers south

*The Battle of the Texel, 11–21
August 1673, by Willem Van de
Velde the Elder.* NMM

must be regarded as the first great tactical book of the Royal Navy.[60] For the first time, the various types of signals and instructions used by the fleet, by day and night, at anchor and at sea, were brought together in consolidated form into a single printed volume. Whose idea this was we do not know, nor do we know its date of publication. James went ashore in September 1672 and never took command of the fleet again. Rupert seems to have been designated his successor even before the Test Act received the Royal Assent on 29 March 1673, though James did not actually resign the office of lord high admiral until 15 June. On 9 July 1673 the office of the admiral was put into the hands of commissioners, with Rupert at their head. Allowing for the time-lag in printing, the book probably appeared somewhere between the early summer of 1672 and the late autumn of 1673.

In detail, it contained nothing very new, and most of the instructions were simple repetitions. It is the form of the book which gives it importance, and the fact that it eventually became the tactical textbook of the Royal Navy in a slightly altered form. It marks the end of instructions written out by hand on odd bits of paper and copied into order books without proper headings or references. When the glorious italic flourishes of the original printing gave way to

60 *NMM*, SIG/A/1 (formerly Sp/78) and *MOD*, NM/104 (formerly *RUSI* 104)

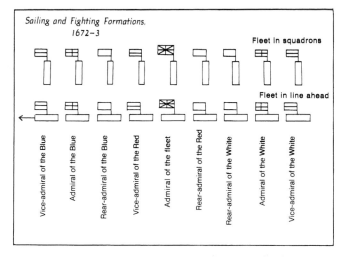

Sailing and Fighting Formations, 1672–3

Fleet in squadrons

Fleet in line ahead

Vice-admiral of the Blue | Admiral of the Blue | Rear-admiral of the Blue | Vice-admiral of the Red | Admiral of the fleet | Rear-admiral of the Red | Rear-admiral of the White | Admiral of the White | Vice-admiral of the White

the more sober roman of the re-issues, the pagination and exact arrangement of words on the pages were carefully preserved. Reprints certainly continued until 1688 and there is some evidence that it was still used officially as late as 1693, that is after Russell's instructions of 1691.

The book begins with twenty-six sailing instructions, which include night and fog signals as well as signals for weighing and anchoring. Then come four instructions for the fireships, signals for the admiral to speak with all of them or with those attached to particular divisions. Fundamentally new is a double-page picture of ships seen from the stern, with spaces for the insertion of names. The format was intended to help scarcely literate seamen recognise where their vessels came in the line of battle. The divisions of the fleet placed the sub-divisions of the vice-admirals of the Blue and White at the respective ends of the line, Blue in the van. The rear-admirals of the Blue and White were placed next to the Red squadron in the centre. This gave the impression that the sailing order intended was always for squadrons abreast, either divisions in single line ahead with the Blue Squadron on the left and White on the right, or with each division in a double column with flagships leading. However, it is unlikely that these rather sophisticated formations, foreshadowing the sailing orders of the late eighteenth century, were used as early as this. Nor, if it was actually in use, is it likely that the order of sailing could have been limited to these formations. The author of the instructions was probably more interested in prescribing the standard order of the ships within each division than in the 'order' in which the fleet was to sail (in the sense of the term in which Tourville and Père Hoste were to use it later). Following four instructions for 'Masters, Pilots, Ketches, Hoys and Smacks attending the Fleet' as auxiliaries, were the twenty-four fighting instructions. The book concludes with four articles of encouragement for the anti-fireship service, which are the same as those of 20 July 1665.

The compiler of these fighting instructions evidently tried to make a selection of the best articles from the Commonwealth onwards. Although the book was issued in the name of the Duke of York, there is no reason to assume that he was personally responsible for the selection, although he must have approved it.[61] The value of the book as a whole was greatly enhanced by separate indices of contents. Each list of contents was by pages, while the signals were indexed under numbers of guns, lights, sails, flags, standards and jacks, with crossheadings: from the admiral, from the chief of any squadron, from other ships. Having brought so many different items together into one volume, the editors evidently felt as much help should be given the users as possible. In fact the whole compilation gives the air of a

standardised codification intended to last for some years. A summary of the contents is given below:

James Duke of York's Instructions of 1672–3

Articles 1 and 2 are those of 1653.

Article 3	is the same as Rupert and Albemarle's no 2 in the instructions of July 1666, 'to divide the enemy's fleet from leeward'.
Article 4	is the same as James's no 3 of 10 April 1665; if to leeward, 'commanders shall endeavour to put themselves in the line close upon a wind'.
Article 5	is no 8 of 1653, to tack or otherwise gain the weather gage.
Article 6	is no 7 of 1653, to provide a signal for use when either to windward or to leeward of the enemy, for the fleet to keep in line with the admiral. An instruction was also included that, if the enemy fled, all the frigates were to try to make captures. The heavy sailers were to be kept in the rear so that they could supply prize crews.
Article 7	is the same as Rupert and Albemarle's first 'Further Instructions' of 18 July 1666, 'to keep the enemy to leeward' when approaching from windward on the opposite tack. The instruction is drafted much more clearly than it was when first issued by Rupert and Albemarle. The only substantive difference is a provision for getting the fleet properly level with the enemy before tacking into battle.
Article 8	is the same as James's no 2 of 10 April 1665. If to windward the leading squadron was 'to steer for the headmast of the enemy's ships'.
Article 9	is the sames as James's no 15 of 22 November 1664, for forming line ahead of the starboard or larboard tack.
Article 10	is the same as James's no 9 and 10 of 10 April 1665, signals for the van or the rear to tack first.
Article 11	is the same as Commonwealth no 9, for flagships to come into the admiral's grain or wake.
Article 12	is the sames as James's no 2 signal of 27 April 1665, for the other squadrons to make more sail.
Article 13	is the same as James's no 3 of 22 November 1664, to engage in the given order of battle.
Article 14	is the same as James's no 4 of 10 April 1665, ships to fight at half a cable distance.
Article 15	is the same as James's no 16 of 22 November 1664, no firing until near enough to do good execution.
Article 16	is the same as James's no 1 of 10 April 1665, ships to keep in one line and 'preserve the order of battle' as prescribed.
Article 17	is the same as James's no 6, no pursuit of small numbers of the enemy until the main body be disabled or on the run.

61 *BM*, Harleian MS 1247 f. 53 (transcript by Sir Julian Corbett); also in the National Maritime Museum, DAR/15; printed in *Fighting Instructions*, pp139–40

Article 18 is the same as no 11 of 1653, no firing at enemy ships 'laid aboard by any of our own ships'.

Article 19: 'The several commanders in the fleet are to take special care, upon pain of death, that they fire not over any of their own ships.'

Article 20 is the same as no 12 of 1653, for small ships to act against enemy fireships.

Article 21 is the same as no 13, for fireships to keep the wind and signal for small frigates to come under the admiral's stern.

Article 22 is the same as James's no 8 of 10 April 1665, disabled ships to be relieved by the sternmost, the rest to press on. Further paragraphs are the same as nos 4, 5 and 6 of 1653 for relieving ships in real danger, etc, and Rupert's no 3 'additional' of 1666, and also included is a further paragraph summarising James's no 6 of 1672, for closing the line if a flagship has to leave it and for allowing an admiral to shift his flag as convenient.

Article 23 is the same as James's no 7 of 10 April 1665, no chasing beyond sight of the admiral.

Article 24 is the same as no 14 of 1653, anchoring at night and the signal to retreat without anchoring.

This book held the field, so far as tactical doctrine was concerned, for less than twenty years. As regards lay-out, however, the book became the model for all subsequent publications of its type until the innovations of Lord Howe, more than a hundred years later. Clumsy and illogical as its arrangements in sections may have seemed even when it was first issued, it inevitably looked clumsier and still less logical as time went on. Yet such was the sense of conservatism in the sea service that it was not until Howe broke completely with the existing system by giving precedence to signals, and making the various instructions no more than 'explanatory of, and relative to' his new signal book, that the Duke of York's system came to an end. Meanwhile, the original book was twice reprinted but without any of the elaborate section inscriptions such as 'James, Duke of York and Albany,' these spaces being instead left blank. Otherwise, with certain

The Duke of York's printed instructions, with an index of signals and double page of ships allowing individual stations to be written in under each picture.
NMM

When ye ADMIRAL would have all ye Ships of ye FLEET to fall into ye Order of Sayling hereunder prescribed, ye RED FLAGG shall bee put on ye Missen-Peek of ye Admiral's Ship; at sight of which, the Vice Admiral & Rere Admiral his own Squadron, as likewise ye Admirals, Vice-Admirals & Rere-Admirals of ye other Squadrons are to Answer it, by putting out their own-coloure FLAGGS in ye same place, and to Sayl in Order, with their whole Divisions.

The BLEW SQUADRON.			The RED SQUADRON.			The WHITE SQUADRON.		
Vice-Admiral	ADMIRAL	Rere-Admiral	Rere-Admiral	ADMIRAL of ye FLEET	Vice-Admiral	Rere-Admiral	ADMIRAL	Vice-Admiral

minute but significant exceptions, the reprints were exactly the same, word for word on each line of each page, though by now the glorious renaissance flourishes of the original italics had given way to more sober roman type.[62]

Narborough's instructions

In 1678 Sir John Narborough issued fighting instructions to his relatively small force at Zante (Cephalonia). He had thirty-two weakly-gunned warships, and was expecting to meet a greatly superior French force. His flag was in a ship of only 52 guns, which was nevertheless the strongest in his force, and this no doubt, apart from other reasons, led him to place himself in the centre of his line of battle. Only five other ships mounted more than 40 guns and these he stationed two at each end of his line, with the fifth in the line on the larboard wing when the fleet was in line abreast. The rest, mostly averaging 10 to 30 guns, were stationed in no particular order of gunpower. Article 2 apparently envisaged a line-abreast formation, and would seem to be the precursor of Tourville's 'order of retreat', with the two wings of the fleet thrown forward, the centre ship being at the angular point

and nearest the pursuing enemy. If engaged with a superior force 'and we see it most convenient to fight before the wind, and the enemy follow us, I would have every commander place his ships in this order of sailing prescribed as followeth, and so continue sailing and fighting, doing his utmost to annoy the enemy, so long as shall be required for the defence

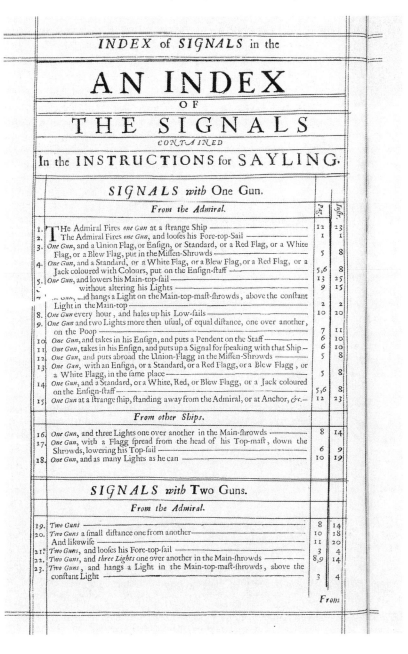

Narborough's Order of Battle

40 gun ships   Flagship (52 guns)   40 gun ships

---

62 *NMM*, SIG/A/5 and 6 (formerly Sp/1, 3); no 5 is marked in pencil '1689' and this may well be the correct date. The form 'His Majesty's' is used in the main part of the text but in the Encouragement 'Their Majesties' is substituted, confirming that it was printed after James's abdication. No 6 is marked in ink '1693', suggesting that it may still have been in use then. R C Anderson connects no 5 with a Navy Board letter ordering 100 copies of the Fighting Instructions to be sent to Lord Torrington at Spithead, dated 12 June 1689 (*The Mariner's Mirror*, 6 (1920), pp130–5).

INDEX of SIGNALS in the

AN INDEX
OF
THE SIGNALS
CONTAINED

In the INSTRUCTIONS for SAYLING.

SIGNALS with One Gun.

From the Admiral.

		Pag.	Inftr.
1.	THe Admiral Fires *one Gun* at a ftrange Ship	12	23
2.	The Admiral Fires *one Gun*, and loofes his Fore-top-Sail	1	1
3.	*One Gun*, and a Union Flag, or Enfign, or Standard, or a Red Flag, or a White Flag, or a Blew Flag, put in the Miffen-Shrowds	5	8
4.	*One Gun*, and a Standard, or a White Flag, or a Blew Flag, or a Red Flag, or a Jack coloured with Colours, put on the Enfign-ftaff	5,6	8
5.	*One Gun*, and lowers his Main-top-fail	13	25
6.	without altering his Lights	9	15
7.	*One Gun*, and hangs a Light on the Main-top-maft-fhrowds, above the conftant Light in the Main-top	2	2
8.	*One Gun* every hour, and hales up his Low-fails	10	20
9.	*One Gun* and two Lights more then ufual, of equal diftance, one over another, on the Poop	7	11
10.	*One Gun*, and takes in his Enfign, and puts a Pendent on the Staff	6	10
11.	*One Gun*, takes in his Enfign, and puts up a Signal for fpeaking with that Ship	6	10
12.	*One Gun*, and puts abroad the Union-Flagg in the Miffen-Shrowds	5	8
13.	*One Gun*, with an Enfign, or a Standard, or a Red Flagg, or a Blew Flagg, or a White Flagg, in the fame place	5	8
14.	*One Gun*, and a Standard, or a White, Red, or Blew Flagg, or a Jack coloured on the Enfign-ftaff	5,6	8
15.	*One Gun* at a ftrange ship, ftanding away from the Admiral, or at Anchor, &c.	12	23

From other Ships.

16.	*One Gun*, and three Lights one over another in the Main-shrowds	8	14
17.	*One Gun*, with a Flagg fpread from the head of his Top-maft, down the Shrowds, lowering his Top-fail	6	9
18.	*One Gun*, and as many Lights as he can	10	19

SIGNALS with Two Guns.

From the Admiral.

19.	*Two Guns*	8	14
20.	*Two Guns* a fmall diftance one from another	10	18
	And likewife	11	20
21.	*Two Guns*, and loofes his Fore-top-fail	3	4
22.	*Two Guns*, and *three Lights* one over another in the Main-shrowds	8,9	14
23.	*Two Guns*, and hangs a Light in the Main-top-maft-fhrowds, above the conftant Light	3	4

From

Pag. 1.

JAMES
Duke of York and Albany,
EARL OF ULSTER,
Lord High ADMIRAL
OF
England, Scotland & Ireland,
CONSTABLE of DOVER-CASTLE,
LORD WARDEN of the CINQUE PORTS,
And GOVERNOR of PORTSMOUTH, &c.

INSTRUCTIONS
FOR

The better Ordering His MAJESTIES FLEET
IN

SAYLING.

INSTR. I.

As soon as the Admiral shall loose his fore-top-sail, and fire a gun, every ship in the fleet is to make ready and weigh, and being come to sail, to follow the Chief of their Squadron.

The Signal for Weighing by Day.

I N-

*George Legge (1648–1691),
first Lord Dartmouth, artist
unknown (English School,
seventeenth century).* NMM

of himself and [the] whole fleet'. In view of the semi-permanency accorded to the Duke of York's instructions, which kept them in use until 1688, Narborough's must be regarded as the earliest known additional fighting instructions. If so, they set the pattern for the future issue of such instructions, valid only for a particular fleet commanded by a particular admiral during the time of his command.[63]

Dartmouth's instructions

In October 1688, James II gave command of a fairly substantial fleet to George Legge, Lord Dartmouth, to prevent a landing by William of Orange. Dartmouth, who might be described as belonging to the third generation of Restoration admirals, had been present at all the battles of the Third Dutch War. On this occasion he issued sailing and fighting instructions which already show a new departure in tactical thought, even before the Revolution had begun.[64]

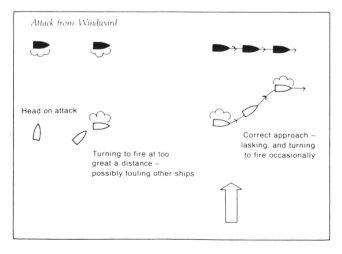

There is little change in the substance of the fighting instructions, but there are important differences in wording. Article 3 seems to suggest that in fighting to leeward, commanders (he does not specify whether this refers to any captain or any divisional flag officer) might divide the enemy's fleet, should they find themselves to windward of a part of it. Article 5, which is supplied with a signal for re-forming the line when to windward of the enemy, indicated that 'each ship or division are not unreasonably to strive for their proper places in the first line of battle given, but they are to form a line, the best that may be with the admiral, and with all the expedition that can be, not regarding what place they fall into or between'. Sir Julian Corbett regarded this as a reaction against formalism, but perhaps it is no more than a repetition of Sandwich's instructions of 29 August 1665, implying that somehow a line of battle must be formed and that difficulty in reaching the precise station must not be an excuse for failing to get into line at all.[65]

Article 6 is an attempt to improve the attack from windward after having approached the enemy on an opposite course and then having tacked when level with them.

They are not to bear down all at once, but to observe the working of the admiral and to bring to as often as he thinks fit, the better to bring his fleet to fight in good order; and at last only to lask away [sailing with a quartering wind] when they come near within shot towards the enemy as much as may be, and not bringing their heads to bear against the enemy's broadsides.

This comparatively sophisticated direction was intended to deal with what had always been, and would become still more so, the bugbear of all admirals attacking from windward. How was it possible to bring a fleet into action simultaneously and in such a way as to avoid ships being crippled while still at too great a distance, thus holding up the whole advance? How, again, was it possible to prevent ships, when coming under fire as they approached, from turning parallel to the enemy to return fire at too great a distance, again holding up the advance and endangering ships on each side of them? These twin difficulties were never satisfactorily resolved and it is clear that with the unwieldy fleets of the later Stuarts, Dartmouth's most praiseworthy notions could only be properly interpreted by squadron commanders and division flag officers on their own initiative. This, however, was not allowed to occur to the prejudice of the battle as a whole, in the sense of allowing it to degenerate into a disconnected series of fights between separate detached squadrons before any general engagement had taken place, as Article 12 makes clear: when the admiral engages or signals to engage,

each division shall take the best advantage they can to engage the enemy, according to such order of battle as shall be given them, and no ship or division whatsoever is upon any pretence to lie by to fight or engage the enemy whereby to endanger parting the main body of the fleet till such time as the whole line be brought to fight by this signal.[66]

63 'Instructions for all Commanders to place their ships for their better fighting and securing the whole fleet if a powerful enemy acts upon us,' *Fighting Instructions*, pp155 7

64 *BM*, Sloane MS 3560, printed in *Fighting Instructions*, pp168 72

65 Pencil note in Corbett's file copy of *Fighting Instructions*, p171

66 Corbett connected this instruction with Spragge's action at the Texel, but is it not equally a commentary on nearly all the battles of all three Dutch Wars? Here Dartmouth directs that the fleet as a whole must engage before squadron or division fights begin. Such fights may occur, and indeed should occur, in a real pell-mell, but the fleet as a whole must engage first.

The Naval Library manuscript

The Naval Library possesses a richly bound manuscript volume purporting to be the Duke of York's sailing and fighting instructions.[67] It is not, however, a copy of the printed book of 1672–3, nor, from internal evidence, can it be dated before the Glorious Revolution of 1688. Dating is difficult because the terms 'His Majesty's' and 'Their Majesty's' fleet are used indiscriminately throughout, and the word 'General' is used interchangeably with 'Admiral'. Its chief interest lies in the observations attached to the fighting instructions and in the watercolour sketches of the battles of the Second and Third Dutch Wars. It contains thirty-three sailing instructions made up of the original twenty-five, consolidated with several insertions and rearrangements, which, with the observations, clearly are of a later date. The observations and sketches are important for their rarity, rather than for what they depict or the observations on them. Indeed, apart from the second-hand tittle-tattle quoted by Pepys, we have practically no contemporary evidence about tactical thought other than that included in this official set of instructions.[68]

The twenty night-sailing signals and three fog signals in this work are separated from the day signals, following the method adopted by Russell's book of 1691 and also undated but official sailing and fighting instructions of a slightly later date. Separate contents lists for the day, night and fog signals were provided, as well as an index of the signals themselves, divided into those for guns, lights and flags, and an abstract of the signals used both in the sailing and fighting instructions. This abstract is arranged in three columns, with painted flags on the left, the place at which each was to be flown in the middle, and the signification on the right. In effect, this list amounts to the first known signal book.[69] The fighting instructions, the encouragement and the fireship signals follow the sailing instructions. These are also provided with a separate contents list for each section, and with indexes of the signals.

The commentator's only real contribution to tactical thought was his advocacy of pressing the enemy's headmost ships, later modified to mean windmost ships. The observation made to Article 7, Rupert and Albemarle's first further instruction of 18 July 1666 for engaging the enemy from windward, was that that part of the enemy's line should be attacked which was most to windward, as this would be the most exposed:

The battle fought in 1666, [Four Day's Battle] the headmost or winderly ships were beaten in three hours and put to run before half the rest of the fleet were engaged. We suffered the like on the 4th of June [Second Schooneveldt] for Tromp and de Ruyter never bore down to engage the whole of the fleet, but pressed the leading shops where Spragge and his squadron were like to have been ruined.

Article 8 reads 'If the enemy stay to fight, His Majesty's fleet having the wind, the headmost squadron of His Majesty's fleet shall steer for the headmost of the enemy's ships for this passage'. The manuscript book substitutes '. . . the windmost ships of the enemy's fleet and endeavour all that can be to force them to leeward', and the commentator observes: 'It may happen that the headmost of their fleet may be the most leewardly, then in such a case you are to follow this instruction'.

The author evidently believed that the enemy's van should be attacked, and from windward. He assumed that, whereas the attackers would always be able to bring down fresh ships to support the leaders, the fleet to leeward would find its headmost and weathermost ships isolated and would not be able to support them with the same ease. This may seem a naive view, later disproven by the tacticians of the American

War, but changes in naval architecture in the intervening century no doubt had a bearing on the capacity of ships to work to windward while fighting their guns. The tactical thinking of the admiralty manuscript does at least represent an advance on the simple alternative of passing, or charging, through gaps between enemy squadrons and back again, or merely engaging the enemy on a parallel course. No doubt, like other tacticians of the time, the commentator was worried by the difficulty of reconciling central and squadron control, as well as by the difficulty of exercising any control at all in the smoke of battle.

Taking the observations and sketches all together, we can see that the author was opposed to any effort to make weathering and doubling the enemy's van, generally from leeward, a frequent and more effective manoeuvre. Article 3, which repeats Rupert and Albemarle's 1666 instruction for dividing the enemy's fleet from leeward, has the observation that this causes disorder and is best omitted unless pressed on a lee shore; ships on either side of the enemy, it was felt, would fire into each other, and furthermore the enemy had only to tack with an equal number and 'then is your fleet divided and not the enemy's'.

Article 5 indicated the signal for gaining the wind of the enemy 'by tacking or otherwise' (1653). The manuscript book added that the ships which gain the wind 'must observe what other signals the general makes', and if they cannot see him must 'press the headmost ships of the enemy all they can, or assist any of ours that are annoyed by them'. In an observation it is noted that,

This signal was wanting in the battle of 11th August 1673 (Texel). The Fourth Squadron followed this instruction and got the wind of the enemy about four in the afternoon, and kept the wind for want of another signal to bear down on the enemy, as Monsieur d'Estrées alleged at the council of war next day. For want of this the enemy left only five or six ships to attend their [the French] motion, and pressed the other squadrons of ours to such a degree they were forced to give way.

Article 6 gave the signal to bear up into the wake or grain of the admiral if he had the wind of the enemy. An additional signal was given for use when any of HM ships had gained wind of the enemy and 'the general or admiral would have them bear down and come to a close fight', thus doubling the enemy. It is noted that 'in the Battle of Sole Bay, Sir Joseph Jordan got the wind and kept it for want of a signal or fireships', the latter presumably to carry a message. Should the admiral be to leeward of the enemy, a signal was provided for ships to leeward of him to come up on a line with him. The instruction adds that if the enemy ran, frigates were to pursue and board them while 'heavy sailers . . . are to keep in a body in the rear of the fleet', and collect and man the prizes.

The 28th May, '73, [it was observed], the [first] Battle fought in the Schooneveld, the rear-admiral of their fleet commanded by Bankart upon a signal from de Ruyter gave way for some time, and being immediately followed by Spragge and his division, it proved only a design to draw us to leeward, and that de Ruyter might have the advantage of weathering us. So that for any small number giving way it is not safe for the like number to go after them, but to press the others which still maintain the fight according to the article following.

67 MOD, Ec/48 and Fighting Instructions, pp139–40, 152–63. It bears the signature of 'James Luttrell 1784.' This was Captain James Luttrell, Surveyor General of Ordnance.

68 Discussed at length in Fighting Instructions, pp120–2

69 See W G Perrin, British Flags, Their Early History, and their Development at Sea; with an Account of the Origin of the Flag as a National Device, Cambridge, 1922, p162.

This is clearly meant as justification for discouraging disorganised pursuit.

The author seems to have been worried by the apparent need for the general to make the determining signal, especially in a case where the commander of the van division could not see his own squadron commander, let alone the commander-in-chief. The first observation, which was by way of an introduction, suggested that the general should have more battle signals to make known his 'pleasure'. It also proposed that 'a person fitly qualified command the reserve who shall by signals make known to the general in what condition or posture the other parts of the fleet are in, he having his station where the whole can best be discovered, and his signals, answering the general's may also be discerned by the rest of the fleet'. Here we have the idea of a repeating frigate, but how could such a ship signal *back* to the general without a signalling code more akin to Admiral Popham's code of the Napoleonic period? And who, moreover, would trust the complete impartiality of such an influential intelligencer? Besides, what is meant by the 'reserve'? That two-way signalling was being recognised as a need is evident from an instruction inserted after the fourth article, which indicated that a close-hauled line of battle should be formed if the enemy had the wind. The additional instruction stated that if any flagship or squadron ahead of the fleet were in a position to weather the enemy, it was to make the signal the general would make for the fleet to tack. Only if the general repeated the signal would it then become an order to the fleet.

Articles 9 and 10 were signals for the 'vice-admiral and the ships of the starboard quarter' to come to the starboard tack, or for the 'rear-admiral and the ships of the larboard quarter', and for the van or rear of the fleet to tack first. The inadequacy of this arrangement had become apparent. The author suggested that there ought to be a separate signal for the different squadrons:

It may happen that by the winds shifting there may be neither van nor rear; then in that case a signal for each squadron would be better understood, so that you may follow the 14th and 15th of the sailing instructions. For in the Battle of August '73 [Texel] the wind shifted and put the whole line out of order.

Accordingly, a new signal was added, to be used 'If the general would have those ships to windward of the enemy to bear down through their line to join the body of the fleet' (Manuscript Article 17).

The note inserted in Article 16 for preserving the order of battle, indicates that the purpose was 'having regard to press the weathermost ships [of the enemy] and relieve such [of ours] as are in distress'.

The observation beside Article 20, describing the duties of commanders of small frigates, ketches and smacks in action against enemy fireships, is that 'The reward of saving a friend [should] be equal to that of destroying an enemy'.[70]

A caution was appended to the provision in Article 22 that a flag officer might shift his flag to any ship in his division if his own were disabled:

In changing ships be as careful as you can not to give the enemy any advantage or knowledge thereof by striking the flag. In case of the death of any flag officer, the flag to be continued aloft till the fight be over, notice to be given to the next commander-in-chief [squadron commander or C-in-C of the fleet], and not to bear out of the line unless in very great danger. It has been observed what very great encouragement the bare shooting of an admiral's flag gives the enemy, but this

may be prevented by taking in all the flags before going to engage. It was the ruin of Spragge in the battle of August '73 [Texel] by taking his flag in his boat, which gave the enemy an opportunity to discover his motive, when at the same [time] he saw three flags flying on board the main topmast-head of three ships which Tromp had quitted.

The first of the accompanying sketches shows two fleets, the one to windward in line-abreast, the other to leeward close-hauled. The author observed that, in 1665 and St James's Day 1666, the English were to windward and pressed the headmost of the Dutch fleet. In 1672 the English lost Sole Bay by being to leeward. On 4 June 1673 the enemy 'came out upon us' and beat the English van 'never minding the rest of the fleet'. 'The same manner the 11th of August we lay to receive the enemy in which we had the same success as ever attended this order of battle.' The second sketch is of the start of the St James's Day battle, with the remark that the English had the wind with which to beat the Dutch. The third is a later stage in the same battle, with the English pressing towards the Dutch in disorder, and Tromp, keeping the Dutch rear well together, weathering some of the English ships and destroying the *Resolution* with a fireship. It is observed: 'This was for not pressing them successively as each ship gave way; for in case of a defeat it is not pursuing the first disabled but to press the next in condition.' The remaining three sketches are of the second battle of the Schooneveldt, 4 June 1673. In the first the Dutch are to windward in line abreast and the allies are close-hauled. The author notes that the French, being in the rear, tacked by order and stood south, and the 'Dutch sent one third of their fleet to attend them'. This appears to be a confusion with the events of Sole Bay. The next depicts a later stage in the battle, with the Dutch pressing the Blue and Red allied squadrons. The observation attached to the last sketch is that Tromp was beaten but nevertheless carried on the fight and was caught to leeward. Spragge, pressing Bankert in the rear, was nearly cut off by de Ruyter. The question is asked that if 'when we have had all the advantage imaginable of [the enemy] and yet do not them the least hurt, whether the same order might not be further improved and kept to'.

A brief reflection on all three of the Dutch wars reveals the huge difficulty of bringing any battle to a decisive conclusion. Though cutting through the enemy's line was held in theory to be dangerous, it happened often and sometimes involuntarily, through the separation of the squadrons from each other. Nor did either side necessarily gain or lose as a result, since the ships which broke through were seldom capable of making an effective concentration against the isolated part of the enemy's line. Cutting the enemy's line could only be really effective if it enabled the cutters to isolate an enemy group weaker then themselves, or at least to place these ships at a tactical disadvantage. A more promising idea was to organise the divisions and squadrons so effectively that they were always ready to cut through and double on an ill-formed line opposed to them, in so far as this action might seem promising. Alternatively, they could present a continuous line of broadside fire to the enemy, and so prevent their own line being dismembered. As long as their own line was preserved intact they were ready for anything and could exploit any situation, compared to a line with gaps in it or with ships not properly aligned with the admiral. Gradually, however, the more defensive aspects of the line theory came to be seen as the most valuable; those who had striven hardest to make the line an effective offensive formation soon came to be regarded as mere formalists, while those who put less stress on warlike order, preferring the old haphazard method of fighting in loose squadrons, came to be regarded as the truly offensive fighters.

---

70 Article 12 in 1653, and no 7 in 1650 to Penn (as numbered in the copy printed in *The First Dutch War*, vol i, p37, and *not* as numbered in *Fighting Instructions*.

HMS Royal Prince *and other vessels at the Four Days' Battle, 1–4 June 1666, by Abraham Storck. NMM*

CHAPTER TWO:

Tactics become a Science

IN THE last years of the seventeenth century, during the wars between England and France under Louis XIV which followed the Glorious Revolution of 1688 in England, naval tactics developed rapidly. Writing in 1780, Kempenfelt observed French progress:

What an extraordinary genius prevailed amongst the French in Louis XIV's reign, to push up everything to the summit of perfection. They had no sooner created a great Navy in that reign (for they never had any before), to the astonishment of all Europe, as at that time they had scarce any commerce at sea . . . than they had the quickness of discernment immediately to see (what we have never been able to see yet) the great advantage that would result in sea fights from a system of naval tactics . . . They judiciously perceived that military tactics might be adapted to naval tactics in the arrangement and evolution of fleets. They set to work about it, and completed it so well as to exhaust the subject, and so were as astonishingly rapid in perfecting the discipline and manoeuvres of a fleet as they were in forming one.[1]

On the English side of the Channel, reform was stimulated by pragmatism and battle experience. During this period

Amiral *Anne-Hilarion de Contentin, Comte de Tourville (1642–1701), by Roger-Viollet.* MM

were published the English 'permanent' Sailing and Fighting Instructions, Tourville's *Signals and Instructions,* and Père Paul Hoste's theoretical study of tactical formations. Tactical practice was developed by the restructuring of the battle line, and the first signal books appeared.

Tourville gave the greatest impetus to the development of tactics and signalling during the wars of the late seventeenth century, mainly because he held the chief command of the French fleet most of the time. Anne-Hilarion de Contentin, Comte de Tourville, was born in 1642 and first distinguished himself against the Turks in 1665. He commanded a ship of the line in d'Estrées' French squadron in the Third Dutch War both in 1672 and 1673, and thereafter commanded a squadron under Du Quesne in the Mediterranean operations against the Dutch and Spaniards.

Tourville did for the French navy, and ultimately for all navies, what *maréchal* Jean Martinet did for the army. His contribution to the science of naval tactics lay chiefly in drilling his very large fleet into a disciplined and controlled force which could deploy from a relatively sophisticated order of sailing to other sailing formations, or into the line of battle. He seems to have been the first naval commander to issue detailed sailing instructions for a large fleet. His order of sailing in six columns was not copied by Howe until ninety years later. He was an adept at fleet organisation, and his aim was to facilitate changes from one order of sailing to another or into line of battle for tactical purposes rather than to create a mere parade-like series of evolutions. His use of printed signals and printed forms for the orders of sailing shows the progress already made by the French in fleet organisation. Similar businesslike methods were not used in the British service for at least another half century. One would imagine also that any fleet commanded by Tourville would be well practised in gunnery as well as in station keeping and seamanship.

As a tactician, however, he does not seem to have striven for anything new, not even to the extent of developing the ideas already initiated by Prince Rupert and the Duke of York. While wisely rejecting the idea that the commander-in-chief should be expected to exercise control over the whole battle, he made no suggestions about squadron initiative. His signal books were in advance of the times in the sense that he exploited the traditional system of 'place were flown' and 'signification' to an unprecedented extent. Apart from all this he was the first of the great French admirals to command a genuinely large fleet with outstanding success in a war against the combined fleets of England and Holland.

Tourville's signed and printed 'Signaux Généraux pour les Vaisseaux de l'Armée du Roy', issued from his flagship the *Conquérant* on 12 May 1689, is a printed book of instructions, in which the signals were somewhat inconveniently

1 Kempenfelt to Middleton, 18 January [1780], printed in *The Letters and Papers of Charles, Lord Barham,* Sir John Knox Laughton, editor, London, 1907 19 (hereafter cited: *Barham*), vol i, pp309 13

expressed in words, embedded in the text of the instructions. However, at the end there was also a separate table of signals. There were twenty general signals by day, seven signals for sailing by day and thirteen by night, thirty-four 'pour le combat' (fighting instructions), eight for chasing, and five for fog. There were only three for calling officers, because Tourville preferred to keep these quite separate from his instructions, and no 'encouragements' for the fireships, their activities being included in the fighting instructions. These, though slightly more elaborate than the contemporary English forms, and certainly better arranged, are equally conservative in character. Seven are instructions only, requiring no signal. The main difference seems to be in the eight signals for chase, which, addressed both to private ships and to the fleet as a whole, suggested possible future developments in tactical initiative. An important manuscript addition on a loose sheet of paper, dated 4 August 1689, stated that when weighing, captains were to proceed ahead to take their station in the line, passing any ship which failed to obey the signal, even a flagship. A separately printed book of night signals, signed and dated from the *Conquérant*, 9 June 1689, required use of guns, lights and fuzées (rockets); they also included compass signals.[2]

Tourville's instructions for 1690, the year of his great victory over the Anglo-Dutch fleet at Beachy Head, have survived in the form of a pocket manuscript signal book. This is probably the oldest pocket size, thumb-indexed, naval signal book in existence. It is thumb-indexed for the positions in which the flags should be flown, not for the flags themselves. Under each position are grouped the various flags flown from it, each hand coloured, and each with its signification and a page and article reference to the instructions. What seems surprising is that in addition to signals made with a flag and a pendant, there were eighty-five signals made with two flags, the second flag hoisted in a different position to the first; no doubt this governs the arrangement of the book.

By noting the page and article reference, it is possible to go some way towards reconstructing the instructions, except for those which required no signal. There were about sixty-seven general instructions, of which five were without signals, covering getting under way, boats, sailing, calling officers to speak with the admiral, making the land, danger, mooring, watering and landing the sick. There were, apparently, two series of sailing instructions, one of about thirty-five articles and the other of about twenty-three, the latter being chiefly concerned with sailing in six columns. There were some nine separate chasing signals. References to the night and fog signals are confused, but there were clearly some thirty-six fighting instructions for which none of the signals actually given are new or specially interesting.

The doubling of the number of instructions included in the 1689 book is mainly accounted for by the increase in the general instructions and the sailing instructions. It is in this latter group that the real significance of Tourville's work is found. For whereas the fighting instructions scarcely mention the enemy, except for engaging and breaking off action, together with doubling (number 8) and boarding (number 33), the sailing instructions indicate elaborate evolutions for a large fleet; hence the instructions for the order of sailing in six as well as in three columns. Evidently Tourville was aiming at the maximum degree of tactical control over his force of more than seventy ships of the line. Wisely, perhaps, he eschewed fighting instructions, being content to ensure a variety of sailing formations and methods of changing from one to another. In this way he could dispose his fleet in the manner best suited to meet the enemy, subject to prevailing conditions of wind and weather, and also the tides, currents and coastal

'Signeaux out'on tire du canon', in Admiral Tourville's 'Signaux ordres de combat, etc' of 1690. NMM

restrictions in the Channel and its western approaches. Flexibility of manoeuvre thus implied quickness and precision in forming the line of battle. Once this was formed and the enemy engaged, little further attempt could be made at detailed tactical control. A fleet of seventy ships in line of battle ahead, even at half a cable interval between ships, would cover a distance of five miles; fighting at a cable's interval, the distance might be nearly ten miles, according to the length of the individual ships. Whatever theoretical assumptions lay behind Tourville's instructions and signals for the year 1690, they were amply justified in practice by the victory off Beachy Head.

Tourville also issued printed 'Ordres et Signaux Généraux', specifically described as being for use, in 1690.[3] These consisted of fourteen printed diagrams of orders of sailing for a fleet of eighty-four ships-of-the-line and thirty fireships, organised in three squadrons. Actually, sixty ships-of-the-line only were shown, with their names inserted in ink. The squadron colours were White, White and Blue, and Blue. In

---

2 *NNM*, HOL/1 and 2 (formerly S(H) 1 and 2). See also HOL/ (formerly S(H)), printed sheets of signals for calling officers, mostly undated.

3 *NMM*, HOL/8 (formerly S(H) 8)

the close-hauled single line ahead, the *Soleil Royal* led the fleet. In three examples of the fleet sailing close-hauled in three parallel columns, each in line ahead, the commanders again led their respective squadrons. Unless otherwise stated, the accustomed arrangement, as in the English fleet, was for the C-in-C's White squadron to be in the centre, the second in command's White and Blue to be to starboard of the C-in-C, and the third-in-command's Blue to be to larboard, thus preserving the traditional order of military precedence. With the fleet in six parallel columns, each squadron in two divisions and the wind astern, the three flagships are shown ahead of and equidistant from the two leading ships of their respective pairs of divisions. Other variants are a wedge formation, close-hauled line of battle ahead with the White in the centre or leading, line of battle abreast with the wind astern, and line of bearing.

Tourville's order of sailing and battle for 1690 contains an unusual item: an 'Ordre de retraitte [sic] sur deux lignes et un front'. Unlike the usual inverted wedge formation, this shows the fleet with the general's squadron in the centre in line abreast, and the other two squadrons each disposed in line ahead directly ahead of the two respective flanking ships of the centre, the formation thus approximating to three sides of a square with the front side open.

The original English 'Sailing & Fighting Instructions'

At some date as yet unknown a new book was officially issued by William and Mary's government with the unadorned title of 'Sailing & Fighting Instrctions' [sic].[4] In general form and layout it looks remarkably like a larger version of the standard folio volume of the eighteenth century, which is indeed what it is. In the main it retains the arrangements of sections first made in the Duke of York's book of 1672–3, but it has a completely new set of fighting instructions. These, moreover, are the instructions which, with only marginal changes, become the official Fighting Instructions of the eighteenth century.

The new book contained nine day, seven night and eight fog signals for weighing and anchoring, and fourteen day and nine night sailing instructions. The day sailing instructions showed a substantial advance on Commonwealth longwindedness, and their separation from the night signals reduced their number. Their form suggests an attempt to exercise more general control over the fleet when at sea, partly by signal and partly by standing orders, thus obviating the ancient, inefficient and time-wasting business of sending ordinary routine orders by boat. Three separate instructions carefully laid down the ceremonial precedence to be accorded to senior captains by juniors, who are not to pass to windward of the latter. There were eighteen separate signals for calling flag officers, captains and officers of various grades on board the flagship, and for the now familiar instructions for the masters and pilots, etc, attending the fleet. The twenty-seven fighting instructions were arranged in an entirely new way, and are the same as those issued, soon after, by Admiral Edward Russell.

Taken as a whole, these instructions tended to concentrate more power in the hands of the admiral, while giving him a wider tactical initiative. Except for Article 17, with its

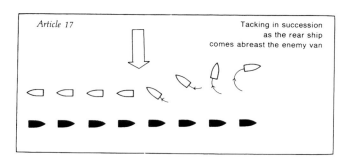

Article 17   Tacking in succession
                                                  as the rear ship
                                      comes abreast the enemy van

notorious later history, the trend was away from the hypothetical cases postulated in earlier instructions and more towards *ad hoc* solutions expressed through a more precise signalling system. The effect of the anonymity of the whole book was to give it greater authority than that of the individual admirals using it. We are no longer dealing with the instructions of the Duke of York for his fleet, or with those of Rupert and Albemarle for theirs, but rather with official sailing and fighting instructions, issued anonymously and not specifically addressed even to His, Her or Their Majesties' Fleet. Even if intended at the start to be no more than a guide book for admirals, it could and did, by entirely natural processes, come to be regarded as an order book for all admirals to obey.

Article 17 of the fighting instructions is a new version of Rupert and Albemarle's further instructions no 1, combined with James's 1672 (OS) no 8, for engaging from windward when approaching on opposite tacks. Here, when the two fleets are level, 'then he that is in the rear of our fleet is to tack first, and every ship one after another, as fast as they can, throughout the line, that they may engage on the same tack with the enemy . . .' Rupert and Albemarle had directed: 'Our whole line is to stand along with them [having first tacked], and same tacks on board [as the enemy]'. James put it more clearly: 'and their [our] whole line is to tack, every ship in his own place and to bear down upon them . . .' The new instructions followed James rather than Rupert and Albemarle in directing an equal number of ships to tack first if the enemy's rear tacks. The importance of this article lies in its uniquely specific direction as to how, under certain given circumstances, the battle was to be fought. Once the instructions had become 'Admiralty', the article became mandatory for any admiral in command of a British fleet in any part of the world. It was in the light of this interpretation of the fighting instructions that Admiral John Byng fought La Galissonière in 1756.

The second printing of the book was different from the first in that it left out the double-page ship plan for the order of sailing. Maybe it was felt that, as these orders were often changed, it was difficult to record them properly on a printed plan. Possibly there is also some connection here with the incorporation of the Dutch in the confederate fleet.

The fact that a copy of the first printing has three further fighting instructions (Articles 28–30) inserted in manuscript, and that these are the same as those added by Admiral Edward Russell to his own printed instructions of 1691, suggests a date somewhere between 1689 and 1691, most probably 1690. We know from Admiral Arthur Herbert, Earl of Torrington's defence of his conduct that at Beachy Head, 30 June 1690, he was still using the Duke of York's book of 1672–3. Russell replaced Torrington in December 1690 and his own instructions of 1691 presumably appeared in time for the summer operations of the following year. It would seem, therefore, that the anonymous Sailing & Fighting Instructions must have been drafted in the light of experience at the battles of Bantry Bay (1689) and Beachy Head (1690), sometime in the late summer or autumn of

4 *NMM*, SIG/A/3 (formerly L52/60); the letter U is omitted in this printing. *NMM*, SIG/A/4 (formerly Sp/2) is a later printing with the word 'Instructions' correctly spelt. R C Anderson makes an interesting comparison between SIG/A/4 and 5 (formerly Sp/1), suggesting the SIG/A/4 was drawn up by Torrington in 1690; *The Mariner's Mirror*, 6 (1920) pp130–35.

1690, and that they remained in force even after Russell's personal adoption of them for the 1691–2 campaign.

How long the book remained authoritative we do not know, but the copy of the first printing in the National Maritime Museum has manuscript insertions showing that it was issued by Admiral Sir George Rooke as late as 12 September 1695, when he was appointed to succeed Russell as commander-in-chief in the Mediterranean. These include the three further fighting instructions mentioned above. Quite probably the original book never completely lost its authority and was the actual parent rather than the ancestor of the eighteenth-century work, Russell's and Rooke's books being only personal and temporary adaptations. The 1690 instructions, printed and issued anonymously under the simple title 'Sailing & Fighting Instructions', are summarised below:

Summary of the 1690 Sailing & Fighting Instructions
Article   1 Signal for the line ahead.
Article   2 Signal for the line abreast.
Article   3 Signal for White squadron to tack and gain the wind of the enemy.
Article   4 Blue squadron ditto.
Article   5 Vice-admirals of the Red, White or Blue ditto.
Article   6 Rear-admirals ditto.
Article   7 If admiral to leeward of fleet or port of it, signal for them to come into his grain or wake.
Article   8 Same as 1653 Article 7, second part.
Article   9 Signal for the van to tack first.
Article 10 Rear ditto.
Article 11 Signal for all flagships to come into the admiral's wake or grain.
Article 12 Signal for the White or Blue to make more sail.
Article 13 Signal to engage.
Article 14 Signal for small frigates to come under admiral's stern.
Article 15 Signal to brace head sails if in line of battle by a wind: ships in rear to brace first.
Article 16 Signal to fill and stand on when with head sails to the mast in line of battle: van ships to fill first.
Article 17 A new version of Rupert and Albemarle's Further Instruction no 1, combined with James's 1672 (OS) no VIII for engaging from to windward when approaching on opposite tacks. The new instructions followed James rather than Rupert and Albemarle in directing an equal number of ships to tack first if the enemy's rear tacked.
Article 18 Restated the simpler position, 'If the admiral and his fleet have the wind of the enemy and they have stretched themselves in a line of battle, the van of the admiral's fleet is to steer with the van of the enemy's and there to engage them.' This replaces Article 8 of James's instructions, which reads: 'If the enemy stay to fight (His Majesty's fleet having the wind) the headmost squadron of His Majesty's fleet shall steer for the headmost of the enemy's ships.'
Article 19 No firing until at point-blank range.
Article 20 No pursuit of small numbers, as in James's additional no 6 of 1665.
Article 21–Signal to be made by ship in distress: Next ship to relieve.
Article 22 Signal that the admiral or another flagship is in distress: fleet to cover him.
Article 23 Next ships to close gap left by ship forced out of the line. This and no 21 have the same sense as James's 1672 (OS) no 12 paras v and vi.

Article 24 If a flagship was disabled, the flag might go on board any ship in his own squadron.
Article 25 Signal for the whole fleet to pursue beaten enemy.
Article 26 Signal for flagship, squadron or division to chase, the same as for tacking to weather the enemy.
Article 27 Signal to stop chasing.

Vice-Amiral *François Louis de Rousselet, Marquis de Chateau-Renault (1637–1716), contemporary engraving.* MM

The Battle of Bantry Bay, 1 May 1689; French and English ships in action during the War of the English Succession, by Adriaen van Diest. NMM

The Battles of Bantry Bay and Beachy Head

On 1 May 1689 the French and English fleets met in fleet action in Bantry Bay for the first time since 1545, except for the inshore operations of 1627–8, when the English were attempting to relieve the Huguenots at La Rochelle. This was the first battle of the War of the English Succession, now merged with the War of the League of Augsburg. Both fleets were supporting their respective land forces in Ireland. Torrington, with a weakly-gunned fleet of eighteen of the line, a 36 and two fireships, sighted a French fleet of twenty-four of the line, also weakly-gunned but with five frigates and ten fireships, under Vice-Admiral François Louis de Rousselet, Marquis de Château-Renault. The French were at a disadvantage, as they were still landing troops and stores when the English were sighted. Torrington was to leeward and when he saw the French at last coming down in good order to attack him, he left the confined waters of the bay so as to form his own line and to try to gain the weather gage. The ensuing battle was a victory for the French despite the fact that Château-Renault was not properly supported by his rear-admiral.[5]

5 Charles de la Roncière, *Historie de la Marine Française*, 6 vols, Paris, 1899–1932 (hereafter cited: de la Roncière), vol vi, pp47–50, where the chief French and English sources are ably summarised.

A Prospect of the Late Engagement at Sea between the English and the French Fleets on Monday the thirtieth of June 1690 the wind from NNE to the ENE with an exact List of ij English Dutch & French Ships Engraven by the Earle of Torringtons Order & approbation

The Battle of Beachy Head, 10 July 1690. NMM

The next main engagement was in the English Channel off Beachy Head on the morning of 30 June 1690. The wind was north-east and the confederate fleet bore down on the French, who could be seen in line of battle on the starboard tack, heading NNW towards the English coast with their head sails aback. Whether by design or not, the French line sagged to leeward in the centre, exactly where the French commander, Tourville, was stationed. He had sixty-eight ships in his line of battle but only sixteen of 70 guns or over. The Dutch, numbering twenty-two, formed the confederate van, but with only six of 70 guns or more. Torrington had the Red, centre squadron, and fifteen of his twenty-one ships carried upwards of 70 guns. Sir Ralph Delavall's Blue squadron, in the rear, consisted of only thirteen ships, of which, however, no less than nine were of 70 guns or more. In each fleet, the squadron commanders were in the centre of their respective squadrons and the division flag officers in the centre of their divisions. Each flag officer, except in the Dutch squadron, was in a major ship but not all were supported by powerful seconds.

The Dutch opened the battle at about 9am by turning into line ahead on the starboard tack to engage the French van on a parallel course. In his defence before the House of Commons, Torrington reported that 'About eight I ordered the signal for battle, to prevent the Dutch steering to the southward, as I did; for by the eighth Article of the Fighting Instructions when that signal is made, the headmost ships of our fleet are to steer away with the headmost ships of the enemy.'[6] This shows that he was still using the Duke of York's instructions of 1672 or one of the later reprints and not the new anonymous official book. Nor was he using Lord Dartmouth's instructions, in which this particular article is numbered 7.

Torrington, fearing that his rear might be overlapped and doubled, avoided steering straight for Tourville, according to traditional precepts, and edged to larboard, so as to engage the rear division of Tourville's centre squadron. This allowed the English Blue squadron to measure its length more favourably with the French rear. His second reason, closely connected with the first, was that, since the French lay somewhat off the wind, their rear ships were their weathermost. If left unengaged, they might have weathered the confederate centre as well as the rear. He also gave as a reason that, being inferior in numbers, he judged it better to leave a gap in the centre rather than be overlapped at his extremities. Vice-Admiral Sir John Ashby, commanding Torrington's van division of the centre, seeing the gap becoming too wide, began to steer to close the Dutch. Torrington, with the remainder of the Red squadron, soon followed him. Meanwhile, Delavall was engaging the centre and rear divisions of the French rear squadron on fairly even terms. When Torrington brought the centre squadron into action he found difficulty in getting close enough because of the sag to leeward in the French line.

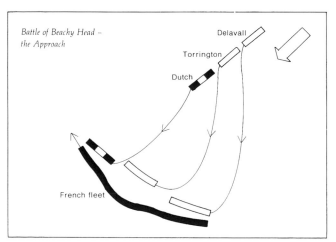

Battle of Beachy Head –
the Approach

Delavall

Torrington

Dutch

French fleet

6 *The Earl of Torrington's Speech to the House of Commons in November, 1690*, (1690) 1710, p32 [NNM, TUN/167]

Meanwhile Tourville had begun to exploit his superior numbers and tactical skill. Finding himself less heavily opposed than he expected, he managed to bring most of his own centre squadron against Ashby. He then pushed ahead to join in the fight against the Dutch, the van of the French fleet having begun to overlap, then weather, and finally double on them. The Dutch were now opposed by the whole of the French van squadron and the van and centre divisions of the French centre. Ashby's division immediately astern of the Dutch were also in a poor way, having several ships towed out of the line. Torrington meanwhile tried to push ahead to come up level with Tourville. By now he could see eleven ships of the French van to windward of the Dutch, and soon likely to weather his own squadron also. Tourville himself, with 'four or five more, [came] pretty near to two disabled ships of the Dutch, and galled them cruelly'.

A French engraving of the Battle of Bévéziers (Beachy Head), 10 July 1690. MM

The crisis of the battle had arrived just as the wind dropped. Damaged ships were towed away by boats, and undamaged ships were towed into better fighting positions. Torrington ordered the nearest damaged ships to anchor and had his own flagship towed between them and the French. The whole confederate fleet then began to anchor with all sail set, thus causing the French to drift away to the south-west on the ebb tide, which had now begun to run strongly. Tourville was not quick enough to appreciate what was happening. When at last he also anchored, the French fleet had drifted out of gunshot to a distance of about three miles.

The battle had lasted eight hours. One small Dutch ship was lost through lack of anchors, but there were no other actual losses on either side, though many Dutch and English ships were severely damaged, and eight or nine of these were subsequently sunk or destroyed.

Russell's additions of 1691

On 23 December 1690, Admiral Edward Russell was named commander-in-chief of the confederate fleet. Sometime in the following year he issued a set of instructions which were an exact reprint of the second printing of the original Sailing & Fighting Instructions, except for three new and one extended fighting instructions.[7] His book is the first to bear an admiral's name and the date on the title page since that of the Duke of York, but Russell himself had no hand in compiling the book as a whole, and there is no clue to the names of those responsible for the changes. The issue of the instructions under Russell's name suggests that they were liable to change when he was superseded, but in fact the National Maritime Museum's earliest copy is endorsed in manuscript as issued by Rooke on 10 June 1702, before setting out for the Cadiz and Vigo operations. Russell by then had retired altogether both from the fleet and the Admiralty, Rooke himself having been appointed an Admiralty commissioner on 20 May.

Summary of Russell's 1691 additions

Article 17 Russell added a second paragraph which provided a signal indicating that the admiral wished the ship that led the fleet in line of battle to make some change of sail. The precise nature of the intended sail change was indicated by the admiral making that change on his own ship. The subordinate flagships were to repeat the signal.

Article 28 A disabled ship may be towed by the ship next ahead, with permission of a flag officer.

Article 29 Signal for a flagship to cut or slip anchor by day, the same as that for tacking and weathering the enemy.

7 'Instructions made by the Right Honourable Edward Russell, Admiral, in the year 1691, for the better ordering of the Fleet in sailing by day and by night and in fighting,' *NMM*, SIG/A/2 (formerly L.58/120). A second printing, also dated 1691, on folio paper but not quite so large and in smaller type, was made at some unknown date. There is also a third printing, in the same type as the second but with slightly different spacing on the title page. There is variation in the references to 'His' 'Her' and 'Their' majesties, but these probably do not help to establish the date. 'His' was sufficient for William III even though Mary exercised the royal power during William's absences in Ireland and the Netherlands.'

Article 30 Signals for the Red, White or Blue suqadrons to form line of battle ahead or abreast, and ditto for vice-admirals and rear-admirals of squadrons.

The Battle of Barfleur

The Battle of Barfleur (19 May 1692) is the first example of a major fleet action in which one side completely outnumbered and outgunned the other. The confederate fleet, commanded by Edward Russell (later Earl of Orford), mustered at least eighty-two ships-of-the-line, and may have included as many as ninety, together with a strong detachment of frigates and fireships. Tourville had only forty-four of the line, including eight mounting between 90 and 104 guns.

In both fleets the order of battle was similar, with commanders-in-chief and squadron commanders in large ships in the centre of their respective squadrons, supported by a pair of powerful seconds. Division flag officers were also in the centre of their respective divisions, also supported by powerful seconds, though in the French fleet this was not possible for lack of large ships. The French squadrons, moreover, were so small, fourteen, fifteen and fourteen respectively, that the division flag officers were very near the extremities of their division lines.

Russell says that by 8am he had formed 'an indifferent line, stretching from the SSW to the NNE the Dutch in the van'.[9] By 9am the French fleet were seen to be formed, and at 10am they bore down to attack. Russell sent orders to the Dutch van to weather the French as soon as possible, and then to tack 'and get to the westward of them', so as to cut them off from Brest. He also ordered the Blue squadron in the rear to close up. A sudden calm, however, stopped this manoeuvre.

Tourville, having the wind, attacked with the best skill he could muster, instructing his van squadron to avoid close action so as to prevent his whole fleet being doubled and completely enveloped. He then brought his centre to close action with the confederate centre, while spacing out his rear squadron with wide intervals between ships so as to hold off the confederate rear. The rearmost French ships kept out of gunshot, presumably on orders, so as to avoid being doubled.

The two centres were heavily engaged for about three hours (times throughout being extremely vague). Tourville in the *Soleil Royal* was towed out of action to windward 'much gauled'. At about 2pm a breeze sprang up roughly north-west by west and, soon after, five ships from the French van ranged themselves between Tourville and the English led by Russell himself and his two seconds. After heavy fighting for about an hour, a thick fog stopped the battle at 4pm. Meanwhile, Sir Clowdisley Shovell, commanding the van division of the centre (Red) squadron, had managed to weather Tourville's centre division of the French centre and double on him. When next seen, Tourville was being towed to the north to avoid being completely enveloped. Russell ordered his own division to head north under tow, and at about 5.30pm, when a light easterly breeze gave him the weather gauge, he 'made the signal for the fleet to chase, sending notice to all the ships about me that the enemy was running'. Shovell's division could be heard firing to westward, though still hidden by fog. Tourville, seeing that the tide was running strongly to the north-east, anchored together with his nearest supporters. When the wind dropped again, Shovell and the ships with him also anchored, but not until they had been carried some distance with the tide, just as had happened to Tourville at Beachy Head.

9 Russell's official letter to Lord Nottingham, *Harleian Miscellany*, vol XII p43

Admiral Edward Russell (1653–1727), first Earl of Orford, by Thomas Gibson. NMM

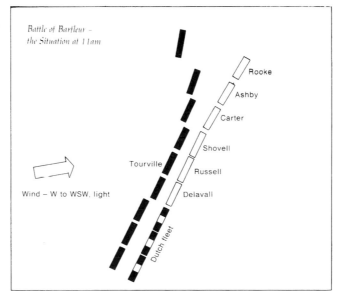

Battle of Barfleur – the Situation at 11am

Rooke
Ashby
Carter
Shovell
Tourville
Russell
Delavall

Wind – W to WSW, light

Dutch fleet

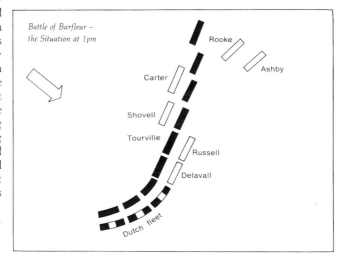

Battle of Barfleur – the Situation at 1pm

Rooke
Ashby
Carter
Shovell
Tourville
Russell
Delavall

Dutch fleet

Fog and smoke made it impossible for either commander-in-chief to direct his fleet. Russell ordered all ships near him to pursue the French to Brest. There was a confused action later in the evening between Rear-Admiral Carter's rear division of the rear (Blue) squadron and some of the retreating French; Carter himself was killed.[10]

Next day the chase continued, but from a tactical point of view the final defeat of the French was no more than a series of widespread piecemeal encounters, with the confederates always attacking and chasing in superior numbers. Nor does the subsequent destruction of part of the French fleet in the Bay of La Hogue offer tactical insight of any significance.

Tourville's signal books, 1691–3

Tourville faced no large-scale tactical encounters in 1691, the year of his famous *campagne du large*, a masterly series of strategic movements at the western entrance to the Channel designed to disrupt Anglo-Dutch trade. His 'Signaux pour le Combat' for 1691, with unnumbered pages, was printed

---

10 The contemporary chart showing 'Les quatre differentes attitudes' of the battle is quite unrealistic as it represents the enveloping of the French fleet in a series of orderly movements; see de la Roncière, vol vi, p112.
11 *NMM*, HOL/3 (formerly S(H)3), dated '1691' on the vellum cover in a contemporary hand
12 *NMM*, HOL/6 (formerly S(H)6)

separately from his general and sailing instructions.[11] This collection of signals contains thirty-six articles and covers much the same ground as those of 1689 and 1690, though the articles are differently numbered and arranged. The only new signals of interest are one for the general in the van wishing to resume his station in the centre (Article 3), and for the van only, or the rear only, to tack in succession (Articles 13 and 14).

The National Maritime Museum has no material for 1692, the year of Tourville's disastrous defeat at Barfleur and later at La Hogue. This may well be due to the destruction of the *Soleil Royal* and also of the *Ambitieux*, to which Tourville shifted his flag. For the year 1693, however, there is plenty of material. Indeed, nothing impresses one more about the resilience of the French navy at this time than the fact that exactly a year after one of the most crushing series of defeats ever inflicted in naval history, Tourville was at sea again with fifty ships-of-the-line.

The year following Barfleur, Tourville issued another signal book containing 229 printed items,[12] by far the most elaborate body of instructions with signals to have appeared up to that time. Unlike all previous and subsequent codes, at least until the end of the Seven Years War, it aimed at covering all the admiral's needs in comprehensive fashion. Even so, administrative signals were omitted, except in the first section of fifty general day signals for use at anchor or under sail. The fourteen signals (of which some were instruc-

The Battle of Barfleur, 19 May 1692, by Richard Paton. NMM

tions only), entitled 'Pour les Ordres de March[e] au plus pres du vent sur une ligne,' covered tacking, wearing and change of distance from the wind. Normally, the general was to lead the fleet, his own squadron, the White, occupying its normal position in the centre. The twenty-seven signals for sailing close-hauled, of which again some were instructions only, allowed highly sophisticated manipulation of the fleet by changing the relative position of the squadrons and forming the line of battle in different orders of columns. Great emphasis was laid on correct station-keeping. Four signals covered the order of sailing with the wind large. The twenty-one signals 'Pour Tous les Ordres de Bataille' when close-hauled, provided for manoeuvring the line and making more or less sail when coming closer to or further from the wind. Twelve more signals dealt with these evolutions with the wind aft. Eleven chasing signals then followed, including the lowering and hoisting of flags to indicate the number of ships sighted.

There were only eight fog signals. The night signals were more elaborate; getting under sail, general and special, when in line of battle or otherwise; sailing, tacking and wearing as a fleet or by squadrons and in three columns; movements in line of battle close-hauled or before the wind. There were fifty-one night signals in all.

The thirty-one 'Signaux de Combat' appear to be tactically limited, cautious and uninspired. However, it would be unwise to dismiss them without knowing more precisely what they were intended to achieve. It is probably more useful to ask what kind of fleet Tourville commanded. Was it a fleet sufficiently well trained both to understand and obey the signals, always assuming that they could be seen by the actual flag officers and captains concerned? If the answer is 'yes', these signals can probably be regarded as sufficient. These signals and instructions are summarised below.

Tourville's 1693 'Signaux de Combat'

Article 1 The ships to fight at half a cable distance, weather permitting. No ship to quit the line unless disabled and then to be assisted by the nearest ship to her. (No signal)

Article 2 Fireships to be stationed sufficiently far on the disengaged side of the line to avoid being disabled but near enough to be ready for action. (No signal)

Article 3 Storeships and merchantmen to keep on the disengaged side. (No signal)

Article 4 Small ships to be ready to stop enemy fireships by boarding them and attacking their boats; such conduct to be rewarded; this was an exactly similar provision to the British 'encouragement'. (No signal)

Article 5 Captains not to open fire except at close range. (No signal)

Article 6 If to leeward, signal for the van to crowd sail and double on the enemy.

Article 7 If to windward, signal for part of the van to double.

Article 8 If to windward, signal by rear-admiral for the rear to double.

Article 9 Signal for squadron commanders to their squadrons to tack or wear together.

Article 10 Signal to general asking approval for Article no 9.

Article 11 Signal for the whole fleet to put about if the van or rear has already done so.

Article 12 Signal by rear-admiral for disengaged ships to come to the wind and form a reserve corps to assist disabled ships.

Article 13 Fireships to perform 'plus grands services' by boarding enemy ships trying to cut the line from leeward, for which they will deserve 'une plus grand recompense.'

Article 14 If the enemy are to windward, the fleet to form a close-hauled line according to the order of battle.

SIGNAUX POUR LE
Combat.

I

Pour mettre l'Armée en bataille au plus prés du vent.

Outes les fois que le Général voudra mettre l'Armée du Roy en bataille au plus-prés du vent, il mettra un petit Pavillon blanc au bout de la vergue d'artimon ; les Vaisseaux portant Pavillons feront le même Sgnal. Le Commandant de l'Escadre blanche & bleüe au milieu de son Escadre se mettra a l'avant garde ; & le Commandant de l'Escadre bleüe au milieu de la sienne a l'arriere garde, les Brûlots seront disposez suivant l'ordre de bataille que l'on a donné.

Un petit pavillon blanc a la vergue d'Artimon,

II

Quand le General voudra prendre l'avant garde.

Quand le Général voudra prendre l'avant garde au milieu de son Escadre, l'Armée étant en bataille au plus prés du vent, il mettra un Pavillon bleu a croix blanche au bout de la vergue d'Artimon, à quoy tous les Vaisseaux portant Pavillons répondront par un Pavillon semblable au même endroit : alors les vaisseaux de l'avant-garde mettront les voilles de l'arriere sur panne, laissant le petit Lunier a porter, & arriveront afin de laisser passer au vent l'Escadre du Général, qui fera force de voille & ira se mettre a l'avant-garde, l'arriere-garde fera aussi force de voille pour gaigner la queüe de l'Escadre blanche & bleüe.

Un pavillon bleu a croix blanche au bout de la vergue d'Artimon.

III

Et lors que l'Armée sera en bataille au plus prés du vent le Général étant a l'avant-garde ou milieu de son Escadre, & qu'il voudra se remettre au corps de bataille au milieu de son Escadre, il en fera le Signal par un Pavillon blanc au bout de la vergue d'Artimon, a quoy les Vaisseaux portant Pavillons répondront par un semblable au même endroit : alors les Vaisseaux de l'Escadre du Général mettront les voilles de l'arriere sur panne laissant le petit Lunier a porter, & feront arriver pour laisser passer

A

Admiral Tourville's 'Signaux pour le combat' of 1691. NMM

L'AUTRE ORDRE DE BATAILLE VENT ARRIERE OU LARGUE.

QUAND le Général mettra le Yac à la Vergne d'artimon, à quoy tous les pavillons & les Vaisseaux qui doivent répéter les signaux, répondrent par le même signal, l'escadre blanche le Général au milieu se mettra au Corps de bataille, & tous les Vaisseaux de son escadre qui seront à sa droite, comme aussi tous les Vaisseaux de l'escadre qui se trouvera à sa droite, feront tous les uns à l'égard des autres, dans l'air de vent qu'il faut être pour se trouver en ligne dans l'Ordre de bataille au plus près du vent, l'amure à stribord, & les Vaisseaux de l'escadre du Général qui seront à sa gauche, & tous les Vaisseaux de l'escadre qui se trouvera à sa gauche, feront les uns à l'égard des autres, dans l'air de vent qu'il faut être pour se trouver en ligne dans l'Ordre de bataille, l'amure à basbord,

One of Admiral Tourville's orders of sailing for 1963.
NMM

Article 15 Signals for ships out of station when to windward or leeward of the enemy.

Article 16 For the fireships to prime and to attack.

Article 17 For calling a particular fireship.

Article 18 For fireships to return to their station.

Article 19 No special ship will be appointed to escort each fireship but any ship whose bow is crossed by a fireship en route for the enemy is to send armed boats to escort her; the two other nearest ships (of the line) to do the same.

Article 20 Fireships to keep a little to windward of their escorts and to lay the enemy aboard at the fore shrouds, and to avoid masking the fire of the escort.

Article 21 Escort to bring to so as to facilitate fireship attack.

Article 22 Further directions for fireships.

Article 23 Signal to engage.

Article 24 If to windward, for van to bear away.

Article 25 For ships following the general to bear away; to board the enemy.

Article 26 To stop the battle.

Article 27 Signal by flag officer, if disabled, for his division to assist.

Article 28 For a ship in danger of sinking to signal for her division to assist.

Article 29 For anchoring in an action lasting until night.

Article 30 For the fleet to close without anchoring.

Article 31 When in the presence of the enemy, for the fleet to form in the same order of battle as on the previous day; and for ships to follow in the same order.

A further batch of documents is an exact repeat of the fourteen printed diagrams of orders of sailings which

Tourville issued in 1690, except that it includes only fifty ships, all with their names printed. By 17 May, however, Tourville had seventy ships, as given in his anchoring list for Brest harbour issued on that day.[13]

The Smyrna convoy action

The attack on the Smyrna convoy, which Charles de la Roncière describes as *le coup de filet*, is part of *la guerre d'escadre* rather than of *la guerre de course*.[14] The tactical details of his immensely important stroke against the economy of the allied states are impossible to establish. At the beginning of June 1693 a huge convoy of some four hundred English, Dutch, German, Danish and Swedish merchantmen left the English Channel for Spanish, Portuguese and Mediterranean ports, escorted by a force of twenty-two English and Dutch warships and eight bombs, fireships and smaller vessels, commanded by Sir George Rooke. The main Anglo-Dutch fleet covered the movement as far as some distance to the south-west of Ushant and then returned to the Channel. French intelligence was good, and Tourville with the Brest fleet gathered at Lagos Bay with twenty ships from Toulon under the younger d'Estrées. Rooke's reconnaissance seems to have been extremely inefficient. He was badly surprised, and in a confused series of piecemeal actions, 17–18 June, lost two Dutch ships-of-the-line and over ninety merchantmen.

Tourville's secret instructions for the attack on the Smyrna convoy are in printed form, 'Fait à bord du *Soleil Royal*, Est & Ouest de Cap de Finisterre le 29 May, 1693'.[15] (This

---

13 *NMM*, HOL./8 (formerly S(H)8), *Ordres et Signaux Generaux de Monsieur le Comte de Tourville*, dated 1693 in pencil on the cover.

14 de la Roncière, *op cit*

15 *NMM*, HOL./8 (formerly S(H)8). This was the new *Soleil Royal* built to replace the one burnt at La Hogue.

suggests the existence of a printing press in the flagship!) Tourville expected two hundred allied and neutral merchantmen to enter the Mediterranean escorted by thirty to thirty-two ships of war. The admiral's ship was expected to be of 70 guns and to have the Union flag at the main. Tourville exhorted his captains: 'Il seroit plus avantageux de prendre cette Flotte que de gagner une bataille ... Les Capitaines ne communiqueront cét ordre qu'aux officiers qu'ils jugeront à propos.'

The next day, 30 May, Tourville issued his *ordre de marche,* fully printed, with the ships' names. His fleet then consisted of sixty-five of the line and twenty-five fireships, but a further printed *ordre de march* in three columns, close-hauled, dated 3 June, gives the names of seventy ships-of-the-line and twenty-five fireships, which shows that Tourville's fleet was at full strength again. His printed order of battle issued on 7 July, immediately after his successful attack, lists eighty-seven ships-of-the-line and thirty fireships, as does his printed set of signals for calling officers, issued on the same day. Even as late in the season as 20 September, his order of sailing and battle included seventy ships, some having already returned to Brest.[16]

Père Paul Hoste

The greatest of all the French tactical theorists was Paul Hoste, born at Brest in 1652. He was trained by the Jesuits and, like so many members of his order, became a leading figure in contemporary mathematics and science. He became Tourville's *aumônier* and was eventually appointed Professor of Mathematics at the Royal Seminary at Toulon, where he died in 1700 at the age of forty-eight. He spent twelve years at sea with d'Estrées and Montemart (Duc de Vienne), and also with Tourville, from whom he learned many tactical ideas. It was on Tourville's suggestion, or command, that Hoste produced the first major work on naval tactics. *L'Art des Armées Navales ou Traité des Evolutions Navales* was published at Lyon in 1697 and dedicated to Louis XIV, who rewarded Hoste generously.[10] Its success was immediate and lasting. It was republished in 1727 in its original form, and fifty years later was still a sound text-book.

Hoste's claim to have described all the chief sea battles since galleys were superseded by *gross vaisseaux* or broadside battleships, is fully sustained by accounts, sometimes first-hand, of all the important actions between the French, Dutch and English navies. In his accounts Hoste is extremely fair in commending the skill of Tromp, de Ruyter, Blake, Prince Rupert and the Duke of York. The book is illustrated with 130 magnificently engraved plates, the ships in each case carrying the sail appropriate for the particular evolution.

In his preface (dated 1696), Hoste declared that without evolutions, fleets were like barbarians who waged war without knowledge and without order, everything depending on caprice and chance. Evolutions provide a framework without which tactical opportunities cannot be seized. *The Art of Evolutions* showed the generals and other officers not only what was necessary but also what was possible.

Hoste's whole system of sailing and battle formations was based on his five *ordres de marche*. These retained their primacy in the French service throughout the age of sail. The basis of these orders of sailing was that each provided a means of forming a close-hauled line of battle, which Hoste strongly favoured. If it was not close-hauled Hoste believed, the windward fleet might lose its position to windward. He also maintained that the leeward fleet should form up close-hauled, because only thus was it able to defend itself effectively and to profit from a sudden change of wind or a mistake by the enemy which enabled it to gain the weather gage.

Hoste's first order of sailing was a close-hauled line of bearing on either tack, one of the most difficult formations both to understand and to execute. The ships were disposed in such a manner that a straight line drawn through their centres in a windward direction would give a close-hauled line. The ships themselves might point in any direction the admiral pleased; it was their bearing from each other which

TRAITÉ
DES EVOLUTIONS
NAVALES.

TROISIE'ME PARTIE,
Rétablir les Ordres quand le vent change.

EXPLICATION DU SUJET.

IL eft fort ordinaire à la mer d'avoir des changemens de vent, & ils font tous capables de mettre le défordre dans une Armée, qui n'eft pas bien difciplinée. Les Ordres étant établis par rapport au vent, ils font troublez quand le vent change, & l'Armée fe trouveroit dans une terrible confufion, fi elle n'avoit pas des régles aifées, pour rétablir l'Ordre que le changement de vent lui a fait perdre. Je fçai qu'en ces rencontres on peut rétablir l'Ordre par les mêmes voyes qu'on le forme, lorfque l'Armée n'en a encore point : mais on voit aifément que ce feroit là une fource de fâcheux accidens ; les Efcadres fe fépareroient, les Vaiffeaux s'aborderoient, toute l'Armée demeureroit un temps infini à fe ranger : au lieu que fi on fuit les régles que nous allons donner, le changement de vent ne dérange nullement l'Armée; chaque Vaiffeau ne laiffe pas de fe trouver dans fon pofte, & un petit mouvement qui fe fait d'une maniere également exacte & imperceptible, remet l'Armée dans l'Ordre qu'elle avoit perdu. Je ne fçai fi je me trompe, quand je me perfuade que le bon Ordre d'une Armée exige que le pofte & la maneuvre de chaque Vaiffeau foient fi exactement

O

Père Paul Hoste SJ (1652–1700): frontispiece of L'Art des Armées Navales, *1697, showing Hoste in his library.* MM

16 *NMM,* HOL/8 (formerly S(H)8). There are two copies of the order of battle, one endorsed on the back *Le Magnifique.* This was Coëtlogen's ship and he was possibly the original owner of the Tourville collection in the National Maritime Museum.

17 *L'Art des Armées Navales ou Traité des Evolutions Navales, qui contient des regles utiles aux officiers généraux, et particuliers d'une Armée Navale; avec des examples itez de ce qui s'est passé de considérable sur la mer depuis cinquante ans. NNM,* TUN/82 Hoste was also the author of *Théorie de la Construction des Vaisseaux,* published at Lyons at the same time, the two works being often bound together in single volume with a composite title page. The latter is a highly scientific work. Unlike the tactics, it has the royal imprimatur, dated 7 November 1696, and also the permission of the Reverend Father Provincial of the Jesuits of the Province of Lyons, dated in Rome, 13 November 1696. In the prefeace Hoste states that he was encouraged to write the book by Tourville. The book concludes with the words 'A plus grande gloire de Dieu'. Evidently ship construction was within the scope of superior religious jurisdiction whereas tactics were not!

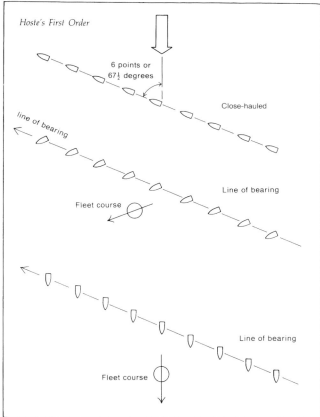

The third order was with the ships in two divisions, ranged along the two sides of an obtuse angle of 135 degrees, exactly bisected by the direction of the wind. The admiral was in the centre, at the angle and therefore at the most leewardly point. When sailing close-hauled in this order the ships of the leading division would already be in line ahead.

The fourth order was for the van, centre and rear of the fleet to be disposed each in two columns with the division commanders between and slightly ahead of the two leading ships in the respective pairs of columns. The centre pair of columns, the admiral's division, would be ahead of the other two, thus reproducing the obtuse angle arrangement of the previous order. Any course before the wind could be steered, and there were variations of the order enabling all three divisions to be level with each other, that is in line abreast as regards divisions.

The fifth order was for three divisions of the fleet in three separate parallel columns in line ahead, close-hauled, with the admiral's column in the centre, led by him. In the case of a large fleet, each division could be in two parallel columns. The formation then came to resemble one of the variations of the fourth order, except that the fleet would be close-hauled instead of before the wind.

A plate from Père Paul Hoste's L'Art Des Armées Navales ou Traité Des Evolutions Navales, Lyon, 1697, showing his third order of sailing, a formation in which the fleet on any point of sailing formed an oblique angle with the flag at the apex and to leeward. NMM

mattered, not their course. The point of the formation was that when the admiral signalled the fleet to come to the wind on the tack in the direction of the imaginary straight line, the ships would automatically find themselves in a close-hauled line ahead, regardless of their previous course. Other lines of bearing could be set by compass direction, as came to be done in the middle of the eighteenth century, but the principle of the relationship was always the same: as long as the bearing was retained, the ships could always be brought into a line ahead on that bearing.

The second order of sailing was what is known as a 'perpendicular of the wind'. This meant that a straight line drawn through the centre of the ships would be at right angles to the direction of the wind. Again the course was immaterial, though generally the second order was used for a fleet with the wind abeam or abaft the beam.

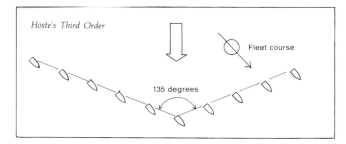

Hoste's system became highly complex when he attempted to show how squadrons or divisions could be changed about when the fleet was in a particular order. The treatment was more theoretical and mathematical, and he felt compelled to defend himself; these evolutions, he claimed, might have little practical application, but it remained valuable for commanders to study them. At the least they would have training value, and at most they could be used seriously to disconcert the enemy.[18] In subsequent chapters he demonstrated more practical manoeuvres to re-establish the line of battle when the wind changed, and to change the order of battle into one of the various orders of sailing, with or without transposing the squadrons. This, in some cases, involved the fleet moving in three columns, which again necessitated a more complicated treatment.[19]

Hoste did not feel competent to give more than examples

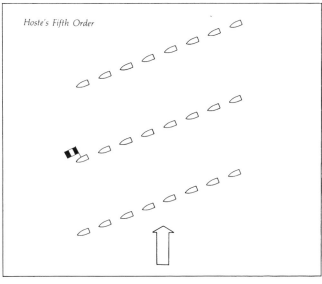

and some general ideas about battle tactics: 'On ne peut presque rien décider dans la théorie: c'est le grand génie d'un Héros avec beaucoup d'expérience, qui en doit faire un grand Général.'[20] He did, however, begin his tactical expositions with a sound practical warning on the dangers of allowing the fleet to be surprised at anchor, with Sole Bay (1672) and Palermo (1676) as examples. He also permitted himself one important generalisation, his first main axiom: if two fleets remained for a long time in sight of each other, either should be able, *régulièrement*, to force an engagement, if it so wished.[21] Later in the book he corrected himself; fleets of equal strength, he admitted, might avoid an engagement (and he described methods of avoiding battle with or without the weather gage), but a fleet of inferior strength caught in the open sea had no recourse but to fight as best it could.[22] Unlike the situation on land, where an inferior force might sometimes occupy a strong defensive position, covered by woods and rivers, an inferior fleet, he argued, was like an inferior army in open country without either defendable ground or the time available for entrenching itself.

The commander of a superior fleet could either detach his best sailers to pin down part of the enemy fleet, and so bring on a general action, or he could afford to divide his fleet into three widely separated squadrons to maximise the chances of catching the enemy fleet. In the summer season, with long days and short nights, forced marches were of little value to a fleet attempting to evade battle, while to try to escape by carrying a press of sail at night only invited dispersion. If the fleet wishing to force a battle were to leeward, and could not win the weather gage, it had only to keep on the same course as the enemy until the wind changed, as Tourville did at the Battle of Beachy Head in 1690. To force a battle from windward was in some respects more difficult, since it was necessary for the attacking fleet to approach level with the enemy fleet and close to them, before actually going down to attack. Otherwise the attacking fleet would lose distance in going down too far before the wind, with the result that the van would fall short of the enemy's van and the enemy would be able to avoid *un combat décicif*.[23] Hoste carefully illustrated navigational methods for ensuring an effective approach.

He summarised the respective advantages of fighting from the windward and leeward positions. The windward position gave a fleet the initiative in approaching, and allowed a superior fleet to double the enemy's rear. It facilitated attack by fireships on disabled enemy ships, attack on enemy ships attempting to escape, and further facilitated attempts to cut off the enemy's van or rear. Finally, the windward position carried smoke and burning matter from the guns towards the enemy, thus avoiding confusion and the danger of fire.

The advantages of the leeward position were that in a strong blow it was safer for the leeward fleet to use its lower-deck guns because the side presented to the enemy would be healed up, clear of the sea. The enemy fleet to windward might not be able to use its lower-deck guns at all without risk of the sea flooding in. The leeward fleet also benefited from the ability of its disabled ships to escape capture by dropping further to leeward, whereas the disabled ships of the windward fleet might not be able to claw to windward away from the enemy. If the admiral of the leeward fleet were obliged to break off action he could signal his entire fleet to

18 Part II, pp104-44
19 Part III, pp145-82; part IV, pp185-330
20 p331
21 p363
22 pp358-63, 375
23 p369

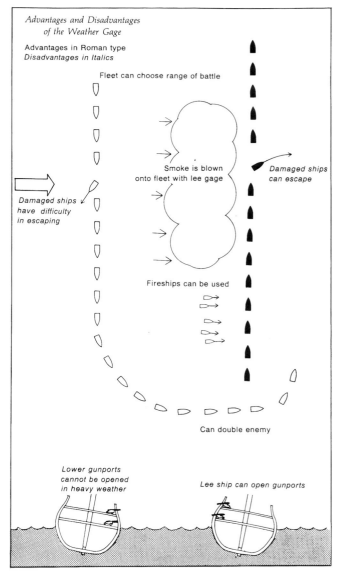

*Advantages and Disadvantages
of the Weather Gage*

Advantages in Roman type
Disadvantages in Italics

Fleet can choose range of battle

Smoke is blown
onto fleet with lee gage

*Damaged ships
can escape*

*Damaged ships
have difficulty
in escaping*

Fireships can be used

Can double enemy

Lower gunports
cannot be opened
in heavy weather

Lee ship can open gunports

bear away. The admiral of the windward fleet could only retreat by breaking through the leeward one, 'Ce qui est d'une tres dangereuse conséquence.'[24] Retreat by the leeward fleet might also be dangerous, but less so in the case of night coming on, the wind freshening, the sea getting up, or the enemy to windward being embarrassed by a convoy.

Hoste's major tactical expedient was doubling. A superior fleet, he advised, should seek to take the enemy between two fires by dividing and coming up on both sides of the end of his line.[25] If the superior fleet were to windward, it could double the enemy more easily; if to leeward, it should still overlap the enemy astern so as to take advantage of a change of wind. The leeward fleet in this position could gradually thin its line so as to build up a mass at the rear for use should the enemy haul its wind.[26]

Many experienced commanders, Hoste noted, preferred doubling the enemy's van, because a disordered van would throw the remainder of the enemy fleet into disorder as it came up. He regarded this notion, however, as erroneous: if

the leading ship were dismasted, the centre and rear could come to her assistance without causing any disorder, and this was easier still when the doubled fleet was to leeward, as the damaged ship would simply drop still further to leeward. Nor could the doubling ships pursue the enemy vessels they had disabled without placing themselves in danger of being cut off. If, however, a fleet were doubled from the rear, their disabled ships could hardly avoid capture either by the enemy's doubling ships or by the rest of the enemy rear. Ships which had undertaken the doubling manoeuvre and were themselves disabled could easily escape capture if they were in the rear, but not if they had doubled the enemy's van. At Barfleur in 1692 Russell succeeded in doubling Tourville's rear simply because he had more ships and so overlapped the French line.[27] This same observation was made by John Clerk of Eldin over eighty years later.

Doubling the rear could only be avoided by preventing the enemy from overlapping the rear. There were five ways of doing this from an inferior position:[28]

1. A fleet to windward should engage the enemy centre and rear only; the enemy van would then be useless. If it tried to put about and take part in the battle it would lose time and be in danger of being cut off from its fleet by the calm which Hoste, in common with many others, believed was the inevitable result of cannon fire. A space might also be left in the centre of the fleet provided care was taken not to allow the van to be cut off. This was the manoeuvre followed by Torrington at Beachy Head in 1690, when the Anglo-Dutch fleet was inferior to Tourville's.
2. A fleet to leeward should leave a wider gap in the centre and allow less overlap to the enemy van, though there had to be ships and fireships ready to stop the enemy breaking through the centre.
3. An alternative was for flag officers to engage their respective opposite numbers in the enemy line (with the same intervals between ships as the enemy). This would mean that gaps were left between the van and centre and between the centre and rear, and thus a number of enemy ships could not engage. The danger here was that the enemy would overlap both the van and rear. The remedy was to station the largest ships at the head and tail of each division, and to ensure that the rear was not overlapped by the enemy rear.
4. A further possibility was for each squadron to attack its opposite squadron in the enemy line, but in such a manner as to leave no enemy ships astern, rather leaving more ships ahead. This amounts to the leeward version of the first tactic.
5. The remaining tactic was to space out the whole fleet evenly along the enemy line, so as to make a line equal in length. This was the least satisfactory method because it allowed the enemy to employ all their ships against the inferior fleet. It might nevertheless prove useful, Hoste suggested, in certain circumstances, for example when the enemy ships were individually weaker. To enable a fleet to get into order to receive an attack from windward, all ships were advised to bear away a little to gain time, and then each captain to attempt to establish his ship abreast the enemy ship indicated to him by the general.

The strongly defensive tone of Hoste's work is obvious. Much attention was given to avoiding an engagement. Defence against doubling was treated at greater length than the art of doubling itself, and seen exclusively in terms of the inferior fleet. Above all, his defensive cast of mind is revealed by the continuous acceptance of the leeward station as a basis for tactical demonstration.

Finally Hoste dealt with breaking the enemy line, a manoeuvre he felt was much used in the Anglo-Dutch Wars. A fleet to leeward seeking to break the enemy line should

24 Part 1, p54
25 p376
26 p376
27 pp381–82
28 pp382–88

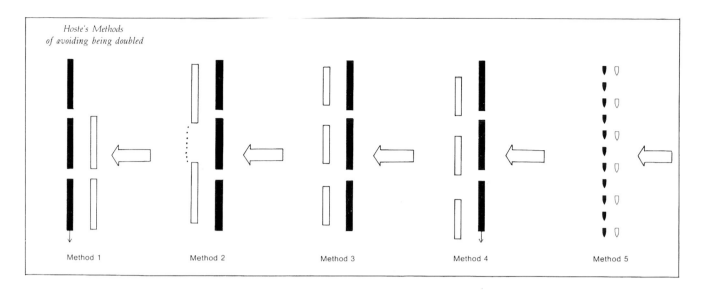

Hoste's Methods of avoiding being doubled

Method 1   Method 2   Method 3   Method 4   Method 5

tack in succession. As the leading ships passed through the enemy line the fleet should tack once more and resume its original course, now to windward of the enemy and roughly parallel. At this point the enemy could be expected to reply in kind, and in this manner the two fleets would pass through each other several times, giving each the chance to cut off, capture and destroy many ships. Hoste argued that this manoeuvre would have to be executed with great technical skill in order to succeed as happily as did d'Estrées in the Battle of the Texel in 1673, in which, according to Hoste, the French squadron passed through the Zealand squadron, weathered it, broke it up, and put the Dutch into such confusion that complete defeat resulted.

Hoste's defence against being cut from leeward was for the windward fleet to tack together as soon as the leeward fleet tacked in succession. If this looked too much like retreat, the windward fleet could wait until the van ships of the leeward fleet were passing through their line, and then tack together, thus putting the leading leeward ships between two fires as they attempted to pass through, while at the same time cutting off those ships which succeeded in doing so.

His view was that there should be little to fear from the enemy's breaking through the line, and it therefore followed that commanders should not attempt the manoeuvre except in special circumstances. Under pressure, an admiral might choose to break the enemy line if, by so doing, he avoided a worse danger. If a wide gap opened in the enemy line, so that a part of the fleet could get into action, it might be in the interest of a commander to steer through the gap. If several of the enemy ships were disabled, the ships opposite them might wish to tack together and pass through the enemy line, followed in succession by the ships astern of them, with the object of cutting off the enemy rear.

It might also be necessary to break through the enemy line to rescue ships which the enemy had cut off. This manoeuvre involved some risk, but several precautions could be taken. The ships cutting through were advised to keep closed up and to carry a press of sail to pass the most dangerous point quickly without troubling to fight the enemy. Once through the enemy line, the ships which had broken through should tack as soon as possible, placing themselves parallel to the enemy, in order to prevent the enemy following on the same tack.

Hoste's 'Projets des Signaux', with an engraved plate of signal flags, is, as he claims, simple, clear and extensive, although it could only be effective in a situation where the flagship had all three masts intact. The system as a whole was, like Tourville's, far in advance of the British equivalent.

It was a true signalling system, and not merely a signalling projection of a series of sailing and fighting instructions. The British were relying on the Royal Standard, Jack, Red Ensign and flags of plain colours including yellow, two striped flags, two part-coloured, three crosses and the distinctive yellow and white stripes (for fireships) for signalling. Hoste preferred to create thirty-six signal flags supplemented by broad pendants, pendants, and wefts. The designs of some of these flags appear to be so much alike as to lead to possible confusion. There are six with crosses, six plain colours with borders, six plain colours pierced with a round blob, and six plain colours with a single fesse-wise stripe. On the other hand, as only three colours were used – red, white and blue – and particular types of signals were restricted to particular

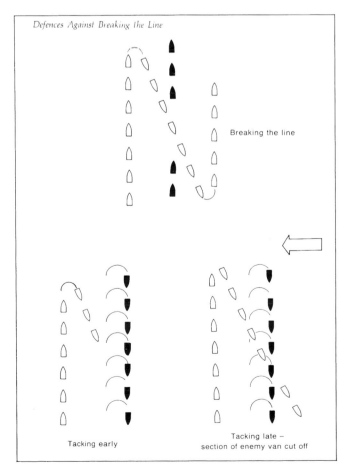

Defences Against Breaking the Line

Breaking the line

Tacking early

Tacking late – section of enemy van cut off

positions recognition would have been straightforward. As Hoste himself claimed, 'On multiple les Pavillons par le mélange des trois couleurs qu'on peut distinguer à la mer: Ce qui fournit un très-grand nombre des Signaux.'

Hoste's first signals were intended to indicate the address to which the message was directed, the whole fleet, a particular squadron, a division, or a single ship. Signals for the various orders of sailing were each indicated by a single flag at the mizzen yard. Those for increasing or decreasing sail and/or for changes of course, etc, were made by a single flag at the main topmast. Battle signals, including to board and to double, were signalled by single flags at the fore topmast. Councils on board the flagship and consultations by signal alone (whether to fight, to put into port, pursue the enemy or retreat), were all indicated by single flags at the ensign staff. Signals for ships to come to the admiral were made by a single pendant at the mizzen yard. Signals to be made to the admiral (whether by squadron or division commanders or by private ships was not specified) were divided into two groups: operational signals were made by a single flag at the jackstaff, and administrative signals by a single pendant at the ensign staff.

Rooke's reissued instructions, 1702

When Rooke reissued Russell's instructions of 1691 in June 1702, William III had only been dead three months. Probably there had not been time to issue new instructions for Queen Anne's navy, and it was the next year before Rooke was able to do so. In effect, his instructions of June 1702 were the final version of the original and anonymous Sailing & Fighting Instructions.[8] There appears to be no further issue of instructions on a personal basis. Eventually Rooke's name was removed and his instructions became the so-called permanent instructions of the eighteenth century.

In so far as Rooke had any hand in it, his intervention was disastrous. Not only does his version follow Russell in leaving out the double-page ship table, which may or may not have been a wise omission, but he also left out all the separate lists of contents and indexes, thus reducing the book to a mere collection of separate chapters. In all the old issues dating from that of 1672–3 to 1691, the indexes enabled the meaning of signals to be read almost at sight. Their removal meant that the signals remained buried in the text of the articles. While it was still possible for the sender, having decided on the article, to discover the signal for it very nearly as quickly as before, the new book was virtually useless for reading signals quickly. The signals therefore either had to be learnt by heart, or each officer had to draw up his own reference book of signals; very soon they began to do just that.

Had Rooke's removal of the indexes stimulated initiative in signalling, his action might not have been so disastrous. On the contrary, however, it had a restricting effect on tactics and tended not only to encourage defensive methods but defensive thinking as well. A book must be judged as a whole: what Rooke handed on was a collection of material which no doubt the Duke of York, Rupert and Albemarle would have recognised as useful but transitory. In detail it

represented the collective experience of the Anglo-Dutch Wars and the War of the English Succession, but as a whole, it turned its back on the slow progress towards efficient fleet control.

'Sailing & Fighting Instructions for Her Majesty's Fleet'

The earliest known copy of the final form of what eventually came to be called the Permanent Sailing & Fighting Instructions was issued in the reign of Queen Anne.[29] The book consists of thirty-one pages divided into ten chapters or sections:

'Signals to be observed at an anchor, in weighing anchor, in sailing & anchoring in the day time', pages 1–7.

'Distinguishing Lights for the Flags', page 8.

'Signals to be observed at an Anchor, in Weighing Anchor, in Sailing and Anchoring in the Night-time', pages 9–12. There are nineteen articles in this section, limited by their form of communication. Up to four lights were used, hoisted in various patterns in various positions in the ship's rigging, or on the poop or ensign staff. Up to eight signal guns were employed for ordinary occasions. Certain emergencies might require 'a great number of lights' and 'gun after gun'.

'Instructions for Sailing in a Fog', pages 13–14. There are eight articles here, mostly listing signals, but containing some instructional material.

'Instructions to be observed by Younger Captains to the Elder', page 15. The four articles in this section mainly deal with punctilios and compliments to be observed when under sail, junior captains giving way to senior.

'Signals for calling the Flag-Officers, and other Officers, on Board the Admiral', pages 16–18.

'Instructions to be observed by all Masters, Pilots, Ketches, Hoys and Smacks, attending the Fleet', pages 19–20. These four articles include the important anti-fireship duties and reward.

'Fighting Instructions', pages 21–27.

'Encouragement for the Captains and Companies of Fire-Ships, Small Frigates, and Ketches', pages 28–9.

'Instructions for the Fire-Ships', pages 30–1.

Once fighting instructions and sailing instructions were bound together and issued as a single book, even though they were paged separately, naval tactics became an expression of all those parts of the book which dealt with the fleet in motion.

However, nearly half the sailing signals were concerned with weighing and anchoring, and there were five signals for various forms of chasing. The admiral had only a very few signals for handling the fleet as a whole when actually cruising; an interesting question is what order of sailing the fleet assumed under ordinary cruising conditions. In most of the major battles of which we have any reasonable account, the rival fleets appear already formed in line of battle ahead or abreast well before the battle began. Where any deployment is mentioned, it is of the simplest kind, from line-abreast, from converging line-ahead or from converging bow and quarter line.

Granted that admirals were skilful at deploying their fleets well in advance, was not this practice, in some ways, disadvantageous, especially if carried out on the previous day? Did not premature deployment make for clumsiness thereafter, quite apart from revealing the fleet's full battle strength and battle area to the enemy? How, moreover, was it supposed that fleets moved between leaving port and encountering the enemy? Did they sail by squadrons, and if so, how? What facilities did the admiral possess, in the way of

---

8 'Instructions for directing & governing Her Majesties Fleet in sailing & fighting by the Right Honourable Sir George Rooke, Kt, Vice-Admiral of England, etc, and Admiral and Commander-in-Chief of Her Majesty's Fleet, etc in the year 1703', BM, Add MS 28126, issued and signed by Rooke, Royal Sovereign, Spithead, 2 May 1703. MOD NM/80 (formerly RUSI 80) is a mutilated copy.

29 MOD, NM/80 (formerly RUSI 80), undated and bound up with Rooke's instructions of 1703

instructions with signals attached, for spreading his fleet over a wide area in search of the enemy, or for keeping it closed up and ready to deploy at any moment? What standard formations had he at his command and how could he change from one of these to another or transpose the relative order or position of his squadrons? To what extent were these orders of sailing also orders of battle and to what extent merely drill exercises, 'evolutions', to improve the skill and promptitude of captains and junior flag officers? These are important questions and the answers to them are central to this study.

Of the thirty-two articles which made up the fighting instructions, only nine, with their respective signals, can be said to have had a truly mandatory character for the admiral, and his subordinate flag-officers and captains.

Article 17: If the Admiral see the Enemies Fleet standing towards him, and he has the wind of them, the Van of the Fleet is to make Sail until they come the Length of the Enemies Rear, and our Rear a-breast of the Enemies Van; then he that is in the Rear of our Fleet is to tack first, and every Ship one after another, as fast as they can throughout the Line; and if the Admiral would have the whole Fleet to tack together, the sooner to put them in a Posture of engaging the Enemy, then he will hoist an Union Flag on the Flag-staves at the Fore and Mizen-top-mast-heads and fire a Gun; and all the Flag-ships in the Fleet are to do the same: But in Case the Enemies Fleet should tack in the Rear, our Fleet is to do the same with an equal Number of Ships; and whilst they are in Fight with the Enemy, To keep within half a Cable's Length one of another; or, if the Weather be bad, according to the discretion of the Commanders.

There can be no doubt that this article is mandatory for an admiral to windward of the enemy and about to pass on an opposite tack. His only discretion lay in choosing the exact moment at which to make the signal to tack and whether to tack in succession, beginning from the rear, or all together. It undoubtedly committed him, if to windward, to a co-linear engagement without chance of variation, except in so far as the two fleets might not be equal in numbers or at the same intervals of distance.

Article 19: If the Admiral and his Fleet have the wind of the Enemy, and they have stretched themselves in a Line of Battle, the Van of the Admiral's Fleet is to steer with the Van of the Enemies, and there to engage them.

This presumably meant that when the fleets were on the same tack an admiral to windward was to force an action by bearing down with his van towards the enemy van, so as to ensure a co-linear engagement.

Article 21: None of the Ships in the Fleet shall pursue any small Number of the Enemies Ships, until the main Body be disabled, or run.

This is a repetition of the Duke of York's instruction.

The remaining mandatory articles are less important. Of the nine mandatory articles, therefore, only two, no 17 and no 19, actually instructed the admiral, in advance, on the conduct of the battle, and even then only under particular circumstances. Article 19, moreover, depended to a considerable extent on the conditions envisaged in Article 17 taking effect. The question remains how far the admiral could arrange matters so that Article 17 did not become applicable. If he was to leeward he was free anyhow, but if he was to windward, was he free to make some other tactical approach, given that the enemy fleet did not make some precipitate move? The answer would seem to be that although he was free enough in theory, the number of signalling options allowed him was insufficient to enable him to deliver a different kind of attack. Apart from signals for tacking (van

or rear) with or without endeavouring 'to gain the wind of the Enemy' and various signals for ensuring the correct formation and regularity of the line of battle, there was nothing in the articles to help him at all. When opposing a markedly inferior force, already in retreat, he could presumably make one or more of the chasing signals given in the sailing signals, combined with the signal for battle, Article 13. In this way he might push home his attack without even forming into line, trusting to his captains to act to the best advantage by exercising individual initiative and providing mutual support. In a full-dress action, however, between roughly equal fleets, with the enemy to leeward and in line ahead, there was little he could do but conform to Article 17.

From all this, it would seem that tactical doctrine favoured the line-ahead formation together with a co-linear engagement with the enemy, mainly because of its defensive value. This being the accepted doctrine, authority, in the shape of the leading flag-officers of the day, took care to keep the system starved of new signals, possibly for fear that some admirals might be tempted to indulge in unprofitable experiments.

The devising and promulgation of additional signals which would increase the admiral's options was always within the competence of the Board of Admiralty or of Prince George of Denmark's Lord High Admiral's Council. We can only conclude that, to the fighting admirals of the period, lack of adequate signals was no more than an inconvenience, and possibly not even that. Torrington, Russell and Rooke, men who had helped put the Instructions into their final form, together with Admiral Sir Clowdisley Shovell, Admiral Sir Stafford Fairborne, Admiral Sir John Leake and Admiral Sir George Byng, who had reached high positions operating them, had all had a hand in Admiralty administration between 1689 and 1714. There is no reason to suppose that the civilian members of the Admiralty Board would have interfered had the admirals wished to extend the signalling system. When in later years the inconvenience became irksome, however, tradition served to put a brake on reform.

There seems reason to suppose that these fighting instructions were regarded mainly as a guide for the admiral, albeit perhaps a strict one. They were of course mandatory for captains and 'inferior officers', since once they were signalled by the admiral they became his instructions. In fact, the authority for these fighting instructions was always derived from a particular admiral, not from the Admiralty. No copy of the instructions as a whole had any authority until it was signed by an admiral and issued to some other admiral or captain. When admirals or captains were tried by court-martial for misconduct or neglect of duty, the charges were always related to the Articles of War rather than to the instructions. Commanders-in-chief, at any rate, seem to have considered themselves entitled to amend the articles of any of the instructions, either by making changes to the printed text, or by issuing separate additional sailing or fighting instructions, or additional signals which, unless cancelled, remained valid in an admiral's own fleet for his own period of command. The fact that a newly appointed admiral was often prepared to take over his predecessor's additional instructions tended to give the more useful instructions a more general acceptance in the service. The earliest known British additional instruction, excluding Narborough's of 1678, date from 1710. The earliest additional signals date from 1711, that is, within a very few years of the issue of the consolidated instructions (of 1703 or later) which for so long remained the accepted pattern. The very fact that changes and additions could be made shows that the system was not as rigid as is frequently claimed. After all, these instructions were drawn up by fighting admirals for fighting admirals and

Battle of Malaga

French attempt to pass between van and centre

French attempt to weather van

were kept in force by a succession of fighting admirals appointed to the Board of Admiralty. With Russell, Leake, Sir George Byng, Admiral Sir John Jennings, Admiral Sir John Norris, Anson, Boscawen, Howe, Admiral Sir Charles Saunders, Keppel and Hawke at the Board between 1709 and 1770, it is unrealistic to argue otherwise.

The Battle of Malaga

At the Battle of Malaga (24 August 1704) two powerful fleets pounded each other for a whole day without either side displaying any tactical initiative and without a single ship being actually sunk in the battle. The action was, however, of immense importance. Strategically it settled the future of Gibraltar, which had just been captured by the confederate fleet, ostensibly in the name of Charles III, the Austrian candidate for the territories of the Spanish Empire. Had the French been able to recapture Gibraltar for the future Philip V as a result of the battle, England would have been without a Mediterranean base and in addition might have failed to capture and hold Minorca.

When the two fleets began to engage off Malaga on the morning of 13 August 1704, each in line ahead, heading south, their paper strength of some fifty of the line apiece made them practically equal. The French ships were comparatively clean, being fresh out from Toulon, but many of the confederate ships had been six months at sea. They were, moreover, very short of ammunition, due to the preliminary bombardment of Gibraltar, which had been far heavier than was actually necessary. They were also short of 800 marines, landed to occupy and garrison the fortress. Furthermore the French were between them and Gibraltar, so that their only advantage lay in having the weather gage with a slight easterly breeze.

Descriptions of the battle are vague, contradictory and highly partisan. There seems no doubt that the French van tried to weather the confederate van, and that later in the day they drew off some distance to leeward. There was also some attempt by the French centre to pass between the confederate van and centre. Shovell was much praised for supporting his own van division early on and for supporting the centre squadron later. Conversely, he was also blamed for not

exploiting the retirement of the French van.

No ships were lost on either side during the actual battle, but a number on each side were towed out of the line disabled; on the confederate side this was in some cases due to lack of ammunition. Both fleets spent the next day making repairs, but, with the wind having changed to west, the Franco-Spanish fleet held the weather gage and maintained its position between the confederates and Gibraltar. At a council of war the confederate flag officers decided to share out their remaining ammunition and try to break through. Rooke's bluff was not put to the test. On the following day, 15 August, the wind was round to the east again and the Franco-Spanish fleet departed for Toulon.

Leake's biographer criticised the French 'manner of firing chiefly wounding the masts and rigging (as if to secure a retreat rather than a victory). Whereas our shot, being levelled at the hulls, must consequently kill many more than they did of the Confederates.'[30] This is one of the earliest mentions of what came to be a standard criticism of French gunnery.

With no major fleet actions for forty years, the legend of Malaga was created. When the War of the Austrian Succession (1739–48) began, no senior officer in the British service except Norris had experience of any major fleet action prior to Malaga. Malaga, it was held, proved the efficacy of the line of battle ahead as the best and indeed the only possible method of engaging. Furthermore, it proved that the maintenance of the line intact was the best and only sure method of avoiding defeat. Victory might come if, in the face of a well formed line, the enemy wilted, but the line must be held intact at the start to avoid defeat. Only when a superior force was faced by an enemy unwilling to engage, as was Sir George Byng fourteen years later, was immediate attack without forming a line the way to victory.

On the French side, Malaga seemed to justify the defensive tactical philosophy expounded by Hoste seven years earlier. Mindful of the fearful defeat of Tourville at Barfleur and the

30 Stephen Martin-Leake, (Geoffrey Callender, editor), *Life of Sir John Leake Rear-Admiral of Great Britain*, London, 1920 (hereafter cited *Leake*), vol i, pp364 5

The Battle of Malaga, 13 August, 1704 (detail) by Isaac Sailmaker. NMM

subsequent holocaust at La Hogue, French naval officers of the eighteenth century saw in Malaga the apparent triumph of defensive methods.

The Battle of Marbella

Whatever influence the legend of Malaga may have had on the thinking of officers in Admiral Thomas Mathews's fleet at Toulon forty years later, the flag officers serving under Rooke must have seen the battle for what it was. None of them was therefore likely to try to apply the tactics of Malaga to an entirely different situation, least of all Leake, who was one of the most vigorous and talented sea officers of the age. In command of the confederate fleet destined for the second relief of Gibraltar, he issued instructions on 4 March 1705, as his fleet was about to meet a French force under Rear-Admiral Jean Desjeans, Baron de Pointis, which was supporting the siege.

Leake had twenty-three English ships, four Dutch and eight Portuguese, and his instructions to them were:

> If they had intelligence that the enemy were in Gibraltar Bay and the wind westerly, they were to sail in a line of battle abreast of each other, after they were past Tangier until they got near the length of Cape Cabarita [facing the extreme end of Gibraltar on the west side of the Bay]; and if the enemy should be at anchor on the west side of the bay [so] they could not weather them, the ship that led with the starboard [larboard?] tacks was, as soon as she got the length of the said cape, to spring her luff and endeavour to get ahead of them, every ship following in a line ahead of each other, till the van got ahead of the innermost [French] ship; and then to make the signal for tacking and to tack accordingly; and as soon as he was about, he was to lay his head sails to the mast, every ship astern doing the same, observing to tack when the ship ahead tacked; and to endeavour all that was possible to annoy the enemy by either laying them athwart the hawse, or otherwise boarding them before they could set their sails; which method was to be observed if they cut or slipped, if any advantage was to be gained before they could form their line of battle.[31]

31 *Leake*, pp257-8

Admiral of the Fleet Sir Clowdisley Shovell (1650–1707), by Michael Dahl. NMM

Leake's problem was how to enter the bay with the west wind which it was assumed would be necessary to carry him through the Straits, and then work to windward of the French who might well be anchored close to the Spanish shore on the west side of the bay.

Four days later, 'Sir John drew the fleet into a line of battle

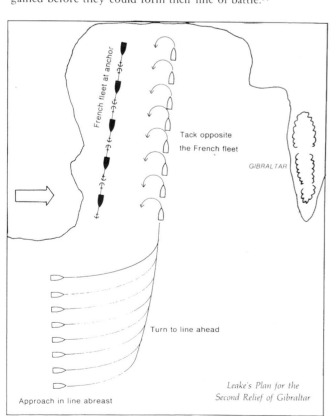

French fleet at anchor

Tack opposite the French fleet

GIBRALTAR

Turn to line ahead

Approach in line abreast

Leake's Plan for the Second Relief of Gibraltar

Vice-Admiral Sir John Leake (1656–1720), by Sir Godfrey Kneller. NMM

to try how the Portuguese would behave, who indeed answered his expectation, he being obliged to bring to for half an hour, to give them time to get into the line; which must have produced fatal consequences had it been to proceed upon action, as he expected very shortly to do.' Luckily for Leake, de Pointis had only five of the line in the Bay of Gibraltar with which to meet him, so that the ensuing Battle of Marbella, fought on 10 March 1705, was a runaway victory with no real tactical features. The interest lies in what Leake was planning to do if faced by a larger force. In neither case could the situation in any way resemble that at Malaga.

Leake's instructions for the projected attack on Cadiz in 1706 include his line of battle for entering the bay with sixteen English and Dutch ships and with his own ship seventh in the line. Each ship's barge was to carry a fire-grapnel to tow off enemy fireships and to be ready to deal with enemy galleys trying to attack confederate fireships. If there was a French squadron in the bay, the fireships were to attack them first. Article no 8 required that 'every ship comes to an anchor the most advantageously he can, for annoying the enemy'.[32] This injunction shows that the doctrine of the line only applied to particular circumstances. As at Marbella, in circumstances different from those at the battle of Malaga, Leake assumed different tactics to be both acceptable and necessary.

The Battle of Cape Passaro

Sir George Byng's defeat of the Spanish Mediterranean fleet in a pell-mell runaway action at Cape Passero on 11 August 1718 is significant less for its tactical features than for its refutation of the idea that the years immediately following the Treaty of Utrecht were years of British tactical decadence. The approach to battle depended on the strategic question of whether or not the Spanish fleet could be treated

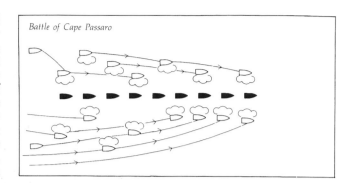

Battle of Cape Passaro

as hostile. Luckily for Byng, the Spaniards opened fire first, largely as a result of the British fleet's threatening demeanour. His instructions ostensibly gave him ample powers, but his position immediately before the battle had nontheless been highly delicate, and even a slight mistake or misunderstanding on the part of his subordinates might have ruined his career. He called no council of war beforehand, being ready to accept full responsibility for whatever happened, on the grounds that 'a commanding officer should only call a Council of War to screen him from what he has no mind to undertake'.[33] He is said to have given all his captains written orders, presumably before fighting began, that they were 'to attack and destroy the Spanish ships'. If so, this seems to assume that an action could and would be provoked.[34]

From the British and Spanish official accounts it seems clear that Byng's force was superior to that of Vice-Admiral Don Antonio Castañeta to the extent of having twenty-one of the line, including a 90 and two 80s, against eighteen much weaker ships-of-the-line and about ten ships which might be classed as frigates. On the previous evening, Byng had ordered his four fastest ships-of-the-line to keep touch with the Spaniards during the night. In the morning he sent a lieutenant to order the leading ship to engage the nearest Spaniard if she opened fire, which she soon did, and she was eventually taken. When the Spanish rear-admiral, the Marques de Mari, became separated from his fleet and led inshore with some six of the line, nine frigates and various galleys, fireships, bomb-vessels and fleet auxiliaries, Byng ordered Captain George Walton in the *Canterbury* to pursue them with eight ships. Meanwhile Rear-Admiral George Camock, an Irish Jacobite and a rear-admiral in the Spanish fleet, who had strongly advised Castañeta against a fleet action, led away with some other ships so that in the end Castañeta was left with only nine of the line against the remainder of Byng's fleet.

Byng appears to have signalled simultaneously for general chase and to engage. As his fleet gradually caught up with the Spaniards, his ships attacked them piecemeal. While some of Byng's ships stayed by their first opponents, others pressed on towards the head of the Spanish line, after delivering broadsides at the Spanish ships astern as they passed. Altogether, six of the line and a frigate were captured, including the Spanish commander-in-chief and another flag-ship. Walton was equally successful, capturing two of the line, one of which was a flagship, and three smaller ships, and burning or causing the Spaniards themselves to burn seven more warships and auxiliaries. Byng, however, was far from satisfied, thinking 'there was great reason to blame the conduct of several captains of the Fleet who instead of following and attacking ships of equal force to them, fell, two

32 *Leake*, i pp322 3

33 J L Cranmer-Byng, editor, *Pattee Byng's Journal*, London, 1950 (hereafter cited: *Pattee Byng's Journal*), p23

34 *Pattee Byng's Journal*, p24

Admiral of the Fleet George Byng (1663–1733), first Viscount Torrington, by Sir Godfrey Kneller. NMM

or three upon one, and so gave to several the opportunity to escape; but as he was crowned with success he would accuse none.'[35] The only captain we can identify for certain as having 'behaved himself very much like an officer and a seaman' was Nicolas Haddock of the *Grafton*. Byng's own second captain in the battle was Richard Lestock.

Byng's dissatisfaction is a tribute to his ardent nature. Aged only fifty-five, he suffered severely from gout, and there had been doubts about his fitness to command the fleet sent to the Baltic the previous summer.[36] However, the Spanish fleet was undoubtedly weaker than the British in every respect, and its commanders seem to have been demoralised from the start, though individual ships fought stoutly. It was this knowledge of superiority, not only in numbers but in fighting organisation and technical skill, and above all in morale, that made Byng look for an even more devastating victory.

This brilliant success belies the idea that during the early years of the eighteenth century the Royal Navy was tactically stagnant. Equally important, what was the source of Byng's tactic of using some ships to attack the enemy rear while other ships passed ahead of them to engage the van? This is exactly the manoeuvre for which Corbett later gave Hawke such high praise. If Byng had been faced by a well-ordered French fleet of equal strength instead of an inferior Spanish one, he would of course have fought the battle differently. But, as Laughton aptly notes, the completeness of his victory, as in the case of Hawke's at Quiberon, masked both his capacity and his determination to exploit a tactical advantage.[37]

35 *Pattee Byng's Journal*, p23

36 Brian Tunstall, *The Byng Papers*, 3 vols, London, 1930-2 (hereafter cited: *Byng*), III, pp241-2

37 Sir John Laughton to Sir Julian Corbett, 19 December 1905, *NMM*, Corbett Papers, Deed Box 'C': 'The entire and easy success of the method has taken away from its credit — just as in Quiberon Bay — but I have often thought that if people would only look, they would find that George Byng was one of our tactical masters'.

Reordering the line of battle

During the Second Dutch War the notion prevailed that divisional vice-admirals and rear-admirals should be stationed close to the outer extremities of their divisions. This was partly to strengthen the extremities, since the flag officers were in powerful ships, and partly to ensure authority as well as skill in leading the squadron (and indeed the whole fleet), especially when it came to tacking and wearing. When, under Article 6 of the sailing instructions, the sternmost and leewardmost ships were to tack first, the captain of the sternmost ship of the fleet assumed a highly responsible position in the tacking movement, quite apart from leading the fleet thereafter. According to the older view, the best plan was to discount any possible inefficiency on the part of the headmost and sternmost captains by ensuring that the division flag officers were themselves near the head or tail of the line. Towards the end of the century, however, opinions changed and it became the established practice to station division vice-admirals and rear-admirals in the centre of their respective divisions, supported when possible by powerful seconds. This was in conformity with what by then was an old established custom of stationing squadron commanders in the centre of their own centre divisions and thus in the centre of the squadron as a whole. The view must have been held that, provided the captains of the ships which were at the head and tail of the line were competent men, there was no need to have a flag officer actually present in these positions.

The new organisation depended on having enough ships of force to allow one for each flag, together with his two seconds; that is to say at least nine for each squadron and twenty-seven for the whole fleet. As this last condition could not always be fulfilled, allowances had to be made. With the increased construction of first rates and second rates in the Royal Navy, however, and their counterparts in the French, the formation of solid blocks of defensive fire in the centres of each division of each squadron became more practicable. Hence the defensive aspects of the line tended to be enhanced by developments in naval construction.

The allied line of battle in 1672, soon after Sole Bay,

The Battle of Cape Passaro, 11 August 1718 (detail), by Richard Paton. NMM

shows that the heavy ship principle had been carried further
in the Royal Navy than in the French navy, though in this
case only part of the French fleet was present.[38] In the French
White squadron, the flag officers were all in heavy ships and
in both the van and centre the admiral had powerful seconds.
Each division was led by a 70, which also meant that the
sternmost ship would be a 70 with the fleet in reverse order.
In the English Red and Blue squadrons the process was more
pronounced. Each of the six flag officers, except the rear
admiral of the Blue, was in a very powerful ship. In the centre
of the Red, the commander-in-chief and his seconds formed

an exceptionally strong combination. In no case were division flag officers stationed as near to the extremities of their
respective divisions as the English had been in 1665. Nor did
there seem any inclination to station ships of force at the
actual extremities of divisions except in the case of the van of
the French White squadron, which would lead the whole
fleet when on the starboard tack.

In the war of the English Succession a new trend appeared
which was to have a strong influence on future English
tactics. This was to station the two most senior captains, if
not themselves flag captains, at the head and tail of the fleet,
no doubt as a tactical compensation for the centre stationing
of division flag officers. This arrangement gave both seniority and superior gunpower to the head and tail of the line

38 *The Third Dutch War*, pp184–6

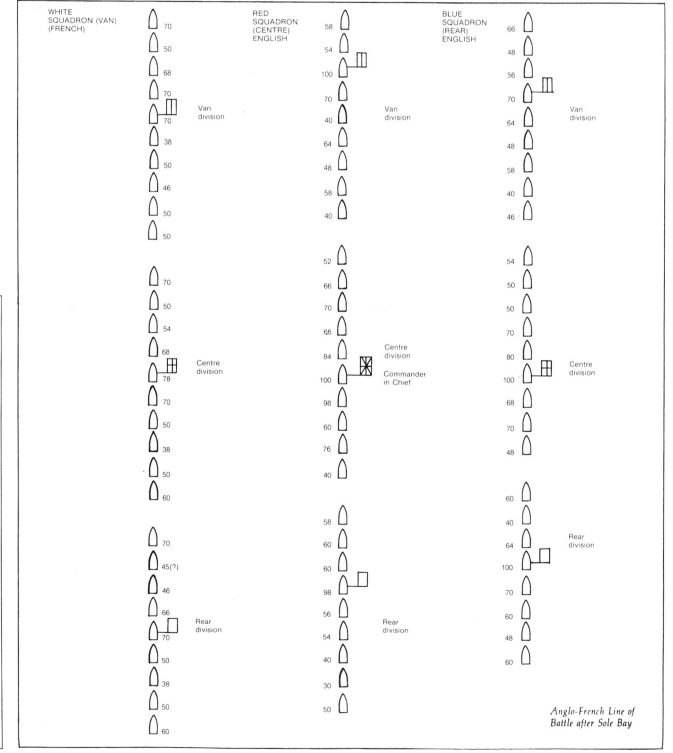

*Anglo-French Line of
Battle after Sole Bay*

because, once the flag ships had been apportioned, the senior captains were more likely to be posted to the heavier ships. The difficulty in tracing this development lies in the lack of reliable lists of orders of battle prior to the war of the Spanish Succession. Even where lists appear in published works, no reliance can be placed on their accuracy.

At the Battle of Bantry Bay in 1689, Admiral Herbert, Lord Torrington, had seven third rates in a weak fleet of nineteen of the line. He stationed these first, third, seventeenth and nineteenth in his line, and positioned the remaining three in the centre with his flag. The six most senior captains were in the first, third, fourteenth, sixteenth (senior captain), seventeenth and nineteenth ships. Hence there was a rough approximation between seniority and strong ships at the head and tail of the line.

At Beachy Head in 1690, Tourville had only eight of his nine flag officers in a ship of at least 70 guns and he did not possess enough first, second and third rates to form a really strong flag officer group anywhere in his line except in the very centre. Here he had a 70, 98 (Le Soleil Royal), 80 and another 80. In the English Red and Blue squadrons there were only four flag officers; all, however, in first or second rates (100 or 90 guns) and strongly supported. The Dutch White squadron was comparatively weak in guns: the vice-admiral was in an 82, supported by a 92, but the centre, 60, 70 (flag), 60, was weak. The English Blue with one flag officers had only a 70 at the head and tail of the squadron. Otherwise there is no sign of heavy ships at the extremities in either fleet. In the Red squadron, the leading captain was second in seniority. In the Blue, the two senior captains were at the head and tail respectively, each in a 70-gun ship.

At Barfleur in 1692 Tourville's fleet, though weak in numbers, was better gunned than previously. All his nine flag officers were in first or second rates and each squadron commander had second rates as his seconds. The English were in overwhelming strength and the Dutch had nine second rates or better. In the English Red and Blue squadrons the most senior captains, though in heavy ships, were not stationed at the extremities of their respective squadrons but in most cases as seconds to flagships.

Sir George Rooke is the only English admiral whose treatment of the line of battle can be traced with reasonable accuracy over any length of time. In 1679 he succeeded Lord Berkeley in command of the confederate fleet employed against France in the Channel and Bay of Biscay, and on 21 June he issued a line of battle for the English part of the fleet.[39] The Red squadron in the centre consisted of eighteen ships-of-the-line, the vice-admiral's division having eight and Rooke himself ten, and there was no rear division. The vice-admiral's division had three first and second rates, and Rooke four, including flagships and in each case they were stationed in the respective centres. Each division was headed (on the starboard tack) by a third rate and tailed by a fourth rate. The head and tail captains were comparatively junior, being in ships of less force. The three divisions of the Blue squadron consisted of eight, nine and nine ships respectively. There were seven first and second rates, distributed two, three and two and in each case in the centre of divisions. The rear-admiral's and the vice-admiral's divisions were headed by fourth rates. All the other head and tail ships of divisions were third rates; as before, the head and tail captains were comparatively junior.

Rooke's Baltic squadron in 1700 was so small, consisting of only ten of the line, that no conclusions can be drawn from his line of battle. He and his rear-admiral, Thomas Hopson, had the two strongest ships, and the three most senior captains were their flag captains and Rooke's first captain. When joined by the Dutch, however, Rooke changed his order of battle so as to have his two captains next in seniority at the head and tail of his own squadron. It is noteworthy here that the Swedish fleet was organised on the old principle, with the division flag officers at the extreme ends of the line, each being next but one to the headmost and sternmost ship.[40]

In the line of battle for Rooke's Anglo-Dutch fleet of 1701 we find the idea of strong centres fully worked out. In the English Red and Blue squadrons there were five flag ships, each supported by strong seconds. In the Dutch White squadron the three strongest ships were the flagships. No clear pattern emerges as regards the captains, several of the most senior being flag captains or flag officers' seconds.[41]

Rooke's Anglo-Dutch line of battle for 1702 shows quite a different pattern. Although the five English flag officers were not over-strongly posted as regards flagships and seconds, the English squadrons when joined together had a 90-gun ship at one end and an 80 at the other. Neither of these captains, however, was the most senior after the first captain and flag captains had been discounted. Three more second rates were stationed in the line at points not apparently having any special significance. The Dutch had only three second rates for their five flagships.[42] For the attack on the treasure ships at Vigo, a composite squadron was formed with four English and four Dutch flag officers.[43]

In Shovell's line of battle in the Mediterranean, 12 May 1704, he stationed his two senior captains at the head and tail of the line, one of them in a 96, though he only had two other second rates, his own flagship and one of his seconds. In his line of battle on 8 June 1704 he again had his senior captains (different from before) at the head and tail of the line in an 80 and 70 respectively, his 96-gun ship (Captain John Jennings) which had previously led now being a flagship.[44]

At Malaga in 1704, the French line was organised in what by now had become the established order. The English van had only two flag officers, but otherwise the line also followed the established order. The Dutch had only two flag officers. Taking the two English squadrons as one, the two senior captains were respectively at the head and tail of the line, though in 70-gun ships.

In Sir George Byng's squadron of sixteen ships for operations against the Jacobites in 1715, three of his captains were senior by ten years and more to all the rest.[45] Of these, one was his flag captain, one his second, and the third had the leading ship on the larboard tack.

In Norris's fleet of twenty-one of the line in 1740, the two senior captains were at the head and tail of the line, each in an 80. Each of the three flag officers was in an 80 and two more 80s were seconds to flagships. A fifth 80 carried the third senior captain whose ship was used to strengthen the centre.

At the Battle of Toulon in 1744, Mathews stationed all his thirteen heavy ships except one in the centres of his three squadrons. The head and tail ships were only 70s and commanded by captains who were by no means amongst the most senior. Apparently, therefore, Mathews did not believe in giving senior captains the leadership of the fleet. Or was it that he wanted their heavier ships to strengthen the squadron

39 NMM, SIG/A/3, a line of battle signed by Rooke loosely inserted in a copy of the Sailing & Fighting Instructions

40 Oscar Browning, editor, The Journal of Sir George Rooke, Admiral of the Fleet 1700-1702, London, 1897 (hereafter cited: Rooke) pp18-9, 24-5, 73

41 Rooke, pp128-9

42 Rooke, pp160-1

43 Rooke, pp231-2

44 Byng, I, pp44-5

45 Byng, III, p138

centres? This seems the most likely explanation, as he does not seem to have had any special reason for choosing either Captain Berkeley of the *Revenge* or Captain Cooper of the *Stirling Castle* to lead the fleet on either tack. Whether he regarded strength in the centres from an offensive or defensive standpoint is uncertain. Writing in 1762, Christopher O'Bryan favoured the earlier view, that where their ships were not too small, the senior captains should be posted either as leaders of the fleet on either tack or as seconds to flag officers.[46]

Fireships

Fireships could be used in two ways, either against shipping in a crowded anchorage, or at sea as part of the general tactics employed in a fleet action. 'Holmes's bonfire', in the Texel in 1666, de Ruyter's destruction of shipping in the Medway in 1667 and Rooke's attack on the French at La Hogue in 1692 are well-known examples of the first. The tactical use of fireships in fleet actions depended on having enough of them ready to hand and enough targets for them to attack, as at Sole Bay, in 1672, which explains why they played very little part in small-scale battles. They were much favoured by the Dutch, no doubt because their shallow draught enabled them to operate in shoal waters.

In battle they were stationed on the disengaged side of the line, generally close to the admiral and other flag officers. The time to set them in motion was when an enemy ship was seen to be disabled and yet still in the line of battle. If the fleet employing the fireships had the leeward position they would probably have to be towed by boats into action. If the fleet was to windward, however, a fireship could slip through the gap between two of its own ships-of-the-line and bear straight down on the disabled enemy.

The fireship was carefully prepared before the attack, filled inside with all kinds of combustible material and with special firing material on the deck. As the fireship approached its target, one or more tenders, generally brigs, ketches, smacks or small 'frigates', stood down with it to take off the crew, who, having set fire to the ship, escaped at the last minute by boat. A few volunteers sometimes remained on board to make sure of steering the fireship right against the enemy, and hooking on with grapnels. They then tried to escape by boat or by swimming.

As the fireship bore down with flames streaming out of it to leeward, the crew of its intended victim often panicked, thus prejudicing last-minute avoiding tactics and attempts to sink it by gunfire. Meanwhile there was even more desperate work for the defenders, whose ship's boats, assisted by their own ketches, smacks and other auxiliaries, attempted to sink the fireship with gunfire, or grapple it and tow it clear, or else board it and divert its course. Even if it struck home there

was still a chance of dragging it clear before its flames really took hold of its victim.

Naturally the attackers were unwilling to allow the action of their fireship to be rendered useless. They, therefore, bore down with their own tenders, ketches, smacks and brigs, not only to cover the retreat of the fireship's crew, but to attack the boats and small vessels of the enemy trying to intercept it. Thus an extensive flotilla battle might take place between the small ships and boats of both sides as a result of an attack launched by only one fireship. The attack might often come as a surprise, the fireship being generally stationed abreast of the flagships or other ships-of-the-line and not opposite the gaps between them. This was to give them protection from the enemy's fire, but it also helped to conceal their presence until they edged through the line. Thus the fireships and their tenders, together with the flotilla opposed to them, had a role analogous to that of the torpedo boats and torpedo boat destroyers under steam. Success depended on the bravery and coolness of the officers and seamen involved, especially those of the fireships themselves, who must not abandon ship too soon, even though the danger of instant death should a lucky shot from the enemy set off all the combustibles and cause the fireship to explode. In the Royal Navy of the Restoration the dangers of the fireship and anti-fireship service were reckoned so great that 'encouragement' was required in the form of the promise of lavish and specified rewards, including a gold medal for successful captains and compensation for relatives of those killed.[47] This was soon afterwards considerably extended, and the rewards increased in value and became incorporated as a separate chapter in the Permanent Sailing & Fighting Instructions.

That the advance of a fireship could be the occasion for a prolonged skirmish between heterogeneous flotillas, with the big ships joining in with gunfire, confirms the slow pace of a naval battle, with ships moving perhaps at only two or three knots. Captain Nathaniel Butler (or Boteler), writing in about 1634, thought fireships were of little value except in a narrow straight with wind and tide or current in their favour, despite their disruptive success against the Armada; he conceded, however, that if disguised as ordinary fighting ships they might score by engaging an enemy at very close quarters and then being fired and abandoned by their crews.[48] Pepys states that the Dutch had proposed to England that both sides should give up using them but that England, on Lord Sandwich's advice, had refused.[49]

Père Hoste, writing in 1697, recommended that the fireships should be stationed to windward of the fleet when cruising. They would then be in less danger from attack, and within easy reach of the admiral. Being better sailers before the wind than on it they would thus have more in hand and would not delay the fleet, while if forced to drop astern through lack of speed they could catch up again whenever the fleet started manoeuvring.[50] Similar views prevailed in the Royal Navy, and a special flag, striped diagonally yellow and white, was appropriated to fireship signals, in contradiction to the general usage of signal flags.

After the Treaty of Ryswick in 1697, fireships fell into disuse. None was used at Malaga, though each side had a number in attendance. Admiral Vernon hoped to make use of

Fireship Battle

Boats and frigates attempt to divert fireships

Boats take off fireship crews

Damaged ship, still in line

Frigate

Ship of the line

Fireship attack

Fireship

Flagship

Boat

Frigate

---

46 *qv* p124 below

47 See the Duke of York's 'Encouragement for the captains and companies of fireships, small frigates and ketches,' *Fighting Instructions*, pp149–51

48 *Boteler's Dialogues*, pp311–2

49 J R Tanner, editor, *Samuel Pepys's Naval Minutes*, London, 1926 (hereafter cited: *Pepys Naval Minutes*) p360

50 Hoste, pp403–4

them while commanding in the West Indies in 1739-42, and issued various orders and signals to their captains and the captains of their tenders to ensure full co-operation and the safe recovery of the fireship's crews.[51] Apart from the abortive attempt by the *Anne* galley to burn the *Real Felipe* at the Battle of Toulon, they were not regarded very seriously again until the time of Howe, who strongly favoured their use.[52] Corbett notes Howe's predilection (as evidenced by his fighting instructions) with some surprise, on the grounds 'that the use of fireships in action had gone out of fashion'.[53] This is true mainly because, for various reasons, the tactical circumstances under which most sea battles were fought after 1744 were unfavourable for the use of fireships. Nevertheless, many British officers seem to have agreed with Howe to the extent that they regarded fireships as having great potential value. Kempenfelt favoured using them at the start of a battle as a means of upsetting the enemy, regardless of whether a particular enemy ship was disabled. He also emphasised the need for frigates to support them.[54]

The first signal books

The National Maritime Museum possesses what would appear to be the earliest surviving English signal book.[55] It is a vellum-bound folio manuscript, bearing the date, 1711, at the beginning of the day signals. It has an index of signals by flags, followed by a subject index. Each signal page has on it a painted picture of two ships, one above and one below, each flying a particular signal in addition to the jack, ensign and commander-in-chief's union flag at the main. The ships are painted accurately and with great spirit and in varied forms. One or more guns are painted above the ships as the particular signals may require. The order of signals starts with flags flown from the maintopmast head and descending down the mainmast; followed by signals flown from the foremast and mizzen in the same manner. Each signal has its meaning written below the picture of the ship in the fewest possible words. There are 134 day signals, six fog signals made with guns, and twenty-one night signals, the entries being shown with the lanterns painted in their various positions. The day signals thus include more than the full total of those given in the Permanent Sailing & Fighting Instructions even counting all the possible variations.

Amongst the extras are eight signals 'made by Sir John Norris in the *Ranelagh* in the Straits 1710', when he was commander-in-chief in the Mediterranean. His first four chasing signals were soon incorporated in the Permanent Sailing & Fighting Instructions under the title of 'Additional Signals to be added to the XVIth Article of the Sailing Instructions by Day and observed as the other Signals'.[56] These included signals to board the enemy; for the squadron to chase to the NE/NW/SE/SW; for the squadron to stand on the contrary tack to that the admiral was on; for the fleet or any particular ship to chase to leeward; and, in case of engaging in the night, for each ship to wear a light (at the mizzen peak).

In 1714 Jonathan Greenwood, a member of the Stationers'

Company, and presumably a printer-publisher with naval connections and a good deal of enterprise, produced a book which he sold to naval officers. It was entitled *The Sailing and Fighting Instructions or Signals, as They are Observed in the Royal Navy of Great Britain*. The book was entirely unofficial, and was dedicated to the Board of Admiralty which took office in October 1714. It is, in fact, a signal book, as stated in the title. Greenwood's intention is clearly set out in his dedication:

May it please your Lordships. My publishing this per-

A page from a manuscript signal book, c1710-11. NMM

For ÿ Fleet to draw into a line of Battle ahead

For ÿ Fleet to Draw into a line of Battle a Brest

51 Brian McL Ranft, *The Vernon Papers*, London, 1958 (hereafter cited: *Vernon*), pp295-6

52 Lord Sandwich to Keppel, 3 Oct 1778; 'Lord Howe, who when he left England left it with me as a sort of legacy that that useful though horrid instrument might not be laid aside.' *Sandwich Papers*, II, p177

53 *Fighting Instructions*, pp248n and 274n

54 See Kempenfelt to Middleton, 5 September [1779], *Barham*, I, p296

55 *NMM*, SIG/B/1 (formerly Sm/2)

56 p33; the exact date when this happened is uncertain.

To Unmoor, if to Cut or Slip two Guns

For the Blew Squadron to tack and gain the Wind of the Enemy.

To Weigh

For the White Squadron to tack & gain the Wind of the Enemy.

A page from Jonathan Greenwood's The Sailing and Fighting Instructions or Signals As they are Observed in the Royal Navy of Great Britain, *London, 1715(?). NMM*

formance is not with any design of derogating from the value and usefulness of the *Printed Instructions;* this being an exact *copy* of them. But what I have endeavour'd at, is the putting of them in such a method, as may make them more easily and more readily found out and likewise to supply the Inferior Officers who cannot have recourse to the printed ones. And as I have disposed matters in such a manner that any *Instruction* may be found out in half a minute; so I have made it a pocket volume that it may be at hand on all occasions. As to the dangerous consequences of mistaking any signall especially by night, I need not mention them to your *Lordships.*

The plan is extremely simple.

For the more immediate finding any signall that shall be made, note that in the first pages are all the signals about the mainmast taking it from the masthead downwards. In the next pages are those about the foremast and after them the signals about the mizenmast and ensign staff. In the latter part are the signals in a fog and by night. The number of guns fired at the making any signals [*sic*] is inserted at the top of each page.

A standardised picture of a warship is printed at the top and bottom of each page, so that when the book is open, four ships can be seen simultaneously. From these ships are seen flying the various signals in addition to the jack, captain's pendant and ensign. The flags are left blank to be coloured by the buyer.

The book makes no pretension to be more than a quick means of reading signals. At the time, it was an obvious boon to all officers, 'inferior' or otherwise. It could not be used with the same facility for sending signals since the arrangement by masts involved lumping together signals having no

particular connection with each other, apart from the signal for a particular instruction to the White squadron being naturally followed by a similar signal for the Blue.

Jonathan Greenwood has never received the recognition he deserves for his pioneer work, his civilian status and unrecorded life causing him to be treated by many writers in a perfunctory and even supercilious manner. He was, however, a real innovator, anticipating the next printed signal book (by Millan – also a commercial publication) by thirty-four years. Moreover, he either invented new signals himself, or at any rate printed and thus publicised additional signals devised by some unrecorded admiral. These signals were incorporated by Millan in his book published in 1748, with Greenwood's name as the authority and included signals for flagships in line ahead to keep two leagues apart; ditto in line abreast; and for flagships in line ahead to keep one league from each, ditto in line abreast.

Mathews adopted these signals while in command of the Mediterranean Fleet, 1742–4. Greenwood also incorporated the six chasing signals of Sir John Norris, which were recorded in the book of 1711, though without any attribution. The use of false fires is indicated in the night signals, though no mention of these had so far been made in the Sailing & Fighting Instructions. Greenwood's book is clearly derived from the manuscript books, both as regards the order of signals by masts and the method of depicting the signals as flying from the rigging of a particular ship. Possibly he did no more than put into printed form what was already being done in manuscript.

It appears from the number of copies which have survived that the sale of Greenwood's book may not have been restricted to naval officers. His remark about its value to 'the Inferior Officers who cannot have recourse to the printed ones [the *Instructions*]' suggests a security element. This, however, may have been only one factor in the situation; the need for marking the superiority of superiors and the inferiority of inferiors may also have been considered important.

Further evidence of the popularity of this type of personal signal book is provided by a tiny pocket signal book which from internal evidence can be dated to between 1719 and 1721.[57] This is in manuscript and organised on the same lines as the Greenwood book, though mainly with painted flags and guns; the place where the signal was flown and the signification were written in words. There are only a few complete pictures of ships with signals flying from them; these, however, are very vigorously painted. It contains very much the same signals, though with a few omissions and additions. Amongst the additions are Norris's additional signals of 1710 'To board the enemy', and 'To stand on the contrary tack the admiral is on'. It also includes the new signals given by Greenwood and afterwards incorporated by Millan, and includes all six of Norris's chasing signals.

All this represents very slow progress compared with the French. Under the Sailing & Fighting Instructions the total number of signals available to the admiral had nowhere near reached that used by Tourville, nor has any example yet been found of a British thumb-indexed pocket signal book comparable with the French book of 1690. By separating signals from instructions at an early stage, the French had gained an advantage which they did not begin to lose until the American War.

57 'R T's' Book: *NMM*, Tunstall Collection, TUN/31 (formerly S/MS/SS/1). The attribution of dates is easily fixed by paintings of the flagships of 'Sir George Byng', (peer in 1721) and of 'Admiral Hopson' as Rear-Admiral of the Blue, (promoted 1719). The book, therefore, must have been made after the battle of Cape Passaro. A possible clue to attribution is supplied by a page entitled 'A View of the *Launceston*' with the opposite page left blank for a picture of the ship.

Convoy tactics

Big battle snobbery and big battle romanticism have long denied convoy actions the full status of naval battles. Our ancestors knew better than this. In fact, they sometimes tended to go to the other extreme, measuring tactical results solely in terms of their broader, strategic influence on seaborne trade. For the merchants, seaborne trade was to a major extent what the war was about, so that in a sense every battle was a convoy battle, whether a convoy was directly involved or not.

Reading the recommendations in *The Seaman's Vade-Mecum and Defensive War at Sea,* published in 1700 by William Mountaine, who was mathematical examiner of the Honourable Corporation of Trinity House, one gathers the impression that by a combination of individual dexterity, mutual support, and sheer weight of numbers, no well organised convoy need fear penetration even by a number of privateers. The section of Mountaine's book dealing with defensive warfare was apparently written by Captain Robert Park of Ipswich in 1704, and was not reprinted; it was deemed relevant seventy-four years later.

If well handled, a merchant ship could, in theory at least, make things very difficult for a would-be captor. Detailed specifications existed for fortifying and barricading the ship against boarders. Equally detailed specifications existed for how to avoid capture either from the windward or leeward position, taking into acount tides, currents and shoals. In addition to sailing techniques there was the extra need for keeping out of range of the chasing warships' guns. Even at close quarters a great variety of strategems were open to the skilful captain to avoid being boarded. Against a single privateer they might well prevail, and indeed the ship might be successfully defended from barricaded positions even if driven ashore.

In theory, a convoy could put up a good fight even if were penetrated by privateers. Convoys generally sailed in three columns, and in more if very large. By keeping well closed up and using such guns and small arms as they possessed the convoy ships could relieve each other by taking a position on the bow and quarter of any privateer attempting to board their next ahead or astern. Warships were a different matter.

In practice, however, convoys were not generally well-organised or well-drilled enough to offer serious resistance to attack, except in the case of the well-armed British, French or Dutch Indiamen. Escort commanders knew this, so that when attacked they generally signalled the convoy to head for safety or scatter. There is no well authenticated case of a convoy or part of a convoy keeping together when attacked and offering organised resistance. Examples of signals for

146   *The Seaman's* Vade-mecum.

IX. *Discharge, but get not in the Cannon open ; both must be done in your Close-quarters, and on the contrary Side to the Enemy.*

BUT if the Enemy sails better than the Ship he attacks, as is evident most Privateers out-go the generality of our Merchant-men, then will they Board him maugre all Opposition, in what Place they please, except thwart the Hawse ; which is not to be done but by Accident, or Want of Conduct in the Ship so boarded.

But before he is on Board, the Commander must order all his Guns in the Waste and upon his Quarter-deck to be discharged without letting them run in ; for if they should be loaded when the Enemy enters, and they should traverse any of them Fore and Aft, they would soon level the Bulk-heads with your own Cannon ; and if they are run in they are soon loaded ; whereas if they are out, and the Tackle-falls moused, or a running and standing Part seized together, their Men will be the more exposed before they can accomplish any such Design ; and left under the Covert of your Smoke, or any other favourable Accident, they should get in a Gun, the designed Advantage may prove their Ruin, by leaving in every Gun when you retire to your Close-quarters, a Piece of lighted Match.

And those on the Side from the Enemy when engaged, in the Close-quarters, must not only be discharged, but got in, that the Enemy do not toss in Hand-Granadoes or Stink-pots, to destroy or suffocate the Men in those Quarters ; and that they should be discharged is necessary, because otherwise an Hand-Granade, Fire-pot, or some such thing may discharge it in your Quarters, and do more Damage to your Ship

The Seaman's Vade-mecum.   147

Ship than the Enemy ; or by carrying away a Port make a Vacancy, where the industrious Enemy may toss in Showers of Hand-Granadoes ; besides, in discharging your Cannon, they run in of themselves, whereas at such a Time the Hands cannot be spared to get them in.

X. *How to act when a Ship comes up your Wake, and lays you aboard upon the Quarter.*

THE Enemy, in his Approach to Board you, comes either up your Wake, upon your Quarter, upon your Broad-side, or lastly upon your Bow.

If the Enemy come up your Wake, ply him briskly with your Chace-Guns loaded with round and Cross-bar ; and as soon as he is within Pistol-shot, give him your Fore-Chace-Guns loaded with a Double-headed-shot and a Bag of Case-shot ; the former may spoil his Masts and Rigging, and the latter destroy his Men : Next let your Guns upon the Quarter be ready loaded with double and Case-shot, that as the Enemy range up your Quarter with his Men ready to enter, they may be discharged among them ; let likewise your Powder-Tubs be ready, and just as the Enemy is going to share aboard, set fire to the Fuze, hoist it up to the Yard-Arm, and then let it run amain among his Men upon the Decks : If he still persist in his Resolution and board you, let all your Ports be lashed in, lest the Enemy wedge them, which is of ill Consequence, as has been before observed : Keep firing your Blunderbusses out of the Look-holes in the Quarter among his Men, as they stand thick and ready to enter ; as soon as he is aboard, spring your Powder Chests upon the Quarter, for then his Men will, mounting your Quarter, be numerous. Let your Men in the Round-house be ready with their Small-Arms to give the Enemy a Volley as soon as they come upon your Quarter-

O 2   Deck,

A page from William Mountain's The Seaman's Vade-Mecum and Defensive War by Sea, *London, 1747. NMM* convoys from 1700 onwards exist in large numbers but they seldom have any tactical significance, either as regards the escort or the merchantmen. *The Seaman's Vade-Mecum* gathers together from various sources what must be the most extensive collection of convoy signals prior to 1790, comprising twenty for day, sixteen for night and eight for fog. However, these are solely concerned with governing the convoy's movements under ordinary conditions. No 17 (by day), which is really an instruction, pertinently remarks:

'Signals made by the Admiral in Chief, the other Flags, the several convoys of private men of war [ie, escorts], and the Captains under those convoys; the care the Trade ought to have in following their respective Commodores and the methods they ought to take when the Convoys are attacked, are peculiar to the Men-of-War, and therefore need not be inserted here'.[58]

The convoy signals issued by Admiral Lord Keith for his Cape of Good Hope expedition, 1795–6, appear rather more sophisticated.[59] Single flags flown from particular positions in the traditional style, provided fifty-eight signals, of four different types (though unclassified as such): signals by the Admiral to ship or ships of the escort, to ships of the convoy, and from warships and merchantmen to the admiral. Although convoy movements are still on a very simple basis, there are hints both of stricter control and of greater responsibility given to individual merchantmen.

Convoys of the ships of the British East India Company were notoriously difficult to handle; their captains often knew the eastern seas far better than did the escort commander, and they treated arrogant naval officers with derision and contempt. If bullied with signals and written reprimands, they risked capture by leaving the convoy altogether. Nor could they be dismissed by the Company as a result of naval complaints to the Admiralty, since they were appointed by the owners of the ships, who treated them with respect and deference. Under such conditions effective signalling depended on tact, politeness and discreet conviviality.[60]

From the tactical standpoint, there were two kinds of convoy battles. The first, such as the Saints, the First of June, St Vincent, and even in a sense the Nile, were those in which convoys are directly or indirectly the cause of the fleets coming together, but no more than that. The second, such as the two battles of Finisterre in 1747, were those in which a

58 p267

59 *NMM*, Tunstall Collection, TUN/17 (formerly S/MS/R/2)

60 See C Northcote Parkinson, *Trade in the Eastern Seas, 1793–1815*, Cambridge, 1937, chapter 10

Convoy Organisation

fleet of merchantmen protected by an escort force were the subject of direct attack by a superior force of warships or privateers. In the first category the battle might take place a long way from the convoy; in the second, it was fought with the convoy actually in sight and with its preservation or destruction the immediate, as well as the overriding, tactical consideration.

Certain general principles governed all French and British convoy tactics under sail. The commander of the escort force was obliged to maintain good reconnaissance directly ahead of the convoy's course and in any quarter from which an attack might be expected. Ships comprising what today would be called the close escort were generally stationed a little ahead and to windward of the weather column of the convoy. The escorts' minor warships were stationed between the columns of merchantmen to regulate their station-keeping and observation of signals. At night the look-out ships ahead closed in so that they did not lose sight of the lights carried by the escort commander. These were counsels of perfection, however, as very often the escort commander had only a tiny handful of warships at his disposal. Further complications arose if he was meeting a home-bound convoy unescorted as far as the pre-arranged rendezvous, or if he was detached with all or part of a convoy hitherto covered by a larger force.

The first duty of the escort commander was to his convoy. If attacked by a force which he was able to repulse, he should never pursue further than the strict protection of his convoy justified. No capture, however attractive, should be attempted if it in any way prejudiced the convoy's safety. In case of a threatened attack by a superior force, the commander of the escort should be prepared, if necessary, to sacrifice his whole force in defence of the convoy. This meant using his own judgment as to how to engage the enemy, so that having once provided for the convoy's safety by completely engaging the enemy's attention, he was free to fight in any manner he chose. If the convoy had no time in which to escape, he might have to form in line-ahead between his convoy and the attackers, or he might choose to retreat slowly in line-abreast so as to prolong his resistance as much as possible. Much depended on whether the convoy had plenty of sea room ahead or whether it was approaching some restricted area or some terminal port.

French requirements were even more exacting, as might be expected. Escort commanders were subject to very severe penalties for deserting their convoys, and merchant ships could be heavily fined for deserting their convoy or for failing to sail in one.

In every convoy action the commander of the attacking force held two advantages, whether in superior force or not. First, the ships of the convoy were forced to sail at the speed of the slowest, while in convoy, and since the speed of even the average merchantman was slower than that of a warship, the attacking commander should have no difficulty in keeping them in sight under normal weather conditions. Second, the commander of the escort, being encumbered by the convoy, was bound to fight under tactical limitations. If the convoy was not near a home or allied port, the attacker could perhaps afford to wait, striving meanwhile for the weather gage and a chance to attack the convoy direct. If, being in superior force, he decided to fight the escort at once, he might detach one or more ships to harry the convoy while the escort was fully engaged. Even when the escort force was beaten or eluded, much skill was necessary to ensure the maximum capture of merchantmen; the available stock of ships' boats and prize crews could be soon exhausted, while to damage a merchantman too much by the usual disabling operation of cutting bowsprits and rigging might mean never

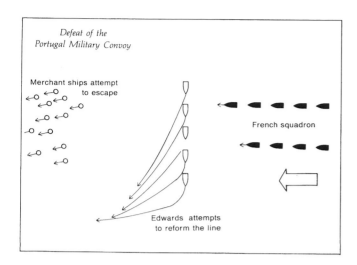

Defeat of the Portugal Military Convoy

Merchant ships attempt to escape

French squadron

Edwards attempts to reform the line

getting her home to the attacker's base.

In the First Dutch War the differentiation between battles brought about by the movements of convoys and those fought in their direct support is scarcely apparent, all the battles being directly or indirectly concerned with convoys sailing or about to sail. With the long series of Anglo-French wars which begin at the end of the seventeenth century, operations were no longer confined to a small area close to the terminal ports of the belligerents. Tourville's success in intercepting the Smyrna convoy in 1693 was gained off Cape St Vincent. Rooke's successful attack on the Spanish treasure ships in Vigo Bay with part of the Anglo-Dutch fleet in 1702 was in a sense a convoy action.

The most important convoy battle of the period, and one quoted as an example more than fifty years later, was the overwhelming success of René Duguay-Trouin and Claude, Chevalier de Forbin against the Portuguese military convoy of some 130 sail on 10 October 1707. The escort force consisted of the *Cumberland* (80) (Captain Richard Edwards), the *Devonshire* (80), the *Royal Oak* (74), the *Chester* (54) and the *Ruby* (80). They were attacked by Duguay-Trouin with four of the line and two frigates, followed by Forbin with five weak ships-of-the-line and a frigate. The setting was simple enough: the French attacked from the east with the wind, the convoy was scattered to the west with the escort interposed between them, forming in line ahead to the south on the larboard tack. Captain Edwards, tried to get his squadron together by turning west into line-abreast and sent a message by boat that they were then to fight in line-ahead, a little from the wind and as close together as possible so as to give mutual support. Heavy seas and the quick approach of the French prevented the *Royal Oak* getting the message. The *Devonshire* sank; the *Cumberland*, *Chester* and *Ruby* surrendered. Only the *Royal Oak* escaped. Twelve valuable transports were captured.

It was a great triumph for French arms. The *Cumberland*, *Devonshire* and *Royal Oak*, all three-deckers, were difficult to board. On the other hand, the *Lys* and the *Achille* were up to British 70-gun standard, while the *Amazone* was as big as a British 50, and the state of the sea prevented the British three-deckers using their lower-deck guns when firing to leeward. The French also had the double advantage of having twelve ships to five and a greater superiority in men, both for boarding and gun crews. More than half a century later Christopher O'Bryan was to make this battle the subject of a searching tactical analysis, urging that Captain Edwards should have made a slow retreat in line-abreast. The real test of the battle, however, seems to lie in its purpose and results. The duty of the British squadron was to save the convoy at all costs and in this it succeeded to a very considerable

measure. The French purpose was to get at the merchantmen, of which in the end they took less than 10 per cent. True, they sank a British three-decker and captured three out of the remaining four of the squadron and this remained a great feat of arms, but in a full tactical sense their success was limited.

When, in February 1708, Leake was preparing to take a large Anglo-Dutch fleet of merchantmen to sea *en route* for Lisbon, he arranged to have an escort ship two leagues ahead of him by day and another two leagues ahead of her:

. . . and in sight of the Admiral's lights by night. Four sail to

Claude, Chavalier de Forbin (1656–1733), by Sergent. MM

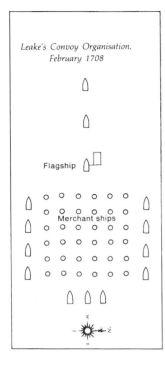

Leake's Convoy Organisation,
February 1708

Flagship

Merchant ships

keep to the southward of the fleet, at a convenient distance from each other; four more sail to the northward of the fleet in a like manner, and three sail to keep astern to secure the rear. No ship was to chase without a signal, nor to lose company as they would answer to the contrary.[61]

So important did the whole business become that a Cruisers and Convoys Act was passed very quickly through the British Parliament in 1708. No doubt arising out of this, Queen Anne's husband, Prince George of Denmark, now Lord High Admiral of Great Britain, issued special instructions for captains of warships and masters of merchant ships 'for the better security of the latter when they shall from time to time proceed under the convoy of the former'.[62] The twenty-three day signals were in fact instructions of a fairly simple kind, including Article XVII which required ships in danger of capture to cut 'jeers' [gears, ie, halyards], or otherwise disable their masts and sails to prevent their being carried off before being relieved by their escort. There are also twenty-four night signals.

Captain Thomas Fox's success against the French convoy from San Domingo in 1747 provides a good example of tactical anti-climax. Very early on the morning of 20 June, Fox, cruising off the west coast of France with a squadron of two 70s, two 60s, a 50, a 40, a frigate and a fireship, sighted a great body of ships to windward. Soon, more than 120 were counted, together with three ships-of-the-line, commanded by Commodore Du Bois de La Motte-Picquet. The wind was

NNE and all the ships on both sides were steering roughly east. After an interval of dense fog, the French escort could be seen drawing into a line-abreast and lying to, but on observing Fox's strength they filled and stood on, keeping about a point off the wind and to leeward of the convoy. Fox, by crowding sail and keeping close-hauled, managed to draw up within five miles of them. During the afternoon the French also crowded sail and steered closer to the wind, and at the same time signalled to the convoy to disperse, at which point the rear ships of the convoy tacked and headed away to the NW while the leading section continued to windward of their escort on an easterly course. So ended the first day.

At daylight on 21 June Fox could see a large part of the convoy about twelve miles away to the north-east but there was no sign of de La Motte's warships. Fox detached one of his 60s to the north-west to look for them. He gradually gained on the convoy, but only managed to capture one ship during the day. On 22 June thirteen more were taken. On 24 June a strong south-westerly gale helped the convoy to make for home ports, but altogether from start to finish Fox's squadron took forty-four of them. An additional four were intercepted by Admiral Warren before they could reach a French port. Thus, through unexplained mismanagement by de La Motte, in a purely tactical issue the French lost a third of their whole convoy without a single shot being fired in its defence.

A typical example of ill-informed criticism of an escort commander is provided by Commodore de Ternay's refusal to engage Admiral Sir William Cornwallis. On 20 June 1780 de Ternay with an 80, two 74s and four 64s was escorting a convoy of 6000 troops bound for Rhode Island. Off Monte Christe he sighted Cornwallis's squadron, which comprised two 74s, two 64s and a 50. De Ternay, seeing his convoy well on its way to leeward, hauled to the wind in line-ahead to block Cornwallis, who was still further to windward. Not only was Cornwallis greatly inferior in strength, but one of his 64s, the *Ruby*, was so far detached and to leeward that the French had a good chance by holding their course to cut her off. Alternatively, they could force Cornwallis to fight at a numerical disadvantage in order to save her. When de Ternay saw that this was what Cornwallis was prepared to do, and that in fact the *Ruby* had managed to rejoin him, he withdrew after a desultory and long-distance cannonade, and rejoined his convoy. He was aware of the presence in the vicinity of another British squadron equal or superior to his own force in strength, and to have been tempted to fight a weaker force, even with the hope of capturing the *Ruby*, would have been contrary to his instructions.

Convoy Action, June 1747 –
Situation on the Afternoon of the 20th

Rear ships of convoy tack and head NE

Forward ships of convoy
continue on course

French escort

Fox's squadron

61 *Leake,* II, p180

62 *NMM,* uncatalogued

CHAPTER THREE:

English Additional Signals

THE period 1714 to 1739 has often been described as one of naval stagnation, but this can fairly only be said of the years after 1728. Until then, at any rate, there was considerable naval activity, including the almost yearly appearance of a British fleet in the Baltic in connection with the Northern War. There was the war against Spain in the Mediterranean, 1718–20, including the Battle of Cape Passaro, the short war with Spain in 1728, involving the British blockade of Spanish ports in the West Indies, and the abortive siege of Gibraltar.

None of this led to a major revision of the Sailing & Fighting Instructions, presumably because no operations lasted long enough to require further development. Thus there was no tactical experience of sufficient importance to modify the system which, having passed through a long formative period, was at last consolidated. However, there was tactical development of a more limited nature. It was open to commanders-in-chief to issue additional sailing or fighting instructions.[1] Alternatively, and more easily, they issued additional signals, devoid of instructions. Such signals and instructions, though official, were only valid for the particular fleet in the particular area concerned, and ceased to operate in the case of a change of command.

How extensive the practice was at this time it is hard to say. *A Narrative of the Proceedings of His Majesty's Fleet in the Mediterranean . . . By a Sea-officer,* published in 1744 in support of Admiral Richard Lestock during the controversy following the battle of Toulon, comments:

Our Signal Book at present, has been found by constant practice in many necessary points to be defective and insufficient. . . . Men in the highest stations at sea, will not deny but that our Sailing and Fighting Instructions might be amended and many added to them, which by every day's experience are found to be absolutely necessary, tho' this truth is universally acknowledged, and the necessity of the Royal Navy very urgent, yet since the institution of these signals, nothing has been added to them, except the chasing signals excellent in their kind by the Right Honourable Sir J[ohn] N[orris].[2]

Norris's additional chasing instructions of 1710 are the earliest known; Vernon's additional signals, issued while he was commander-in-chief in the West Indies, are the most important. In this way, by piecemeal experiment and innovation, the now established British tactical and signalling system gradually threw out fresh shoots. Eventually a selection from these was published by the Admiralty, but individual admirals still continued to give out official instructions and signals of their own with temporary and local force.

Vernon's additional fighting instructions

Unlike some naval officers at that time, Edward Vernon was a man of education and property, and still full of life at the age of fifty-five, when he accepted the West Indies command. He not only issued the first substantial body of additions to the corpus of sailing and fighting instructions, but saw them generally accepted by the navy, if not incorporated in an Admiralty publication. Possibly his additions were no more than a compilation of various individual additions, examples of which are now lost, emanating from Rooke and Shovell, as well as Norris, in all of whose flagships Vernon had at one time served.[3] Their piecemeal issue, however, suggests personal ideas worked out gradually, rather than the re-issue of an existing body of lore, and the language in which they are phrased is highly personal. The 'sea-officer' was willing to acknowledge Vernon's exceptional contribution to tactical development:

Mr V[ernon], that provident great admiral, who never suffered any useful precaution to escape him, concerted some signals for so good a purpose, wisely foreseeing their use and necessity, giving them to the captains of the squadron under his command.[4]

Vernon was critical of the refusal of British seamen to study tactics:

Admiral Edward Vernon (1684–1757), by Charles Philips. NMM

---

1 In 1735 Sir Charles Wager issued Captain Swale of the *Rippon* with the Sailing & Fighting Instructions which include Norris's additions, but without any additions or changes in ms. The printing is slightly different from that of earlier and later forms. *MOD,* Ec/39

2 pp74 5

3 *Vernon,* pp285 86. Brian McL Ranft's conclusions as editor on this point seem very reasonable.

4 Quoted in *Vernon,* pp74 5

Meanwhile the French van ships stood on unengaged and were soon followed by the *Terrible*, her seconds and the sternmost French ship, after their preliminary fight with Rowley's ships. These sternmost Frenchmen edged away a little to leeward and easily cleared their British opponents. In rear of the *Barfleur* and her seconds, Rowley's three remaining ships and two of his 50s at first found only two French ships to oppose them. They engaged the Spanish van ships as they began to come up, but nothing decisive happened for some time.

In the centre, Mathews and his seconds, together with the two ships ahead of them, engaged their five opposite numbers very closely, damaging the *Poder* (60) severely and driving Navarro's two seconds out of the line. The fifth was forced to make sail to catch up with the rest of the fleet. Even the *Real Felipe* was badly damaged, though she fought back magnificently and inflicted severe damage on both the *Namur* and the *Marlborough*. With the enemy dropping to leeward in disorder, there was a splendid chance for the British ships to press them harder and make at least one capture. No captain, however, was prepared to pass to leeward of the admiral with the signal for the line still flying. The remaining four ships of the British centre could equally have joined in and smashed the *Real Felipe* (only one had so far opened fire and they were now practically unopposed, as the five remaining Spaniards were still straggling into action). Similarly, the British centre ships, so far not seriously engaged, had the chance to concentrate on the remaining Spaniards as they came up, being able to see that Lestock's rear squadron was at last beginning to close up to support them. Yet, apart from the *Dorsetshire*, they were ineffective, holding off to windward and making little attempt even to engage the Spaniards at long range.

It was now nearly 4pm and, as Lestock came up with his

squadron, his two leading ships opened distant fire on the two sternmost Spanish ships. His flagship, the *Neptune*, stationed third in his squadron, fired a few shots at the retreating Spaniards ahead. He made no attempt whatever to bear down and cut them off; on the contrary he kept his whole squadron well to windward and failed completely to join the battle. Meanwhile, the leading French ships were still in good order ahead of the British van and about a mile to leeward of it.

Captain Hawke of the *Berwick*, stationed two astern of the *Barfleur*, had been very active all through the fight. As the *Poder*, much damaged, emerged from the pell-mell round the *Real Felipe*, making sail to catch up with the French, he bore down and, after a battle of two hours, forced her to strike. This was the only capture of the day. There was a gap of a mile and a half between the *Poder* and the *Real Felipe*, which was lying to leeward with another Spanish ship and no doubt waiting for the remaining Spaniards to come up. Mathews tried to organise a fireship attack on the Spanish flagship, but the fireship was not properly covered by the ships astern of the *Namur*, though it was blown up by the Spaniards only a few yards from its goal.

De Court had already seen that only three Spaniards were following him and easily divined that the rest were in danger of capture by the British centre and rear. Sometime between 3 and 4pm he signalled his van to tack in succession. The three leading ships at once obeyed and began to steer north, with the wind still at about ENE. The three leading ships in Rowley's division did the same, so as to protect their main body from being doubled. The rest of the French ships failed to see the signal and made sail as before, on which the three leaders returned to their stations, though keeping a little to windward. De Court then signalled the whole French division of the fleet to tack together.[25] This time they saw the signal, and at once went about, hauling fairly close to the wind on the starboard tack. Rowley quickly signalled his own squadron to tack. Although his three leading ships actually anticipated his signal they were nearly caught by the French owing to their inferior speed. De Court, however, did not waste time on them, nor powder and shot, and, passing under their sterns within musket shot, bore up and steered large to cover the Spaniards.

If Mathews's intention had been to 'bring on a general engagement', his wish seemed about to be fulfilled, though not in the manner he had originally intended. So quick were the French that they retook the *Poder* before Hawke could even withdraw his prize crew. Closing the gap between her and the *Real Felipe*, they re-established close contact with the Spanish squadron. Mathews made no attempt to force a close action for possession of the *Poder;* and nor did he try to draw de Court into a pell-mell round the *Real Felipe*. Nor, with damaged ships and his general tactical disarray, could he be expected to accept the risks of what must become a night action. Instead, he signalled his own squadron to tack. Owing to the heavy swell, he and most of the centre had to wear. They then stood away to the north under easy sail with Lestock's squadron also in retreat, immediately ahead of them, and Rowley's astern. Some of the British centre and van engaged for a time the rear Spanish ships pursuing their original southerly course, with the two forces passing each other on opposite tacks, but it was now dark and both fleets brought to for the night and repaired damage.

*Battle of Toulon –
Situation at 4pm, 11 February*

Lestock's squadron arrives

Real Felipe

Namur

Berwick

Poder

Rowley's squadron

Leading Franco-Spanish ships
begin to tack in succession

25 *A Narrative of the Proceedings of His Majesty's Fleet in the Mediterranean*, 1744 (hereafter cited: *A Narrative of the Proceedings . . .*), p102. (*NMM*, TUN/179).

The courts-martial

The court-martial which sentenced Mathews to be cashiered listened to fifteen separate charges against him, the most important of which were that he made the signal to engage before the fleet was ready and formed in line of battle; that he failed to steer a proper course so as to bring his van and centre against those of the enemy, which it was possible for him to have done; that he failed to support the *Marlborough* sufficiently when engaged with the *Real Felipe;* and that he failed to direct the battle effectively after it had begun, especially as regards those captains directly astern of him who neglected their duty. His conduct on the days following the battle was also heavily censured.

It would seem that the attack Mathews devised was the best possible under the circumstances, but, being irregular, it had left some of the captains in his own squadron puzzled as to what they should do, with the result that they did very little. Ironically, it was the straggling of the rear Spanish ships (all five of which should have been taken) which left the puzzled British with nothing to fire at.

Although the court-martial found no fault with Mathews for keeping the signal for the line flying during the battle, this really does seem to have been a major mistake, since it invoked the taboo on breaking the line enshrined in Article 21 of the Fighting Instructions. If he had hauled down the signal for the line and kept the signal to engage flying as soon as the *Namur* came to close quarters with the Spaniards, his captains would have been free to fight pell-mell. There was no signal ordering the fleet to come to a closer engagement when to windward, though Rowley later added one for the 'Leewardmost ships to come to a closer Engagement', the precise meaning of which is itself unclear.[26]

When Lestock was tried he was charged with all the shortcomings that might have been expected from the way in which he hung back on 10 February and during the battle, and the evidence against him was voluminous, well-attested and absolutely damning. He was, however, acquitted on all charges. His defence statement was extremely well-argued, and based on a literal and pedantic adherence to established principles. The evidence he called was, in Sir Herbert Richmond's view, not only inferior in authority and character, but was supported by seemingly perjured statements, wholesale falsification of journals and the admission of arguments and statements which should have been rejected by the court as contradictory and inadmissible. It would seem that the nature of the men who composed the court inclined them more towards condoning superior backwardness than superior hastiness. For inferior backwardness, that is in the case of captains, they had no mercy, nor for the display of individual initiative by the three leading captains of the van.

Today it is easy to see where Mathews went wrong. Neither he nor Lestock, while in temporary command, seem to have done enough, or indeed any, exercising of the fleet in forming the line of battle ahead and abreast, at varying distances between the ships, and in bearing down together on a supposed enemy.[27] His relations with Lestock were hopeless from the start. According to Beatson,[28] he had made Lestock's early recall a condition of his taking the command and had publicly rebuked him at their very first meeting in the Mediterranean. On the evening of 9 February Lestock had gone on board the *Namur* to ask if Mathews had any commands for him in the coming battle, to which Mathews is said to have replied that it was a cold night and wished him good evening. If this story is true it shows an unparadonable attitude on Mathews's part. Here was someone he disliked, whom he had himself described as a cripple, making the tiring ship-to-ship visit in an apparent spirit of humility. No wonder Lestock felt bloody-minded next day and the day after, though it was equally unpardonable on his part that His Majesty's Service should suffer thereby.

The second charge against Mathews at his trial was that, although it was his duty 'to appoint necessary and proper signals, for the better conducting the said fleet under his command, as well by night as by day, according to the various exigencies of His Majesty's Service; yet the said Thomas Mathews did not direct and appoint necessary and proper signals'. In particular, he had made no provisions for a night signal ordering the fleet to bring to in line of battle. Instead,

On the tenth of the said month of February His Majesty's fleet bearing down on the said combined fleets of France and Spain, then laying-to in a regular and well-formed line of battle, in full sight, and within four, five or six miles distance of His Majesty's fleet, the said Thomas Mathews, in the night, did make the night-signal for the fleet to bring to; by which signal the windermost ships of the fleet [Lestock's squadron] were to bring to first; and did not appoint and make a night-signal to form the line of battle, and to bring to, and keep in the line.

To all this Mathews answered that he had given the necessary and usual signals in the same manner and form as he himself had received them from Leake, Admiral James Berkeley and Torrington (that is, George Byng, Viscount Torrington); that if Lestock had not shortened sail during the afternoon of 10 February, he would have been well up with the centre and in line abreast by nightfall; and that while it was true that the night signal to bring to directed the weathermost ships to bring to first, (and rightly so, to prevent the weathermost ships colliding with ships to leeward), nevertheless if Lestock had been in his proper station in line abreast, there would have been no ships to leeward of him except the enemy's.

The exact meaning of the charge, and Mathews's answers, are not at all clear. The charge seems to imply that Mathews, as commander-in-chief, should have issued his own set of additional signals and instructions to be used by the ships under his command. In particular he should have been ready with additional night signals. Mathews seems to have contended that this was unnecessary, and that if Lestock had behaved properly everything would have gone satisfactorily.

The court decided in Mathews's favour on the grounds that he was not compelled to issue any additional signals. Nevertheless he should have contrived things better by making a night signal for the line of battle. As, however, no such signal officially existed, he should have sent officers in boats through the fleet, the night being fair and calm.

Asked about this by the court, Mathews replied with remarkable candour that he had had

very little rest for the two nights and days before, and was thereby much fatigued; therefore as soon as I had brought the ship [*Namur*] to, I left the deck in order to rest myself, that I might the better be enabled to go through the next day's expected fatigue: not imagining that an officer of Vice-Admiral Lestock's rank, and very great experience, should require a message sent to him to do his duty.[29]

---

26 A yellow flag at the main topmast head combined with a Union at the mizzen topmast head. Trelawney's Mediterranean Signal Book, *NMM*, TUN/12.

27 *A Narrative of the Proceedings . . .*, p83

28 R Beatson, *Naval and Military Memoirs of Great Britain, 1727–83*, London, 1790, 3 vols, I, p153

29 *Admiral Mathews's Remarks on the Evidence given and the Proceedings Had, on his Trial and Relative thereto*, 1746 (hereafter cited: *Admiral Mathews's Remarks*) pp50 1. (*NMM*, TUN/191).

We know that Mathews had in fact made at least one addition to the Fighting Instructions. This was to Article 1, concerning station-keeping 'save only in time of fight.' It reads: 'Every ship is to observe and keep the same distance those ships do which are next to the Admiral'; meaning that the distance between the admiral and his seconds was to be the determining distance for all. Mathews also issued signals for flagships in a line of battle or abreast to be a league or two leagues apart.[30] He may well have issued more, and the whole situation thus remains obscure. What the court was really saying was that there was a need for more signals to cope with particular needs; this was a fact apparent to all.

The charges, answers, counter-charges and exculpations made in the controversy reveal exactly the attitude of mind satirised by Vernon when he wrote, 'Our enquiry since [the Dutch Wars] has not been whether a Commander be bold, daring and intrepid, but whether his ship lay in such and such a disposition; and what he said and she said; whether a line of battle was properly formed, and if such a signal did not represent this or that.'[31]

Mathews's case was that Lestock failed to support him. Lestock's case was that Mathews had rashly committed the British fleet to battle before the line was properly formed and that it was impossible to support him as long as he kept the signal for the line flying.

The main interest of the controversy lies in the issues raised about the line of battle and the methods of engagement. On the line itself the pro-Lestock History of the Mediterranean Fleet[32] is superbly authoritative:

A line of Battle is the basis and foundation of all discipline in sea fights, and is universally practised by all nations that are masters of any power at sea; it has had the test of a long experience, and stood before the stroke of time, pure, and unaltered, handed down by our predecessors as the most prudential, and best-concerted disposition that can possibly be made at sea.

The History continues with the observation that 'no temptation, no allurement of any sudden advantage to destroy the enemy, no exploit, let it be ever so considerable, and well executed, can extenuate the guilt of disobedience' involved in leaving the line without permission (Article 24) or to 'pursue any small number of the enemies ships until the main body be disabled or run' (Article 21).

Was there, however, anything to prevent the admiral from signalling 'general chase' together with the signal to engage, as Anson and Hawke did later in their respective battles of Finisterre? This question was not fully considered at the time, but the answer would seem to have been that, where the

enemy were of equal force and in a well-formed line, it would have been imprudent and even culpable to try the experiment merely because the attempt to form the line of battle had broken down. Against a weaker enemy, already in retreat, a pell-mell irregular attack might be necessary as well as justifiable.

Mathews also accused Lestock of failing to repeat his signal to engage; in his answer, Lestock contrived to raise wider issues than the comparatively simple one of accepted practice. He argued, first, that the signal, being immediate and executive, required no repetition by other flag-officers, nor was any provision made for repetition in the article itself (no 13), 'for where repeating is necessary, every article expresses it shall be repeated.'[33] Lestock had been in two general battles and the signal to engage was not repeated in either. At Malaga 'he was Lieutenant to the Admiral of the Namur [Shovell] who did not repeat it, nor did any other flagship, either English or Dutch'. If, continued Lestock, Mathews had really intended the signal to be repeated, he would have taken the opportunity to say so at the time of his addition to the article itself, which read '. . . and strictly charged not to fire before the signal be given by the Admiral'.[34] At Malaga Rooke did not make this signal before he was within gunshot of the French admiral; in the Battle of Beachy Head it was not made until just before the action began, long after the fleet had been drawn into a proper order of battle. At La Hogue (Barfleur) Russell did not signal to engage until 'within three quarters musket shot'.[35] To make the signal to engage as Mathews did, with his fleet at a distance and his line unformed was clearly reprehensible. If, however, Mathews had hauled down the signal for the line, and instead signalled to Rowley 'to keep his wind' then the provisions of Article 19 would have become operational, that is, 'the van of the admiral's fleet is to steer with the van of the enemies, and there to engage them'.[36]

Mathews was questioned at his trial as to why he did not change the signal for the line ahead to the line abreast when he bore down to engage the Real Felipe, since the manoeuvre required what was practically a line-abreast approach. To this he answered that at the time of attacking he consulted Captain Cornwall of the Marlborough, who was killed in the battle, and they agreed that the signal for the line abreast 'would not be so proper as that which was then flying', and that had he hoisted it 'confusion and disorder' would have been 'unavoidable'.[37]

The trials and the resulting pamphlets also raise interesting questions about station-keeping in battle. Mathews had made an addition to Article 1 of the Fighting Instructions which is in fact no more than the signal for forming the line of battle ahead. It reads: 'And every ship is to observe the same distance those ships do which are next the Admiral, always taking it from the centre'. Though well-intentioned, this addition must have been difficult to follow with a large fleet. How could ships at the very head of the line judge if they were under or over the distance between the admiral and his seconds? Signals for particular distances would have been better. Armed with this extra fact, Lestock denied Mathews's accusation that he had failed to obey the signal for Article 12 'where the Admiral would have the Admiral of the White and his squadron make more sail, though himself shorten sail [waiting for Lestock to come up], he will hoist a white flag at the ensign staff'.[38]

As to Lestock's failure to cut off the rear ships of the Spanish division, why did not Mathews make the signal for Article 27, 'If the Admiral would have any particular flag ship and his squadron or division give chase to the enemy . . .'[39] As the signal for line ahead was flying he could not possibly have got properly into action by running more to

30 Trelawney's Mediterranean Signal Book, Tunstall Collection, NMM, TUN/12 (formerly S/MS/AS/1)

31 Quoted by Richmond in The Navy in the War of 1739–48, III, p254

32 The History of the Mediterranean Fleet from 1741 to 1744, 1745, p.48 (NMM, TUN/180). See also Admiral Mathews's Charge Against Vice Admiral Lestock, 1745 (hereafter cited: Admiral Mathews's Charge), pp11–3 (NMM, TUN/181).

33 Admiral Mathews's Charge, p9

34 Admiral Mathews's Charge, p10

35 Admiral Mathews's Charge, p12. See also Vice-Admiral Lestock's Defence to the Court Martial, Giving a short View of the Nature of his Evidence, 1746 (hereafter cited: Vice-Admiral Lestock's Defence), pp38–9 (NMM, TUN/182).

36 Admiral Mathews's Charge, p35

37 Admiral Mathews's Remarks, pp54–5

38 Admiral Mathews's Charge, p37

39 Admiral Mathews's Charge, p21. See also Vice-Admiral Lestock's Defence, pp45–6.

leeward, nor could he make more sail than he did. Yet, as Sir Herbert Richmond so cogently pointed out, Lestock only two years before had reprimanded various captains for not making sail quickly enough to join the line of battle irrespective of their divisional commanders. When one of them, Captain Barnett, asserted that he ought to take station on his divisional commander, Lestock had strongly denied it, stating that it was the duty of captains to join the battle as quickly as possible. 'Is it your duty to see two-thirds of the squadron sacrificed to the enemy when you could and did not join the battle?' Here speaks a more vigorous Lestock, even if the general tone of his letter is 'passionate and ridiculous'.[40] Besides, as Richmond again points out, all this was in conformity with the traditional doctrine of the duty of all captains 'to keep up with the admiral of the fleet'.[41]

An interesting suggestion appears quite casually in the *History of the Mediterranean Fleet*, that in battle the admiral should hoist his flag in a frigate, with a boat from every ship ready alongside 'to wait his immediate commands'.[42] He would thus avoid being exposed to the enemy's fire and avoid having his view obscured by smoke.

Tactical development following the Battle of Toulon

The battle of Toulon began a period of tactical development in the English fleet, carried out by successive admirals and commanders of squadrons, trying to find a means of obtaining decisive results from the fleet actions. It seems at least arguable that the battle of Toulon was a providential happening, in that it revived that earlier spirit of continuous innovation which had long been in decline in the Royal Navy, largely due to the static condition of naval architecture and gunnery and the static relationship of the chief maritime states. Scandal provoked reform, and the Battle of Toulon can now be regarded as a great sea-mark in the history of British tactics. From it stemmed new ideas about bringing a fleet into action as quickly and effectively as possible while at the same time avoiding confusion and misunderstanding. Vernon's ideas were taken up and exploited by Anson and Hawke and the course was laid for the triumph of Lagos and Quiberon.

In August 1744 Mathews handed over the command of the British Mediterranean Fleet to Rowley, Lestock having already gone home. Morale was low. A number of captains and other officers had already been ordered home or were expecting to go home, either to face trial by court-martial or to give evidence as witnesses. Rowley rose to the occasion. On 17 September, within a month of taking command, he issued 'Observations and Instructions in Relation to Engaging the Enemy'. These were accompanied (presumably at the same moment) by some fifty new signals (new that is to the Mediterranean fleet).[43]

Rowley apparently decided that decisive action could be obtained only if the traditional concept of station-keeping were modified. After quoting Mathews's additional instruction on station-keeping, he gave it as his opinion that when forming line of battle each ship should get into station in her own division, and that it should be the responsibility of division and squadron commanders to get their own ships into station and align them with the centre of the fleet. He revised Article 29, which required ships to keep their station in action, laying it down that where 'any ship shall be overmatched or pressed by the enemy, that the next ship to her breaks the line and goes to her assistance'.[44] If the next astern failed to act, then the next but one was to do so, after first sending an officer to ask the captain of the ship ahead of him why he had not supported the ship in distress. 'All ships are to keep their barges alongside, manned and armed, for this or any other service during the time of an engagement.' Mutual support, in effect, was to take precedence over line-discipline. Attacks by enemy fireships were to be prevented by the barges of the ships directly ahead and astern of the ship endangered. If the fireship was already too close to her intended victim, the next ahead and astern were 'to close the line between her and the enemy so as to cover her from the fire [from the enemy ships of the line, presumably, not the fireship] while her own people are endeavouring to save her'. Frigates and small tenders were to assist.

As well as these injunctions to break the line of battle, under particular circumstances, Rowley sought to improve the degree of tactical control. He ordered that, if a flag officer were obliged to shift his flag because of a fireship attack, his second was 'to have his [that is, the admiral's] flag ready to hoist the moment the Admiral comes alongside'. Repeating frigates were ordered to fire a gun at ten minute intervals to call attention to any signal not observed by the private ships. Every ship was to have a midshipman on the poop 'to watch and write down all signals, etc, as pass in the engagement'. The master was to be responsible for insuring this was done; he was to report every quarter of an hour, and to replace the midshipman if he were killed or wounded.

Rowley's new day, night and fog signals included all the Additional Signals issued by Vernon in the West Indies operations of 1739-42. Their introduction into the Mediterranean, however, reflects well on Rowley's capacity as commander-in-chief. These signals include those for closer engagement, for closing up scattered squadrons, for small ships to leave the line and form a reserve corps, for the slowest ships to drop astern, and for the same ship to lead even if the fleet tacked, to prevent confusion such as was suffered by Mathews at Hyères Bay. His night signals include those for forming a line of battle, and for identification of friends. They foreshadow the possibility of night action.

The gap between Millan's summary of Vernon's and Rowley's innovations and the later stages of the war in the Mediterranean is covered by a manuscript signal book signed W Trelawney.[45] The middle double page is entitled, 'The Sailing & Fighting Instructions, As they are observed in the Royal Navy of Great Britain, with the Additional Signals of Admiral Mathews, Rowley, Medley and Byng'. The additional signals are of great interest, especially as none of the admirals left much reputation as a tactical innovator. Rowley is credited with twenty-two signals, together with variations to cover different squadrons or points of the compass; Byng has nine, Medley five and Mathews four.

---

40 The full correspondence is printed in Charnock's *Biographia Navalis*, LV, pp213–17. Barnett refused to be put down and answered Lestock in a good-tempered and carefully argued letter to which Lestock did not reply.

41 Richmond, *The Navy in the War of 1739–48*, II, pp18-19. An example of Barnett's interest in tactics is provided by a comment in Christopher O'Bryen's translation of Hoste, *Naval Evolutions . . .*, p17, following Hoste's method for chasing a ship from leeward (p35, Remargue no 3) 'We now practise this method in the navy, when chasing an enemy, as it is found to be mathematically true. It was first put in practice by Admiral Barnett in the Mediterranean, upon reading the treatise of L'Hoste.'

42 pp21-2

43 Richmond, *The Navy in the War of 1739–48*, II, pp261-3

44 Rowley wrote article '29' in error.

45 *NMM*, Tunstall Collection, TUN/12 (formerly S/MS/AS/1). This is a vellum-bound pocket book which may well have belonged to Captain William Trelawney, later a baronet and Governor of Jamaica (1767–72) and not to be confused with Edward Trelawney, also Governor of Jamaica (1736–52). Since the additional day signals of all four admirals are worked in amongst the rest, the book could not have been compiled until Byng issued his signals on taking over from Medley.

Rowley's are by far the most important; they include all those already issued by Vernon and later correctly credited to him in the pro-Lestock booklet *A Narrative of the Proceedings of His Majesty's Fleet in the Mediterrean.* Those of Rowley's signals not derived from Vernon are either variations or extensions of existing signals, or signals of a less important nature, such as for taking tenders in tow, and for victuallers going into port. There is a separate list of Rowley's night signals, fourteen in all, including Vernon's. Rowley's fog signals are listed separately.

Peyton and La Bourdonnais

On 25 June 1746 an action took place off Negapatam between a British squadron under Commodore Edward Peyton and a French squadron under that redoubtable and colourful figure, Mahé de La Bourdonnais, already something of a St Malo corsair cum East India nabob, and now Governor of Mauritius and Bourbon, as well as commander of the French navy in those waters. He was, moreover, as far as we can tell, the first sea officer to devise and use an absolutely unqualified numerical system of signals. Peyton had had no real fighting experience, and he had been only two months in command as successor to Commodore Barnett.

At daylight, with the wind south-west and Peyton's ships in loose order, the French squadron was sighted to the SE. Peyton at once steered a little south of east to cut across them to windward, signalling for the line ahead as he went. The French meanwhile were steering north, and on sighting the British, La Bourdonnais also began to form his ships in line ahead. The interesting feature of this battle was the unusual balance of the two squadrons. Peyton had his broad pendant in the *Medway* (60) with three 50s and a 40. As the *Medway* was leaking at the rate of 30in per half hour, Peyton was in serious danger if heavily attacked. La Bourdonnais had his flag in a 70, and commanded a squadron of eight Indiamen of from 36 to 28 guns. In both gunpower and manpower, he

Amiral Bertrand François Mahé de La Bourdonnais (1699–1753), by Lepaulle. MM

outnumbered Peyton very considerably.

Neither commander showed much inspiration. La Bourdonnais stood so far north before hauling his wind that by the time he brought to and formed a line, tightly closed up pointing NW, Peyton had been able to pass to the SW of him, and so gain the weather gage. La Bourdonnais' *Achille* (70) was the centre ship. Peyton now tacked in succession and began to edge down so as to attack the French van. At about 4.30pm the British closed in, starting to engage more or less ship to ship with the French centre and rear, and passing on towards the French van. The two leading French ships were quickly driven to leeward, on which they wore round the lee side of their own line, then tacked and formed up again in the rear. At 5.30 Peyton signalled his two leaders to tack, and then the remainder, all in succession. By 6.30 the five British ships had tacked, and were passing the French line on the opposite tack. By 7pm Peyton had cleared the French line and the battle had ended.

Next morning the French were still in sight to the NE. Peyton made sail and formed in line of battle. However, the wind dropped by evening without the action being resumed. A council of war aboard the British ships decided in favour of steering south to Trincomalee for repairs. Although La Bourdonnais claimed a victory, he made no attempt to renew the fight, having only provisions for twenty-four hours, and fearing that he might be carried to leeward of Pondicherry by the strong northerly current.

The first convoy battle of Finisterre

The two battles fought off Cape Finisterre in 1747 have markedly similar features. In both battles a stronger British force completely defeated a weaker French force seeking to cover the retreat of a threatened convoy. In both battles the French were overwhelmed by a pell-mell attack from the rear.

In the first battle, that of 3 May, the British admiral, George Anson, profited from having worked up his squadron very efficiently after leaving Portsmouth barely a month before. He had been continually exercising his ships in various tactical and scouting formations, and had them constantly ready for battle. As in the previous year when in

Peyton and La Bourdonnais

Two leading French ships wear, then rejoin end of the line

French stand on too far – lose weather gage

Peyton tacks in succession

French form line of battle

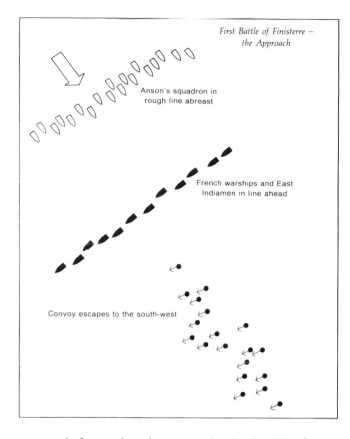

First Battle of Finisterre – the Approach

Anson's squadron in rough line abreast

French warships and East Indiamen in line ahead

Convoy escapes to the south-west

Admiral George, Lord Anson (1697–1762); print by J M Ardell after a painting by Sir Joshua Reynolds. BM

command of a squadron, he was continually signalling for the line ahead and abreast, and for exercising the guns. All this meant additional signals, and at least some reasonable degree of mutual confidence.

Early on 3 May his squadron was spread out in a loose line abreast covering an area of about nine miles, steering large before a NNW wind, when he sighted straight ahead of him the French convoy of which he already had had intelligence. He at once signalled 'general chase', and at 11am called in his cruisers. By noon he was near enough to distinguish details. The French convoy was in flight to the south-west. Astern of it was the escort, about twelve French warships commanded by the Marquis de La Jonquiere, now forming in line abreast. By 1pm the British squadron was only three miles from the French warships, which were now brought to with their heads to the north-west in line ahead, directly barring the British approach to the convoy. Anson signalled for the line of battle abreast, and called his second-in-command,

Rear-Admiral Sir Peter Warren, to come on board the flagship. Warren is said to have urged an immediate pell-mell attack, the weakness of the French both in size and numbers being now clearly observable. Anson apparently rejected this proposal and signalled for the line ahead on the starboard tack, so as to bring his ships roughly parallel to the French. This manoeuvre took two hours to complete. When, at about 3pm, Anson was at last able to signal his ships to fill and

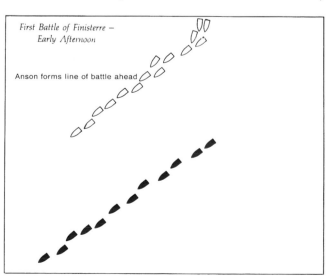

First Battle of Finisterre – Early Afternoon

Anson forms line of battle ahead

Admiral Sir Peter Warren (1703–52), by Thomas Hudson. NMM

*The Battle of Toulon, 1744
(see pp 83–90).*
NMM

The Situation of the ENGLISH, FRENCH and SPANISH FLEETS, when they begun the Engagement

The English Commanded by Admiral Mathews, Vice Adm.ᵗ Lestock and Rear Adm.ᵗ R

This Plate is Humbly Inscrib'd to the Right Honourable the Lords Commissioners of ÿ Admiralty by their Lordship's most Obedient, and most Dutiful Servant, James Main

The Combin

The British Fleet

Squadron

Rowleys

Admiral

Rear

When the French Tack'd at Night they steer'd this way to assist the Spanish Admiral who was in great Distress

When the French Fleet Tack'd, the English steer'd this way

The representation of the Disabled Ships of

When we Engaged we had a fine Top Sail Breeze at E.S.E.

Cape Siae is made to Bear a little to the West ward upon the Account of Shewing the Bay of Hieres

The first Sixteen Ships in their Line are French, and the other Twelve Spanish.

English Line of Battle						
The Revenge to Lead with ÿ Starboard & ÿ Sterling Castle with ÿ Larboard Tacks aboard.						
	Ships	Comanders	Guns	Men	Ships	Comanders	Guns	Men
1	Sterling Castle	Cooper	70	480	Vice Adm.ᵗ Lestock			
2	Warwick	West	60	400	24 Dunkirk	Purvis	60	400
3	Chatham	Hughes	50	300	25 Cambridge	Drumond	80	600
4	Nassau	Lloyd	70	480	26 Neptune	Lestock	90	770
5	Chichester	Dilk	80	600	27 Torbay	Gascoign	80	600
6	Boyne	Frogmore			28 Russel	Long	60	600
7	Barfleur	Ripley	90	765	29 Buckingham	Towry		
8	Prin.ᵗ Caroline	Osborn	80	600	30 Elizabeth	Lingen	70	480
9	Berwick	Hawke	70	480	31 Revenge	Berkeley		
10	Guernsey	Cornish	50	300	32 Romney	Cockshal		
11	Kingston	Lovett	60	400	33 Oxford	Pandot	50	300
12	Salisbury	Osborn	50	300	34 Mercury Fire Ship	Pedel	8	45
13	Bedford	H.G.Townshend	70	480	35 Feversham	Watkins	44	250
14	Dragon	Watson	60	400	36 Nonsuch	Strange	50	300
15	Somerset	Sclater	80	600	37 Diamento repeat Signals	Hodsdel	40	250
16	Princessa	Pett	70	450	38 Dursley D.ᵒ	Vanburgh	22	130
17	Norfolk	H.J Forbes	80	600	39 Winchelsea D.ᵒ	Marsh	22	130
18	Namure	Mathews	90	780	40 Ann Galley Fire Ship	Mackey	8	45
19	Marlborough	Cornwall	90	750	41 Enterprize Barco Longo	Fleming	6	60
20	Dorset shire	Burrish	80	600	42 Sutherland Hospit.ᵗ Ship	Lord Colvil	18	75
21	Essex	Norris	70	480	43 A Tender			
22	Rupert	Ambrose	60	400	45 A Victualler			
23	Royal Oak	Williams	70	480	46 D.ᵒ			
Ja.ʳ Main Delin.ᵗ W.H. Toms Sculp.ᵗ					Total		2550	18022

*Admiral Mathews in the Namure, after consulting with Cap.ᵗ Cornwall of the Marl both bore down upon the Spanish Admiral about One a Clock, the Namure on her Bow, Marlborough on her Weather Quarter, and began the Engagement.
At the same time the Norfolk bore down upon the Constant / the Spanish Admiral and in an Hour and a Quarter beat her out of the Line, upon which the Royal Phi a Stern fired her fore Chase Guns at her to make her come back, but she run to L and never more returned.
About two a Clock Rear Admiral Rowley came pretty near the French Admiral lay sometime without Action; but at last upon his Firing at the Van of the Spa the French Admiral pour'd a Broadside in upon him, which he returnd; and the Caroline coming to his Assistance, they together bore the Fire of four French an Spanish Ships till the French Admiral and his Seconds bore away. After the and Marlborough had Silenced the Firing of the Royal Philip, he order'd Ann Galley Fire Ship to go down and Burn her, but by some Accident or other Sh up within thirty yards of the Spanish Admirals Weather Quarter without d Execution. However this unlucky accident gave new Spirits to the Enemy Royal Philip, now recover'd from the Consternation She was in, employ'd her boat and a stern to take up her Men that had leapt into the Sea to avoid the Conflagration they every moment expected.
The Ships a head of Admiral Mathews Engaged those a head of the Philip; and the Poder a Spanish Ship of 64 Guns Struck to the Berwick, after ged with the Princessa, Somerset, Bedford, and Kingston, and was boarded claiming the Honour of taking her; but the Capt. pointing out the Ship he had submitted, deliver'd his Sword to the Lieutenant of the Berwick. Then French Fleet Tacking to assist the Spanish Admiral, the Berwick had no m than to take the Cap.ᵗ and four of the principal Officers out of her, leaving his Lieuten twenty three Men behind, but the French doubling so fast retook the Ship & after taking out abandon'd her in the Night. next Evening Adm.ᵗ Mathews sent ÿ Essex to Burn her, & she blew up at which time the Enemy were got about Four Leagues to Leeward having bore away in the Night*

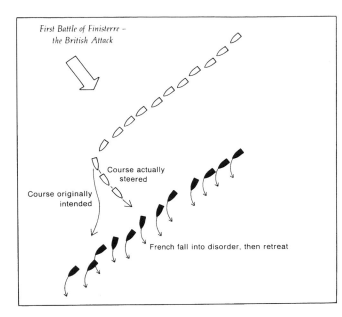

First Battle of Finisterre – the British Attack

Course actually steered

Course originally intended

French fall into disorder, then retreat

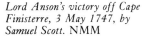

Lord Anson's victory off Cape Finisterre, 3 May 1747, by Samuel Scott. NMM

stand on, he also signalled for the van, Warren's division, to lead large (in succession), that is to steer more to port and so more towards the French van, as otherwise he might have run too far ahead of them. Warren and several of the leading ships responded by steering individually towards the French, thus breaking the line and chasing, instead of bringing the line round in a steady curve. Anson, though no pedant, liked precision. He at once signalled Warren to re-form the line and for the leading ships to lead large, thus allowing an almost perpendicular approach towards the centre of the enemy line.

All this time La Jonquière had been lying to awaiting the

British attack. Having forced Anson to deploy he had gained valuable time for his convoy's escape and as the British bore down, apparently to cross his bows, he signalled his squadron to fill and stand on. The moment, however, that he saw the leading British ship bearing up to lead more directly towards his centre, he bore up himself and took his squadron away as fast as they could go to the south-west, following the convoy. Being inferior in force, he had no wish to fight an ordinary line of battle engagement in which he was certain to be doubled, as well as to have his weaker ships immediately overwhelmed. Very soon his ships became strung out in an irregular formation more like a triple line ahead than a single line ahead or abreast.

Anson hauled down the signal for the line of battle ahead, and hoisted 'general chase' and 'engage the enemy,' both together, thus leaving his ships to get up with the enemy in the best order their respective sailing qualities permitted. Whether La Jonquiere deliberately shortened sail to allow Anson to come up is uncertain. On balance it seems that he did not; in any case, by about 4pm, he was overhauled and brought to action, which showed the immense tactical value of superior speed in pursuit. As the leading British ships closed the French rear, they came under heavy fire, often from two ships at once. After an hour's fighting only six British ships were actually engaged, but by the end of another hour the sternmost French ships were beginning to surrender.

The leading British ships had aimed at crippling the nearest enemy they could find and then making sail to catch up with the next unengaged enemy ship ahead, leaving the damaged enemy astern to be dealt with by the heavier sailing ships as they gradually worked their way into the battle. So great was Anson's confidence that, soon after the battle began, he

First Battle of Finisterre – the Pursuit

British come up with French in ones and twos

detached two of his weaker ships and a sloop to pursue the convoy and shadow it during the night. By sunset he had captured every one of La Jonquière's squadron and, though his own ships were badly damaged aloft, he was able to detach three of them to chase the convoy, of which eighteen out of a total of thirty-eight were subsequently taken.

The most remarkable features of the battle were Anson's refusal to be hurried at the start, followed by his quick decision to chase as soon as the French retreated. La Jonquière's ships fought well individually, several with great gallantry, but having lost their formation and taken to flight they lacked tactical cohesion and acted only out of a sense of self-preservation. In the end a 74, a 66, a 56, two 52s, a 46 and three East Indiamen of 30 guns each, surrendered to a 90, a 74, a 66, four 64s, five 60s, a 54 and a 50. The odds seemed heavy but the difference in comparative strength between British and French ships, mounting the same number of guns, must be kept in mind. It was because of the strength of French ships that some of them were able to hold out so long.

'Anson's' additional fighting instructions, 1747

When, on 1 August 1747, Sir Peter Warren arrived back at Plymouth, too ill with scurvy to continue at sea any longer, the newly appointed Rear-Admiral, Edward Hawke, was named as his replacement. On 12 August Hawke issued a set of additional fighting instructions which are of tremendous importance in the history of naval tactics. Probably they originated with the commander-in-chief, Anson, and apparently there once was a set of additional sailing and fighting instructions in the Naval Library, attributed to Anson, and probably dated 1747 or even 1746.[46] This may have been the first appearance of the new instructions. There is also a manuscript signal book in the National Maritime Museum, dated 19 April 1747, which credits Anson with the idea of forming an emergency line when in pursuit.[47] Hawke himself issued various individual additional instructions, but never again a complete set comparable with that of 1747. Anson's authorship of the latter is not firmly established, and Hawke continued to issue them, but these additional instructions can be attributed to Anson with some reasonable certainty.

There is no doubt that Articles 8, 9 and 10 of Anson's instructions are of the highest importance. They envisaged the maximum degree of offensive effort, and they codified

Byng's tactics at Cape Passaro. Article 8 ordered that, if engaged with an inferior enemy, the overlapping ships either ahead or astern were to leave the line without waiting for a signal, and rake the enemy's van or rear ships. Article 9 ordered that, when chasing an enemy, the five or seven ships nearest the enemy were to draw into a line ahead of the main force 'and to engage the ships in their rear endeavouring at the same time to get up to their van till the rest of my squadron can come up with them'. Article 10 completes the picture by instructing the leading ship in a chase to form an *ad hoc* line of battle, with other ships joining the end of it as they came up. These dynamic instructions, it must be remembered, were only for use against an inferior, beaten or flying enemy. The difficulties of lines tactics where fleets were equally or nearly equally matches were still unresolved.

A summary of the additional fighting instructions of 1747 is given below.

Summary of Anson's additional instructions, 1747

Article 1 Signal for smaller ships to quit the line, to hold themselves in readiness to assist disabled ships, the strongest of those withdrawn to be ready to fill a gap in the line caused by a ship having to withdraw. (This is a repeat of Vernon's instruction in a slightly different form.)

Article 2 Signal for frigates to attack a convoy and for ships to leave the line and do so. (Richmond thought this to be original. Both 1 and 2 are referred to in May 1748 as the first and second articles of Warren's additional fighting instructions.)[48]

Admiral Lord Edward Hawke (1709-81), by Francis Cotes. NMM

46 Letter from Richmond to Corbett, 21 January [1914]; Corbett, *Fighting Instructions*, Corbett's file copy (*NMM*, MS 81/143, Deed Box A)

47 *NMM*, CLE/2/19

48 D Bonner-Smith, editor, *The Barrington Papers*, 2 vols, London, 1937

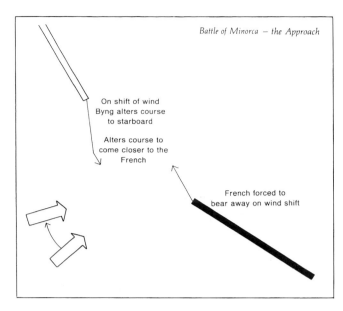

Battle of Minorca – the Approach

On shift of wind
Byng alters course
to starboard

Alters course to
come closer to the
French

French forced to
bear away on wind shift

of the enemy's rear', thus his signal to tack could be no more than a time signal for an evolution which in theory he was bound to execute. He did, however, have two choices as to the method of tacking. He could either signal the fleet to tack in succession, the rear to begin, or he could signal to tack together – as circumstances seemed to require. Of course he could have ignored the article altogether and tried something different, which, provided it had proved successful, would probably have received approbation. The difficulty was that in doing so he would almost certainly have confused his captains; in any case, he did not have a sufficient number of signals to express his orders. Yet Byng was no pedant and, when his secretary, standing beside him on the quarter-deck of the *Ramillies*, 'took the liberty of observing that agreeable to that Article [17] the fleet should tack', the admiral answered 'that he should stand rather beyond their rear before he tacked, as it would give an opportunity to every

ship to lead slanting down on the one she was to engage, and they would not be so liable to be raked by the enemy's shot'.

Here Byng reveals himself as a thoughtful and intelligent tactician, anxious to avoid the typical disadvantage of a perpendicular attack from windward. When, a few minutes later, he signalled the fleet to tack in succession, beginning with the rear, La Galissonnière seems to have interpreted Byng's delay quite differently, fearing a concentration against the French rear. He, therefore, signalled the French fleet to throw all aback, so checking their forward movement. It seems that La Galissonnière had some justification for his fears. He had twelve of the line and two junior flag officers. Byng also had twelve of the line and a 50, which he ordered to withdraw during the battle so as to equalise with the French. Instead of taking station in the centre, he had his only three-decker, his flagship the *Ramillies,* fourth in the line when on the starboard tack and in the exact centre of his own division of seven ships. The remaining five were formed in a rear division under his only other flag officer, Rear-Admiral Temple West. Therefore, when Byng began to tack, Galissonnière saw that the movement would place Byng near the rear of the British line, where he would be well disposed to lead an attack on the French rear.

Byng, seeing that the French had thrown all aback, decided it would now take too long to reach beyond their rear. Wishing to hurry on the battle, he signalled for the British fleet to tack together, thus bringing them on to the same tack as the French.

The two fleets were now on slightly converging courses on the same tack, the French at first with only steerage way on, but soon beginning to pick up speed. Byng's action was now governed by Article 19 of the Fighting Instructions, 'If the admiral and his fleet have the wind of the enemy, and they have stretched themselves in a line of battle, the van of the admiral's fleet is to steer with the van of the enemy's, and there engage them.' Left to themselves, the British van would

*Battle of Minorca – British
Fleet Tacks in Succession*

Temple
West

Byng

Defiance turns one point,
then another, towards
the French

Defiance steers on parallel
course to the French

Fleet follows *Defiance*

*Battle of Minorca – the
Failure to Engage*

French fleet

Byng's intended attack

*Battle of Minorca –
Byng's Attack*

French fleet

Intrepid, damaged

Rearmost ships, further
away, fail to come close
to the French

not to bear down all at once, but to observe the working of the
admiral and to bring to as often as he thinks fit, the better to
bring his fleet to fight in good order; and at last only to lask
away when they come near within shot towards the enemy as
may be, and not bringing their heads to bear against the
enemy's broadsides.[28]

His fleet was now exposed to full French broadsides in its
approach, each vessel effectively having its T crossed by the
French line.

Whether Byng could really have brought off his plan is
doubtful, mainly because of his own distance from the van
with the line of battle in reversed order. Had he stationed the
Ramillies in the centre of the fleet and appointed one of his
captains a commodore, to balance Temple West, he could
both have seen better what was going on and exercised more
control. Had he engaged on the opposite tack and with the
Deptford (50) out of the line, the *Ramillies* would have been
the third ship and he could have controlled the angle of attack
himself. As it was, his tactical approach was a fiasco.

When he signalled to engage, his leading ship was just
under a mile from the leading French ship, with the *Ramillies*
about two and a half miles from her opposite number and the
sternmost British ship over three miles from the sternmost
enemy. The British attack thus became a straggling piecemeal
affair, of which the French took full advantage, especially as
their ships mounted a far higher proportion of heavy guns
than British ships of the same rating. The French moreover
were clean ships, fresh out from Toulon, and could easily
outsail the British.[29]

Having opened a heavy and accurate fire on the five
leading British ships as they approached, the ships in the
French van skilfully filled their topsails and stood on a short
distance so as to upset the aim of the British gunners as their
ships turned into line ahead one after another to engage. The
French then began to bear away to leeward one at a time, but
not in their order in the line, thus giving the impression that
they were beaten. The British van, however, made no
attempt at pursuit, having received heavy damage in masts,
yards and rigging.

Further astern things went badly for Byng. In terms of the
battle as a whole there is much of interest here; in terms of
tactical analysis, however, this part of the battle is no more
than an illustration of the lack of wisdom of approaching an
efficient enemy fleet from windward in a straggling bow and
quarter line, which is what Byng's attack had now become.
In order to get into action, the centre and rear ships of his
fleet had had to keep away a little to larboard, otherwise in
attempting a right-angled approach they would have fallen
too far astern of the enemy. When, therefore, the sixth ship in
his line was damaged aloft and became unmanageable, a gap
opened ahead of it and the ships astern were held up. A
similar hold-up took place when the eighth ship was
damaged. In the confusion caused by smoke obscuring the
scene, collisions were only just avoided, as was the danger of
ships firing at each other. Such was the delay that the
Ramillies, tenth in the line, was never properly in action at all.

There was much subsequent debate as to whether the
captain of the seventh ship should have passed the damaged

eventually come within range of the French and the two
fleets gradually close in battle.

This accepted plan of attack was entirely upset by Byng's
leading captain on the larboard tack, Thomas Andrews of the
Defiance, who kept on a strictly parallel course with the
enemy. Seeing the leading French ship a little ahead of him on
his starboard bow, he seems to have thought it necessary to
make up distance before leading down to attack. Thomas
Andrews was an experienced captain of eleven years standing, though by no means the most senior in the fleet. Byng's
idea of a slanting attack was quite lost on him, and Byng had
no signal with which to express his full intentions. Following
Andrews's lead, Temple West's division, now the van in the
reversed order of sailing, also kept on a parallel course to the
French. As there was no appropriate signal for leading down
to the enemy, Byng 'recollected that the fifth article of the
Additional Fighting Instructions directed the leading ship to
lead to starboard, which he said would answer the purpose he
intended.'[27]

The signal was made and a gun was fired to indicate a turn
of one point, it being impossible from the *Ramillies* to see
exactly how the French van bore from the British. Byng
intended the signal to be a general indication and no more,
but Andrews interpreted it literally and bore away exactly
one point. Seeing that Andrews had failed to understand that
this was really a signal to bear down slantwise and engage,
Byng made the same signal again, with one gun, and
Andrews turned one more point to starboard and towards the
enemy. This was still not enough, as he remained a mile to
windward of the leading French ship. Instead of repeating the
signal yet again and firing, say, three guns, Byng gave up
trying and hoisted the signal to engage, on which the fleet as
a whole turned to starboard and steered what each captain
deemed a direct course for the enemy.

So ended Byng's attempt at a subtle tactical approach. By
his original plan of approaching from slightly astern, he had
hoped to edge down towards the French, as advocated by
Lord Dartmouth in 1688,

---

27 This refers to the Admiralty Additionals, not his own, and is evidence
of their issue as early as 1756. Article V is the same as Hawke's Article V
of 1747.

28 Article VI. 'Lasking', ie sailing with a quartering wind, was not quite
the same as 'bow and quarter' line, which implied an angular relationship to
other ships as well as a special relationship to the wind.

29 Of Byng's thirteen ships only six had been cleaned during the year
1756, one not since May 1755.

Intrepid according to Article 24 of the *Fighting & Instructions*, 'ships astern to close a gap in the line'. All this, however, is beside the point, which is that the British made a poor approach, and the French exacted the appropriate penalty.

When advised by his flag captain to order more sail to be set so as to clear the confusion around him and get into action quicker, Byng made a very significant answer:

You see, Captain Gardiner, that the signal for the line is out [pointing to the signal] and that I am ahead of the *Princess*

---

30 Evidence of Captain Arthur Gardiner at Byng's trial.

The Battle of Quiberon Bay, 20 November 1759, by Dominic Serres the Elder (see pp 115–6). NMM

Louisa [eighth in the line] and Durrell [Captain of the *Trident*, ninth in the line], and you would not have me as Admiral of the Fleet, run down as if I was going to engage a single ship. It was Mr Mathews's misfortune to be prejudiced by not carrying his force down together which I shall endeavour to avoid.[30]

Byng's quasi-pedantry, and the very clear pedantry of his captains, shows the inhibiting effect of the signal for the line of battle. As long as this signal was flying, no-one seemed capable of any obvious or useful initiative, even to the extent of closing gaps in the line as required by Article 19 of the Fighting Instructions. Had Byng hauled down the signal for line and hoisted the signal for closer action, he might have succeeded at least in carrying the *Ramillies* into action against the French rear, though distance was against him and the French ships could always outsail him.

equivalent of the British Sailing & Fighting Instructions, with separate tables of signals.[31] Though officially printed, it appears to represent La Clue's own ideas.

La Clue's signals and orders, 1757

Admiral de La Clue Sabran commanded the French rear under La Galissonière in the Battle of Minorca, and succeeded him as commander-in-chief of the Toulon fleet. In 1757 he was involved in a series of ignominious operations which led to the blockade of his forces in the neutral port of Cartagena by Admiral Osborn, and eventually returning to Toulon. His 'Signaux et Ordres' for that year are the

31 *Signaux et Ordres qui seront observés par l'Escadre du Roi, commandée par M De La Clue, Chef d'Escadre des Armées Navales. De Sa Majesté; Armée à Toulon en mil Sept cens cinquante Sept A Toulon Chez Jean Louis Mallard, Imprimeur ordinaire du Roi, & de la Marine M.DCC.LVII; NMM*, Tunstall Collection, TUN/46, (formerly S/P/F/1). This particular copy is inscribed 'M Le Chev De Fabri cap & état major de l'escadre de M De la Clue.' [sic] The Chevalier de Fabri had been a lieutenant in Galissonière's *Foudroyant* at the Battle of Minorca, and thereafter commanded the frigate *Gracieuse* under La Clue until after the Battle of Lagos Bay. After the war he was associated with Morogues in the reorganisation of the Navy.

The lay-out of the book is clear and simple. Down the centre of the page are the articles, headed with Roman numerals, in the form of 'Lorsque le Géneral . . .' or 'Si le Géneral . . .', giving the order followed by the signal and then by any instructional words necessary. In the left-hand margin is the short title, and in the right margin the signal, written in words. There are 240 day signals and seventy night signals; folding tables of signals are provided for each. These are ruled in squares, each for a signal, in which the words and figures are clearly indicated. Down the left-hand column are the seven positions from which signals could be flown, and along the top the various flags, ten on each page. Reference, therefore, to the square indicated by the flag and its position is easy. The signals were numbered serially from page to page, taking the top line of each page first, then passing to all the second lines, and so on. Each square has the number of the signal at its top, followed by its short title, and the page reference and number of the article of the instructions to which it relates. Eighty-six out of the 280 squares are left blank for extra signals, since only 194 of the 240 articles of instruction required signals. Although the book, a thin folio, could be easily handled on deck, the same cannot be said of the folding tables of signals, on which all reading of signals would depend. Hence the book can really only have been suitable for cabin use.

Thirty-four flags were used and six pendants, compared with the thirty-six flags, three broad pendants, three pen-

---

32 *NMM*, SIG/C/2 (formerly Sm/170); contemporary English translation and copy
33 Sir Julian S Corbett, *England in the Seven Years War*, two vols, London, 1918 (hereafter cited: Corbett, *Seven Years War*), I, p347

Admiral Sir George Pocock (1706-1792); engraving for Hervey's Naval History. BM

dants and a weft employed by Hoste in 1697. The number of flags of the same pattern (though of different colours) had been reduced, and new patterns had been introduced. La Clue used only four crosses, four bordered flags and four *facié* (single stripe or heraldic fesse across a plain colour) in place of six of each as before. All six of Hoste's *percées* (plain colour with a circular blob in the centre) were gone, no doubt because of their obvious confusion with the *bordées*. In place of these were four plain yellow *sautoir* flags (plain colour with a saltire), four *tranchées* (parti-coloured fessewise), and two chequered flags. The night signals appear to be much too simple, consisting of lights in various positions, with or without guns, or of gun-signals alone. Apart from the difficulty of differentiating between the various positions in the ship from which the lights were to be shown, some of the signals by gun were duplicated. It would seem very difficult to operate this system without previous knowledge of the course of action contemplated by the general.

D'Aché's signals 1758-9

D'Aché's signal book, which he used in his three East Indies battles against Sir George Pocock, shows a higher degree of tactical sophistication than anything displayed by Galissonière, Conflans or La Clue.[32] His twenty general orders included an instruction, number 10, that the amount of sail carried by the admiral was not to be made a rule for others. If the admiral shortened sail this was a chance for others to recover lost distance. When the signal was made for ships to correct station, it was not sufficient to set a little more sail; ships were instructed to crowd on sail until they had gained position. Any ship noticing that another ship, especially when chasing, had not seen the admiral's signal, was instructed (Article 16) to repeat it (presumably the entire fleet would have been provided with the necessary flags). Article 18 instructed that in battle no ship was to fire until the admiral had made the signal. No ship was to run a greater risk than the rest, but all should go down to engage together. If the enemy were defeated, the best sailers should go in pursuit, but should take note of the admiral's movements and signals. Several signals were devoted to the possibility that the admiral might wish to avoid close action, either by bearing away (number 7), or by making a running fight (number 18). On the other hand, provision was also made for surrounding a fleeing enemy (number 11), for attacking the enemy's rear (number 24), for the squadron to engage all together (number 26), to close the enemy to within musket shot (number 27), to board (number 28), and even to continue an action into the night (number 29). Flag officers were directed to repeat the admiral's signals, and to take the initiative themselves in battle when necessary by making signals to their own divisions.

Pocock and d'Aché

The three battles in the East Indies fought between Sir George Pocock and the Comte d'Aché in the years 1758-9 show the limitations placed on tactical skill both by internal and external circumstances. In the first battle, off Cuddalore on 29 April 1758, Pocock had only five of the line and two 50s against d'Aché's own 74 and seven Indiamen, of which the strongest was a 60. The battle began with Pocock chasing d'Aché, who eventually formed a line and lay to, waiting. 'Under these circumstances,' wrote Corbett, 'seeing that he was in numerical inferiority, Pocock had practically no choice either by the Fighting Instructions or the tactical science of the time, but to form line himself and attack van to van.'[33] The qualification implied here is important: had he been superior in numbers he might have tried a concentration on the French van. Pocock, therefore, signalled for the line of battle ahead at half a cable distance. Even so, his well-understood form of attack was a failure. For various reasons the three near ships were never fully engaged, and their captains were severely dealt with after the battle. D'Aché also was not properly supported, and after an explosion in his flagship and the failure of several of his ships to stay in action when signalled to do so, he bore away after them. Pocock now signalled general chase but his four leading ships, which had done practically all the fighting, were so damaged aloft that he had to desist as night closed in.

In the Battle of Negapatam, 3 August 1758, Pocock was able to force action because of a sudden change of wind in his favour. He had the same fleet, less one 50, while d'Aché was minus the wrecked Indiaman but plus a 74. Pocock won this battle by hard fighting. The leading French ship lost her mizzen; D'Aché's flagship lost her wheel, left the line for repairs and later fell on board another ship. Two other French ships left the line and d'Aché himself now bore away. Pocock signalled for closer action and then for general chase as the French crowded sail. By nightfall they were out of range, having suffered 850 casualties, against only 200 in Pocock's ships.

In commenting on Pocock's failure to obtain a decisive victory in either action, Corbett points to the general success of his main strategy, adding that it is sounder to dwell on this 'than to disparage him, as the custom is, because he knew no other way of bringing an unwilling enemy to a decisive action than by attacking in general chase'.[34]

In the Battle of Ponchicherry, 10 September 1759, Pocock had seven of the line and two 50s. D'Aché had four king's ships plus a 68 and six 54s of the Company's – over 100 guns more than Pocock. Despite his inferiority, Pocock closed from windward, with the French waiting to receive him in line ahead. Bearing down in line abreast, he seems to have had little choice as to how he should attack except by forming line ahead on a parallel course. As it was, he made an excellent approach. His next ahead and second-in-command, Rear-Admiral Charles Stevens, ran up the French line and drove d'Aché's next ahead out of the line. In this way four leading British ships more than dominated the five leading French vessels, though their respective total armament was 242 to 320 guns (121 to 160 in terms of broadsides), exclusive of the higher proportion of heavier guns mounted by the French ships. Pocock and his two next astern now came into action against the French centre, but, through poor sailing capacity, the two rearmost British ships came very late into the action. After five hours' fighting only six French ships were still left firing in the line, and soon afterwards two of these quitted it. D'Aché was badly wounded and his flag-captain killed, and the officer in command wore the French flagship. The three remaining ships still engaged soon followed, crowding sail. Of Pocock's fleet, only the two rearmost ships were capable of pursuit, the rest having received heavy damage to their masts, yards and rigging, so that he could do nothing more than lie to. The French casualties were 1500 and the British 570.

D'Aché had preserved his fleet and kept it mobile, but only at the cost of heavy casualties and hull damage. Had Pocock attempted a more ambitious form of attack in any of his three battles, it is unlikely that he would have succeeded as well as he did, the French being not only superior in numbers but far more adroit in specialised manoeuvres.

The Battle of Lagos Bay

The Battle of Lagos Bay, 18 August 1759, presents few tactical features but raises some interesting tactical and

signalling questions. The French fleet of ten of the line, two 50s and three frigates, commanded by La Clue, was attempting to reach Brest from Toulon. They had already succeeded in passing Gibraltar on the night of 17 August 1759, pursued by Admiral Boscawen in command of a more powerful fleet issuing from Gibraltar in a great hurry and in no special order. La Clue had given out Cadiz as his first rendezvous, apparently thinking his ships would not be able to keep together over the whole course, which would involve at least 1700 miles of sailing. At midnight, being near Cadiz but having his whole fleet with him and a favourable wind at a little south of east, he decided to lead on round Cape St Vincent.

French authors are in doubt as to what signal he made, but Corbett guessed it was 'sail large on the starboard tack'.[35] His printed signal book issued in 1757 suggests that he made night signal number 27, 'Prendre chasse les amures à tribord, au plus-près au largue'. The signal for this was for the admiral to extinguish his lights and fire three guns or three rockets. If he actually made this signal, it is likely that he fired the guns as these were less likely than rockets to give away his exact position to the pursuing British. The signal continues 'Les vaisseaux seront la même route que le Général, et auront grand soin qu'il ne paroisse aucun feu'.[36]

At 2am he could still count his fleet; soon after, however, Boscawen's ships began to creep up and fire on his rear. According to some French accounts, he then put out his own poop lanterns and so continued till daybreak. Meanwhile three of his line ships, the two 50s and all three frigates (including Captain de Fabri's Gracieuse) had lost contact and worked their way into Cadiz, following their original orders; no doubt they had missed his signal.

There seems some confusion here, as the putting out of the poop lights should have been part of the earlier signal. Moreover, the first article of the 'Advertisement' to La Clue's night signals says 'Lorsque le Général éteindra ses feux de poupe, on redoublera d'attention dans tous les vaisseaux pour ne laisser voir aucun feu'. If he made signal number 27 he must have put out his lights anyhow, which would have indicated that a change of course was being signalled. It seems almost certain from this that the ships which broke away to Cadiz did so because they had already lost contact, not merely with La Clue, but with the ships immediately ahead of them who dowsed their own lights in conformity with the admiral's signal.

The running fight on 18 August occurred, first, because La Clue mistook the British for his own ships which had run into Cadiz and so allowed them time to come up, and second, because the British were able to come up on a strong easterly breeze while the French lay almost becalmed. Beatson says

34 Corbett, Seven Years War, I, p350

35 Corbett, Seven Years War, II, p36n

36 Copy belonging to the Chevalier de Fabri; NMM, Tunstall Collection, TUN/46

that when La Clue saw the British going for his van and centre he

> made his fleet luff up as much as they possibly could, so as to form a sort of crescent, by which position, the whole of his ships in their van and centre were enabled by their fire not only to assist the rear, but each other, in their endeavours to repel the attack which they looked for every moment from the British Admiral. By this manoeuvre of M de la Clue's, such of our ships as first got up with the enemy's rear, and to leeward of their line, were thrown cut of action; while, for want of a sufficient breeze of wind [which had dropped again], they could not get into it again[37]

Boscawen's flagship, after engaging La Clue's flagship for half an hour, lost her mizzen and both topsail yards and fell astern, so that Boscawen had to transfer his flag. This success, however, was of no avail. The French were being clearly overpowered and knew it. La Clue signalled a retreat on a freshening breeze and left his rear ship, which had already been heavily hit, to be partially dismasted and taken. La Clue's difficulty, just like that of Conflans and later Villeneuve, was that he had been ordered to go from one place to another. Had he been instructed to break Boscawen's aggressive blockade of Toulon by boldly attacking him as he lay refitting at Gibraltar, or as he tried to come out, things might have been different.

Boscawen later complained that his leading ships had concentrated on the French rear instead of 'attacking in inverted order, by which each ship as she came up passed on under cover of the one already engaged and got alongside the next ahead, until all the enemy from rear to van were tackled in succession'.[38] As he had not issued any instruction for this, he could only signal his leading ships individually to make more sail, the real point of which they failed to interpret. He had, indeed, just issued additional instructions (numbers 9 and 10) to the effect that when chasing a fleeing enemy he might signal for five or seven of his leading ships to start engaging the enemy by forming a line without regard to seniority or the order of battle but according to their distance from the enemy.[39] This, however, as Corbett points out, was not what Boscawen now required. It was merely a provision

for an *ad hoc* line of battle when in pursuit, instead of a means for trying to get at the enemy as far ahead as possible. In any case, these instructions were merely a repeat of those issued by Hawke in 1747. The anonymous sea-officer who translated Morogues, suggested in a footnote that Boscawen would have done better to have divided his fleet 'into three small divisions, the better to surround the enemy'.[40] He noted that Mathews and Haddock ran right through the retreating Spanish fleet at Cape Passaro (1718) until they were able to engage the two leading Spanish ships, thus holding up the Spanish retreat until Sir George Byng came up with the main body of the British.

The obvious conclusion to be drawn from all this is that Boscawen had doubtless made adequate plans for fighting La Clue off Toulon, but not for a long running fight up the Atlantic coast of Spain and Portugal. When the moment came, he was caught unawares, being at dinner several miles away with the Spanish Governor of San Roque, together with some of his senior officers. He had failed to make known to his captains in advance what he wished done in a running fight, and in the event it was too late. Doubtless some of his captains behaved badly, or at least unimaginatively, but no blame attaches to the official Sailing & Fighting Instructions. Boscawen could either have added pursuit signals to his repeat of Hawke's additional instructions of 1747, or he could, like Nelson, have worked out plans with his captains to cover pursuit contingencies. He apparently had done neither and his captains merely acted according to the more generally accepted 'custom of the service'. Boscawen was no innovator, and in relying on Hawke's twelve-year-old instructions he suffered the inevitable consequences.

---

37 Robert Beatson, *Naval and Military Memoirs of Great Britain from 1727 to 1783*, six vols, Boston, 1972 (hereafter cited Beatson, *Naval and Military Memoirs*), II, p318.

38 Corbett, *Seven Years War*, II, pp37-8

39 Corbett, *Fighting Instructions*, p221

40 p16

The Battle of Lagos, 18 August 1759, by Thomas Luny. NMM

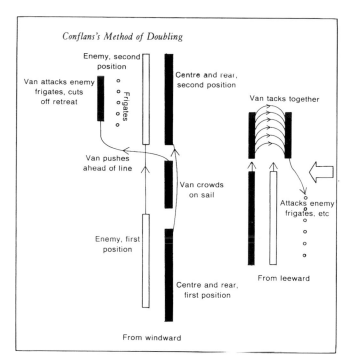

Conflans's Method of Doubling

During the night, two French 74s slipped away, one eventually making Rochefort and the other the Canaries. Next morning's destruction of La Clue's remnant of four on the neutral shore of Lagos Bay and within range of Portuguese forts merely demonstrates the final demoralisation of the French after their navigational failure to weather Cape St Vincent.

The Marquis de Conflans's signals, 1759

Conflans's various sets of signals, bound together in a single volume, mark yet another stage in the attempt to provide instructions and signals for as many eventualities as it seemed appropriate to envisage.[41] The volume might be likened to a series of textbooks for a course of instruction rather than a handy guide to action. Throughout the volume, a standard form is followed, with the order on the left of the page, the signal in the middle and the observation on the right. Each group of signals is preceded by some *avertissements,* a *plan général* or an *order général.* At the end of the book are twenty-three pages of tactical diagrams.

In the day signals, under sail, Conflans reserved the foretopmast head for chasing signals, the mizzen topmast head for simple navigation, and the ensign staff for all non-navigational signals. Forty-four flags and three *flammes* were used. Six bore saltires, six were plain colours with a border, six were striped, eight had crosses and eight included yellow. Reserved for the *général* were the red, white and blue and all flags of these three colours. Yellow mixed with any of the three colours was reserved for signals by private ships.

Apart from the means of tactical control, a great deal can be learned from this volume about Conflans's tactical system. Half a cable was his standard distance between ships in line of battle, weather permitting. No ship was permitted to quit or disorder the line unless disabled, in which case she was to be assisted by ships not in the line. Ships were not to open fire until *à la petite partée du canon.* If the enemy tried to break the line from leeward, ships at the threatened point were to close up so that the enemy was obliged to run aboard and foul one

or more. If the enemy was outnumbered, his van might be doubled from windward or leeward. If the fleet was to windward, and presumably if the attack was launched from windward, the overlapping ships of the French van were to pass to leeward of the enemy van, and then put about in succession so as to regain their original course, now parallel with the enemy van. The leading ships were to shorten sail until the whole detachment was properly formed in line. Meanwhile, these detached ships should not approach too close to the enemy because of the danger of being hit by gunfire from ships on the enemy's windward side, or of hitting them. The mere presence of the detachment was considered sufficient to worry the enemy and make it possible to cut off the retreat of any of his ships driven to leeward. In the meantime the detached ships were expected to attack the enemy frigates, fireships and storeships, stationed on the unengaged side.

If the enemy were to windward, Conflans wanted the French van to crowd sail so as to be sure of weathering theirs, and then put about together, thus coming on a parallel course with the enemy's van, and to attack their frigates, fireships and storeships. To avoid the danger of the enemy doubling the French van from windward, the French van was to attack the enemy's as they came round, ship for ship. To double the enemy's rear from windward, the overlapping ships were required to pass as close as possible across the stern of the enemy's rearmost ship and crowd sail so as to draw up level with them on their leeward side. To double from leeward, the ships had to tack and pass the enemy's rear in bow and quarter line. Once to windward, they could tack again and crowd sail to draw up level.

The mere recital of these recommendations raises the image of an accommodating and docile enemy. Conflans's *signeax de combat,* however, show a great deal of tactical sophistication. Number 6, for instance, is a signal requesting advice about the nature of the enemy as seen by scouts. The night signals include number 40, to prepare for night action, and number 41, to begin a night action.

Conflans retained the traditional system under which the order of sailing in line ahead was the same as the order of battle, except that the *général* was to lead the whole fleet. Orders of sailing in two and three columns followed, the latter described in great detail, as also were the usual single line, line of bearing and wedge formations. Great emphasis was laid on precision in executing all evolutions, and on correct station-keeping. A strong mathematical bias was evident in the evolutions, apparently requiring great skill in steering and sail-handling.

The Battle of Quiberon Bay

A tourist's view of Quiberon Bay from Carnac Plage, St Gildas or le Croisic suggests a huge expanse of sea, bounded only very distantly by the Quiberon Peninsula, Belle-Ile, and La Pointe de Croisic, with some tiny islands of no particular account dotted here and there, and scarcely visible from the mainland. Yet this was the scene of one of the greatest battles in sea history, the special feature of which was heavy fighting in heavy and chaotic seas with the wind against the tide, amidst dangerous shoals and rocks.

When, on the morning of 20 November 1759, a frigate of Hawke's fleet sighted the French fleet off Quiberon Bay, the whole British fleet of twenty-three of the line knew that they were at the end of their long chase from Torbay and also at the end of their long blockade of Brest. What the French knew or thought is uncertain, but their admiral, Conflans, appreciated the situation clearly enough. He had twenty-one of the line. His first duty, having broken out of Brest while

41 *NMM*, HOL/16 (formerly S(H)16): Brest, Chez à Malassis, Imprimeur du Roi et de la Marine, 1759

Hawke was driven to raise his blockade by extremely bad weather, was to make contact with the large assembly of transports and victuallers gathered in the great land-locked gulf of Morbihan within Quiberon Bay. These forces were for the invasion of England, and Conflans's fleet was to provide the escort. Like La Clue only three months earlier, he found himself committed to getting from one port to another, with fighting the enemy only a secondary consideration.

At dawn on 20 November Conflans had sighted a small British force, under Commodore Robert Duff, keeping watch on the invasion forces in the Morbihan. The wind being NW, Conflans decided to attack them, signalling 'close on the general' (the usual dawn signal), 'clear for action', 'pay attention to battle signals', 'prepare for action', and 'general chase'. When he saw Duff's force splitting up as it retreated inshore, he sent his van after one part, took his centre after the other and left his rear close-handed to observe some distant ships seen to be approaching, and which turned out to be Hawke's fleet. When this was realised, Conflans at once signalled his whole fleet to 'abandon chase', 'close', 'order of sailing in a single line', 'pay attention to battle signals', and 'prepare for action'. He then took station at the head of his line and steered for the entrance to the bay, which lay roughly ENE of him. The wind was now WNW, so he could make it quite comfortably, though surprising as it may seem when viewed from the shore or a pleasure boat, the safe entrance for sailing warships was by then only seven miles wide. To starboard, between Conflans and Le Croisic lay the Four shoal, and to port between him and the Île Hoëdic an irregular scattering of sharp-toothed rocky islets, known as Les Cardinaux. Conflans calculated that once inside the bay he could haul his wind and use the local knowledge of his pilots and masters to assume a strong defensive position protected by rocks and shoals. Should Hawke dare to follow, he had only to shift his own position to the centre of the centre squadron to put his fleet in 'the natural order of battle'.

Hawke meanwhile was in loose order south of Belle Île. On sighting the French, he signalled for line abreast so as to close up his fleet, having earlier sent the *Magnanime* (74), commanded by Lord Howe, ahead to make the land. Seeing the French withdraw towards the bay, Hawke made his 1747 additional signal (Article 9) for the seven ships nearest the enemy to draw into a line ahead to engage them until the rest of the fleet could come up. He could see Conflans 'under such sail as all his squadron could carry, and at the time keep

together; while we crowded after him with every sail our ships could bear'. This is really the key sentence in Hawke's despatch, denoting an absolute determination to engage the enemy despite a rising gale, sudden squalls, and heavy seas at the entrance to the bay. Even so, it was not until 2.30pm that a few shots were fired at the French rear, and Hawke made the signal to engage. The shots stopped and the chase continued. Presently Conflans was seen to have led round Les Cardinaux and into the bay. In theory he had had plenty of time to get inside with his whole fleet; his rear, however, not only failed to do so but equally failed a proper line. He had never imagined that the British would carry such a press of sail under such dangerous conditions, with the light of the winter afternoon dimming in the squalls. A very severe rain squall brought the three leading British ships into collision, but they managed to clear each other after some damage and continue the chase.

What happened next no-one really knows, except that the *Magnanime, Warspite, Montague* and other ships which forced themselves into the lead attacked the French rear, the *Magnanime* going as far ahead as possible to get up with the French centre.

The supposition that Howe and others pushed ahead without a special signal is proved by the eye-witness account of Patrick Renny, MD, surgeon of the frigate *Coventry*, stationed ahead of the fleet:

> Lord Howe, in the *Magnanime*, had led the chase the greatest part of the day, but was now passed by Sir Peter Dennis in the *Dorsetshire* and Captain Patrick Baird in the old *Defiance*, who ran along the French line to windward, receiving the fire of every ship that passed without making any return, intending to stop and engage the van of the enemy. Lord Howe followed in the same glorious career, but coming abreast of the Rear Admiral [St André du Verger] in the *Formidable* was disabled from going further, his foreyard being carried away in the slings. He immediately bore down upon the rear admiral.[42]

In this kind of fighting, devoid of tactical order or control, British superiority in fighting power was decisive. The wind shifted suddenly to the NW, and the French line of battle fell into confusion. At about 4pm the French rear-admiral struck; two French ships were sunk by gunfire, and at about 5pm another ship struck but could not be boarded. As Hawke rounded Les Cardinaux in the *Royal George*, he met Conflans in the *Soleil Royal* trying to lead his fleet out to sea again. They exchanged broadsides, and amidst a general pell-mell and a double collision Conflans fell to leeward of the Four shoal and anchored off Le Croisic.

It was now almost dark and Hawke had no pilot. With the wind blowing hard he was on a lee shore with the Pointe du Castelli and Rade du Croisic just under his lee, the Four shoal to his south, and the dangerous Ile Dumet just ahead. Some of the French had retreated towards the Vilaine Estuary in the extreme north-eastern corner of the bay, beyond Le Croisic. More British ships were entering the bay and the wind was kicking up terrifying seas against the tremendous ebb tide fed by the Morbihan and the Rivers Crach, Auray and Vilaine. Hawke made the signal to anchor, with his own position two to three miles west of Ile Dumet.

The chaos of the night and the further destruction inflicted on the French next day and subsequently, are without tactical significance. For the price of two British ships wrecked on the Four shoal, Conflans's fleet was completely destroyed as a fighting organisation. The French invasion scheme was

Battle of Quiberon Bay

Quiberon

French transports at anchor

French ships trapped in the Vilaine estuary

River Vilaine

Quiberon Bay

Wind veers to north-west

Hoëdic Is

Belle Isle

French fleet

Le Four

Main body of British fleet

Leading group of British ships

Eight French ships escape to Rochefort

---

42 From the transcription by General Henry Renny. An incomplete and inaccurate version of Renny's memoirs was printed in W H Long's *Naval Yarns* in 1899.

ruined and, with the defeat of Lagos and the loss of Quebec, the Seven Years War began to change in character from a war fought for the command of the sea to a war in which British naval pressure began to be applied directly to the centres of French military power.

When I consider the season of the year [wrote Hawke in his despatch], the hard gales on the day of action, a flying enemy, the shortness of the day, and the coast they were on, I can boldly affirm that all that could possibly be done was done. As to the loss we have sustained, let it be placed to the account of the necessity I was under of running all risks to break this strong force of the enemy.

Hawke was no longer a tactical innovator. In signalling for his seven leading ships to draw into a line ahead to engage the enemy's rear, he was merely following his own additional instruction of twelve years before. Nevertheless it was enough, partly because the French could retreat no further into the bay, so that their whole fleet was involved, and partly because Hawke's captains had developed such a highly aggressive spirit through the long blockade that they did not require special signals to urge them on.

Additional signals, 1760–2

In the spring of 1760, after his return from the Quebec campaign, Vice-Admiral Sir Charles Saunders was appointed to command the Mediterranean Fleet. On 12 May 1760 he issued the new version of the Admiralty's additional sailing and fighting instructions, and again on 12 July.[43] Saunders's signals, 'extracted from the General Printed Sailing & Fighting Instructions', were formed into a special signal book 'so as to dispose all signals that a Commander-in-Chief may see what Signal is to be made'. The book is in two parts of folio size, separately bound, beautifully written and painted. The first volume is labelled 'Admiral Saunders's Signals for Commanding Officers', possibly indicating that it was specially prepared for his divisional flag officers or commanders of detached squadrons, though the phrase might also apply to captains.[44] In effect, however, the first volume serves as the index to the second volume, which contains most of the signals. References to the general printed Sailing & Fighting Instructions were printed in black, and those to additional signals in red. The second volume was the signal book itself, arranged in the usual pattern, with references also in black and red. The two volumes provide the best example of signal book layout certainly up to the end of the American War, and possibly even until the appearance of the first official Admiralty Signal Book in 1799.

On 8 March 1762 Commodore Howe, having succeeded to the title Lord Howe, issued a printed book of 'Additional Signals & Instructions to be Observed by the Ships of War' to his cruiser squadron in the Basque Roads watching Rochefort.[45] This book provides the link between his earlier essay of 1756 and his epoch-making Signal Book & Instructions of exactly twenty years later. The additionals of 1762 contain sixty-one signals for day, fifteen for night and seven for fog. No attempt was made to separate sailing from fighting signals, with all lumped together under the day signals. A special and distinguishing feature of the book is that it is laid out in signal-book form, except for the purely instructional articles. Each page is ruled into three columns,

the first two headed Flag and Place, the signification being on the right with its respective number.

There were no startling innovations; Howe's main purpose was to increase the efficiency with which his squadron could carry out its immediate task, stopping enemy ships leaving and entering Rochefort. For this he needed flexibility and simplicity of manoeuvres, and a more efficient method of reporting a strange sail by private ships. Combined with this energetic direction of effort were characteristic signs of that obscurity of statement which were to make his later tactical reforms so difficult to understand (particularly to modern readers). His signal number 59, for instance, seems dangerously ambiguous, considering that it was intended for use in a sudden emergency:

When the Commander-in-Chief upon making the Signal for Line-ahead or abreast, though meaning to form on the Starboard Tack, would nevertheless have the Ship appointed to lead on the Larboard Tack, and that Division then to lead: Or on the contrary the Ships of the Starboard Division to lead though forming on the Larboard Tack. The same change of Divisions is meant to be made when this Signal is shewn with the Signal for forming in two separate Lines of Battle.

To set against these ambiguities of drafting, there were some striking advances, clearly expressed. Number 12 was the long overdue proviso: 'For the Squadron when in chase to engage the Enemy in Line of Battle according to the Order in which they may successively arrive up with them', which amplified Anson's additional number 9 establishing an *ad hoc* line of battle. Number 50 was an instruction, not a signal: Howe instructed his squadron that, when tacking to windward, the division which happened to be to windward when the signal was made was to stay there until a further change of wind or course enabled it to regain its designated station in the order of sailing, 'The intent of this distribution of the Ships in Sailing Order being only to prevent accidents by their running through the Squadron and crossing one another in the night time.'

Article 1 of the night signals and instructions, which are

Admiral Sir Charles Saunders (1713?–79), detail by Sir Joshua Reynolds. NMM

Admiral of the Fleet Richard Howe (1726–99), first Earl Howe, by John Singleton Copley. NMM

---

43 *NMM*, CLE/2/8

44 *NMM*, Tunstall Collection, TUN/34 and 35 (formerly S/MS/SY/3 & 4). *NMM*, HOL/17 (formerly S(H)17) is a pocket version of this volume.

45 *Barrington*, I, pp326–56

otherwise of no special interest, reads: 'As it is necessary in cruising that all signals should be made [at night] with as few guns as possible, they will therefore be made in fair weather with lights only, as directed in the Printed Instructions.' Howe specifically indicated which signals should be made using guns. The reduction of gun signals at night as well as by day remained one of his constant preoccupations; Hawke's experience at the Battle of Quiberon Bay, when the gun signal for anchoring was confused by many ships for the distress guns fired by enemy, may have been the origin of Howe's determination to depend primarily upon light signals alone.

Five weeks later, Howe issued three orders of battle. In the first he stationed himself in his 74 in the centre of the first division of five ships. He gave the central position to two 90s by placing them at the rear of the first division and the head of the second. His second order of battle provided for 'being myself opposed to the Commanding Officer of the Enemy placed in the Centre of their Line as is their usual practice'. He took station fifth in the line, and placed his 90s third and eighth on the assumption that numbers were equal and the enemy were to be engaged ship for ship.

His final order of battle provided for the possibility of 'finding them composed, as is possible, of a number of ships greater than that of which the British Squadron consists'. In this case, the order of battle was to be the same as number 2, but, assuming the enemy to number fourteen ships, their fourth, sixth, ninth and eleventh ships were to be left unopposed. Howe's two 90s were to fill the respective gaps.

The idea of trying to contain the whole of the enemy's line instead of concentrating on part of it may seem retrogressive. Nevertheless the mere notion of making any kind of tactical distribution for a specific purpose implies a higher level of tactical thinking than anything so far sanctioned by the 'custom of the service'. His tactical creativity is further revealed by his observation that, in the event of the enemy outnumbering his squadron, 'I may perhaps attempt by studied delays to conceal from them my purpose of bringing them to action, until later in the day, and that opportunity offers of doing it more to advantage.'

What appears to have been Hawke's last signal book of the Seven Years War is entitled 'General Printed & Additional Signals delivered out by Sir Edward Hawke 1762'.[46] Twenty-four flags were used, and two pendants. It incorporates the traditional signals from the Sailing & Fighting Instructions with those in the 1758 version of the Admiralty additionals by means of two pairs of reference columns. An interesting additional day sailing instruction, for ships when not otherwise appointed, is 'to keep the stations in respect of each other (the several Admiral's seconds ahead and astern upon their different quarters respectively) as prescribed by the order of battle'. This suggests the beginnings of a more formalised cruising organisation than had yet been adopted.

Undated Admiralty signals and instructions

The Naval Library possesses a printed set of 'Signals and Instructions for His Majesty's Fleet' of exceptional interest, but to which no certain date can be assigned.[47] There are forty-five fighting instructions in this set, in which are continued the efforts begun after the Battle of Toulon to produce decisive results when the British fleet was superior. Article 38 was for particular ships to quit the line, and to distress the enemy by raking their van or rear, or to assist disabled or hard pressed ships by taking their place in the line; this was a fuller and combined version of existing instructions. Article 34 ordered that, if the admiral hauled down the signal for the line during battle, every ship was to engage as close as possible, and drive the enemy out of their line 'without having regard to the situation which was prescribed to themselves, by the line of battle, before the signal was hauled down'. The intention of this was apparently to allow captains a free hand to act to the best advantage should the course of the battle necessitate such action. Article 43 made the instruction automatic, if British numbers exceeded those of the enemy. Article 40 was a signal for engaging as close as possible, with the British ships at a distance of half a cable or less, should the weather permit, together with the usual warning against firing over the top of British ships.

Sir Julian Corbett conjectured that these instructions were 'possibly never issued to the fleet'.[48] They certainly seem too progressive, as well as to individualistic both in content and arrangement, to have been issued by the Admiralty. The decorative printed designs suggest affinities with the undated and hitherto unknown instructions issued by Howe.[49] Nevertheless the title, text and general arrangement are unlike anything produced by Howe either then or later. They would seem to have emanated from a commander-in-chief, and from someone other than Anson, Hawke, Boscawen, Saunders or Keppel. Apart from Sir George Pocock's, it is difficult to suggest a name. They are quite unlike Keppel's of 1756, and when Rodney sailed for the Capture of Martinique in 1761 he issued the new version of the Admiralty additionals.[50] The date and authorship of this printed set, therefore, remains a mystery.

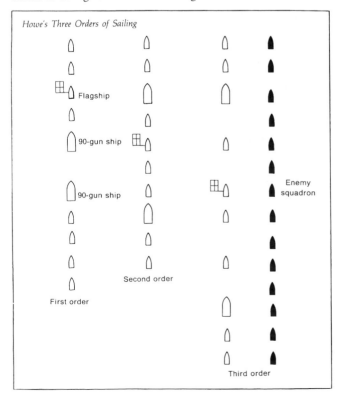

Howe's Three Orders of Sailing

Flagship

90-gun ship

90-gun ship

First order

Second order

Third order

Enemy squadron

46 *MOD*, Ec/35

47 *MOD*, Ec/42

48 *Signals & Instructions*, p383

49 *NMM*, Tunstall Collection, TUN/111 (formerly FI/5)

50 *NMM*, (formerly Sp/130 and Sm/168)

CHAPTER FIVE

Signals Development following the Seven Years War

ONE of the main features of the revival of the French navy under Étienne François, Duc de Choiseul, Louis XV's minister of marine, was the advance made in tactics and signalling. The leading figures were Sébastien François de Bigot, Vicomte de Morogues, and Jean François de Cheyron, Chevalier du Pavillon. Jacques Bourdé de Villehuet was a lesser figure, but he publicised the work of Mahé de La Bourdonnais, who anticipated the important development of the numerical signalling system. Although Morogues and Pavillon succeeded in rescuing tactics from the civilian mathematicians and scientists, they followed Hoste in tending to substitute evolutions for fighting, and in failing to emphasise the need for offensive action. This is not really surprising, for just as Hoste wrote in the shadow of Barfleur and La Hogue, so they wrote in the shadow of Lagos and Quiberon. They were also influenced by the *Ordonnance du Roi*, published on 25 March 1765, which was an unrealistic and theoretical exercise allowing captains no freedom of action. They were required to fight 'yardarm to yardarm', and to 'give more attention to defending the flags of the Generals than to preserving their own ships, letting themselves be sunk rather than deserting the flag'. However, when the French intervened in the American War in 1778, the commander of the French main fleet, the Comte d'Orvilliers, was able to exercise tactical control through a signal system devised by Pavillon. Though baroquely complex, it greatly surpassed anything previously available.

Amiral *Sébastien François de Bigot, Vicomte de Morogues (1706–81)*. MM

Christopher O'Bryan, and possibly his brother Lucius, helped to spread French tactical ideas across the Channel, and they and their father found a market in England for essays on naval discipline and tactics. The printing by Lord Howe of a new set of additional instructions, possibly in 1770, is a milestone in the development of English tactics, and indicates that the victories of the Seven Years War did not have their usual stultifying effect. The Admiralty itself published a companion volume for the Sailing & Fighting Instructions, and, in the opening phase of the American Revolution, Howe had the opportunity further to develop his tactical system.

Morogues's tactical system

As in Hoste's time, naval matters such as ship construction and hydrodynamics, navigation and the determination of longitude, hydrography, chronometers and voyages of discovery, were considered in France as worthy of free discussion between mathematicians, scientists and naval officers. Naval science was something in which civilians could, and indeed were encouraged to, play an active and fruitful part. The Vicomte de Morogues was already eminent in this highly civilised naval-scientific society; he had been chosen as the first director of the Académie Marine when it was founded in 1752. Much of his work *Tactique Navale ou Traité des Evolutions et des Signaux* had been circulating for some years amongst naval officers at Brest before he submitted it in book form to the judgment of commissioners appointed by the Académie Royale des Sciences. Only a month after the signing of the humiliating Treaty of Paris, three members of the Académie signed a laudatory report on his book, published the same year in Paris. It was dedicated to Choiseul, the minister being pointedly invited to give reforming attention to the navy.

Although in some respects modelled on Hoste, whose work Morogues declared in his introduction to be now 'un peu rare' and to contain 'beaucoup de manoeuvres inutiles' though 'excellent dans plusiers parties', his own book was quite different in structure. Whereas Hoste wrote as a mathematical expert with a strong historical bias, Morogues wrote as a naval officer anxious to give the service a complete system both of tactics and signals together with all the theoretical, explanatory and administrative guidance necessary to make the system effective.

During the Seven Years War, Morogues had been active in proposing various strategic measures, and in the disaster of Quiberon Bay he commanded the *Magnifique* (74). This experience must have reinforced whatever prejudice he already had in favour of the over-riding importance of defensive tactics. In his introduction, Morogues wrote that tactical order and discipline were decisive factors in battle. A small well-disciplined force, whose captains kept their line well closed up and gave each other mutual support, could overcome a more numerous but less disciplined force, quite apart from the fact that the smaller the fleet the easier it was to handle. In comparison with fighting on land, the weaker force at sea had more opportunities to escape. Tourville's *campagne au large* showed how a force could make up for its inferiority by *vivacité de l'action*. Morogues's thoughts on tactics were doubtless influenced by his belief that sea fighting was not decisive, at least as far as the war as a whole was concerned; for eighteenth-century France this was indeed the case.

Morogues stressed the importance of large ships, as compared with ordinary two-deckers. They mounted a higher proportion of big guns, they sailed better in stiff winds and heavy seas, and they were less easily boarded. In bad weather they were more likely to be able to use their lower-tier guns than the two-deckers, while if conditions would not allow this, they had the advantage of being able to use two upper tiers of guns against the two-decker's one. In battle, the fleet with most large ships should win even if inferior in numbers. A small fleet was more easily handled than a large one, especially if it were not too closed up in formation, though in fine weather one-third of a cable between ships was considered sufficient. A fleet too greatly extended risked having some of its ships fired on by two enemy ships at the same time.

Morogues's approach to the problem of defeating an inferior enemy by doubling his battle line was much more practical than anything in Hoste's academic exercise. To double the enemy's van from windward, the leading ships should crowd sail so as to draw ahead of the enemy van.

Having doubled, they should then drop to leeward and resume their course as soon as the enemy van drew level; this would place the enemy's leading ships between two fires. For the superior fleet to double the enemy van from leeward, the leading ships should tack right across the enemy's line of advance, each giving the enemy's leading ship a broadside as it passed. They should then resume their course, now to windward and abreast of the enemy van. To double the enemy rear from windward, the rear ships should turn to leeward and crowd sail so as to sweep round the enemy rear and bring them between two fires. To double the enemy rear from leeward, the rear ships were required to tack so as to pass astern of the enemy and so gain the wind. They were then required to resume their course and crowd sail so as to draw up level with the enemy rear. On balance Morogues, like Hoste, favoured doubling the enemy rear rather than their van, and for the same reasons. It was easier in this way to cut off the enemy's doubled ships from support, and it involved less risk that one's own ships, if damaged, might be caught by the enemy as they came up.

All this was highly theoretical, in that it assumed a complacency on the part of the enemy which might not be forthcoming even if they were weaker in numbers and hence more easily overlapped at each end of their line. Nevertheless, these were positive ideas recommended for actual use in battle, rather than manoeuvres debated in theoretical terms on the basis of historical examples.

Below: A page from le Vicomte de Morogues's Tactique Navale ou Traité des Évolutions et des Signaux, *Paris, 1763, showing ships signalling with skyrockets. NMM*

Below right: A plate from Morogues's Tactique Navale, *showing the points of sailing. NMM*

Morogues's Methods of Doubling the Van

Van drops back to double enemy

Van crowds on sail and gets to leeward

From windward

Van tacks ahead of enemy

Van resumes its course on the other side of the enemy

From leeward

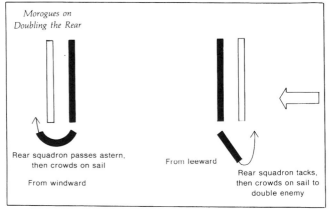

Morogues on Doubling the Rear

Rear squadron passes astern, then crowds on sail

From windward

From leeward

Rear squadron tacks, then crowds on sail to double enemy

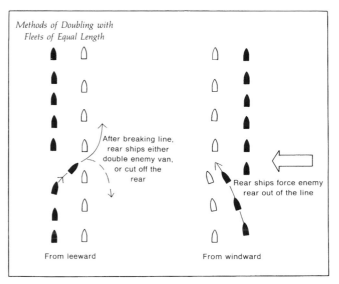

Methods of Doubling with Fleets of Equal Length

After breaking line, rear ships either double enemy van, or cut off the rear

Rear ships force enemy rear out of the line

From leeward

From windward

Morogues suggested another, novel, approach to the tactic of doubling. If the superior fleet were well closed up so that it only equalled the enemy's line in length, it might double the enemy rear from leeward by cutting right through them with its own rear ships. These ships could then fan out, so as either to resume their course to windward of the enemy or to surround the enemy rear completely by passing round them in succession from windward on the opposite tack. Similarly, the superior fleet, if well closed up, might double the enemy rear from windward by bearing down hard with its own rear ships and driving the enemy rear to leeward and right out of their line.

To avoid being doubled when inferior in numbers, Morogues adopted the traditional expedients of spacing the fleet either by individual ships or by squadrons in such a manner as to prevent the enemy from overlapping either end of the line. He also introduced the idea of bringing up the fireships to fill gaps which the enemy might otherwise be tempted to exploit. Large ships should be stationed at the centre and rear of their respective squadrons. He admitted, however, that it would be difficult for an inferior fleet to leeward to avoid being doubled. Much depended upon its bravery and experience.

The second half of the book, entitled 'Des Signaux et

Ordres Généraux', is the more important. It consists of over 200 pages of orders with signals annexed, written in the form of sailing and fighting instructions, and containing over 300 administrative and operational articles in addition to explanatory matter. Embedded in these instructions is the real essence of Morogues's new tactical thinking, and both in number and in detail they go far beyond the scope of the standard French system of the Seven Years War. Although the appropriate signals are annexed to the instructions, there is an eighty-page table of signals, for day, night and fog. This forms the draft lay-out of what a signal book should be. The engraved plates at the end illustrate the work as a whole, including both parts and showing the ships in diagrammatic form. In the middle is a forty-page index with detailed references to both parts of the book and to the table of signals and the diagrams. Each part of the book is heavily cross-referenced, so that the instruction, signal and diagram dealing with a particular evolution can be found at once and vice-versa. The diagrams all bear references back to the text.

Morogues proposed employing thirty-three flags with a choice of six positions, and nine pendants with twelve positions, as against what seems to have been the official system of thirty-four flags and six pendants, all flown from seven positions. The flags and pendants were numbered by Morogues in two separate series. Signals for the whole fleet were to be flown from the mizzen yard. Signals for the van, centre and rear squadrons were to be flown from the fore, main and mizzen topmasts respectively, except for the blue and white, white, and blue flags or pendants which indicated the van, centre and rear squadrons respectively and were supernumerary to the numbered flags. Morogues left the colours blank for all his flags or pendants; except for the three flags mentioned, they all were given numbers only. This was done on the grounds that the colours of the flags should be constantly changed for security reasons.

As in the British system, a flag only had a meaning when flown in a particular position, but Morogues gave each meaning a number which corresponded with the number of the instruction to which it applied. This number was quite distinct from the number given to the flag, which indicated its colour. A true numerical system was superimposed on the main signal system, for use when it was wished to draw attention to a particular article in the instructions, or in cases where figures had to be given, as for soundings and latitude and longitude. The flags numbered 1 to 9 were to be flown from the fore topmast head for the hundreds, at the main for tens, and at the mizzen for the digits. Nought was signified by having no flag flying at the appropriate masthead.

In some cases a signal might have more than one meaning according to particular circumstances. In such cases, and indeed in many other cases, the signal was accompanied by signal guns. Morogues devised an ingenious system to increase the possible vocabulary of gun signals, by varying their rate of fire: *lentement,* at twelve to fifteen second intervals, or *coup sur coup* at four to five second intervals. If the two methods were combined in the same signal, an interval of thirty seconds was to be made between each group. Not more than six guns were used for any one signal. Morogues provided a separate table of gun signals, with cross-references to the flag signals. Each grouping of gun signals was given a number which covered all the flag signals, night and fog signals, to which it applied.

The night signals were expressed in exactly the same form, the numbering of the articles being continuous, but the system employed for transmitting signals was extraordinarily complicated. Each night signal had a number printed in the margin which differed from the number of the article itself. In the next column was printed one, two or three letters of the alphabet, ranging from A to Q. These letters were purely notional: they were simply symbols indicating the use of from one to five lanterns at the fore, main or mizzen shrouds. The lights themselves thus indicated the original number of the signal. Several of the signals had more than one meaning, and, as in the day signals, these were differentiated by signal guns. Numbered references to the table of gun signals were in a separate column opposite those covering the lights. After the title of the signal, a final column provided references back to the numbered articles. Under conditions where the respective masts of the flagship could not be distinguished, rockets *en pluie,* to the right number, might be fired to indicate the foremast, rockets *en étoiles* for the main mast, and rockets *en serpentaux* for the mizzen. Thus for signal number 135 two star rockets, four rain rockets and five serpentine rockets might stand for the same number of lanterns at the main, fore and mizzen respectively. This would indicate the purely notional letters B I Q. With the help of gun signal number 32, six guns *coup sur coup,* captains would then know that the signal (number 135, referring to Article number 292) was to anchor. Lanterns could be used in the same manner as numeral flags by day to indicate numbers.

The fog signals were much more restricted. Even so, nearly two hundred separate signals could be made by an ingenious use of hailing by speaking trumpet (for recognition), various bugle calls, drum, bell, trumpet, fife, musket and swivel gun. Some of these were combined with ordinary signal guns and a large number were made by signal guns alone.

From all this, it is apparent that the book was comprehensive, both in its lay-out and in the system it served, but by no means simple to handle. Whereas the official French signal book was a modest folio of just over 100 pages of large print, Morogues's guide book had 480 pages and 49 engraved plates depicting 133 tactical diagrams and weighed 5lb. There is, however, no real evidence that it ever had any immediate practical usage, even though its influence was very considerable. It was reprinted in Paris with the original date of issue, 1763, and a Dutch edition was published in Amsterdam in 1779.

Morogues's reputation must certainly rest more on the ingenuity of his signal system than on the originality of his tactical thought or his desire to discuss battle tactics at length. Despite his own suggestions to the contrary, he strengthened the notion already advanced by Hoste that tactics are mainly evolutions, and that skill in evolutions is almost a substitute for fighting efficiency. He encouraged the idea that greater tactical control, exercised in terms of evolutions, would be the future answer to the British, with their improvised tactical methods, and their tied signal system.

Bourdé de Villehuet and La Bourdonnais

In 1756 a book was published entitled *De la Manoeuvre des Vaisseaux ou Traité de Méchanique et de Dynamique.* It is not surprising to find that the author, Pierre Bouguer, was a member of the Académie Royal des Sciences and sometime hydrographer at the ports of Croisic and le Havre, and had already written books on observing the height of the stars (1729), compass declination (1731), ship construction and movement (1746) and the shape of the earth (1749). Nor is it surprising that Bourdé de Villehuet, writing on tactics in 1765, remarks of Bouguer, 'mais, pour l'entendre, il faut beaucoup de Géometrie et d'expérience: d'ailleurs, combien d'opérations ne peuvent s'apprendre qu'à la mer!' Bourdé acknowledged the value of the theoretical ideas obtained from earlier writers, but his own work was based on his personal sea experience.

Bourdé described himself as an officer in the French East India Company's service, and he entitled his book *Le Manoeuvrier, ou Essai sur la Théorie et la Pratique des Mouvements du Navire et des Évolutions Navales.* It was dedicated to Choiseul and contained a warm eulogy from the Académie Royal. Compared with Morogues's work it is rather disappointing to the historian, being to some extent a restatement of Hoste in simpler form, lacking the lavish historical examples. Nevertheless it is a distinctly more useful and more practical book, produced in handy form, measuring only 7⅞ x 4⅞in. After a section of theory, Bourdé explained simply and clearly how to handle a single ship, and he presented new ideas about chasing which were later adopted by Howe and Kempenfelt.[1]

In Part III Bourdé wrote about the organisation of the crew for fighting purposes and about battle discipline. He strongly enjoined silence so that orders could be heard. He very clearly explained gunnery and training in gunnery, and stressed the importance of getting in the first broadside, 'quand on est de bien près'. 'Si le vaisseau que vous attaquez tiraille de loin, laissez-le faire; il ne vous sera jamais beaucoup de mal.' The enemy, he argued, must be brought to close quarters as quickly as possible and overwhelmed by rapid fire.[2] He provided a good description of the organisation and conduct of boarding.

In Part IV Bourdé discussed evolutions. On the relative advantages of the weather and lee gages he followed Hoste, adding by way of summary that the weather gage offered fewer disadvantages than the lee gage, though it was correspondingly the more dangerous. On balance, the weather gage was best, especially for the stronger fleet which could detach its surplus ships to double on the enemy rear. In fact, a fleet ought not to accept battle from leeward 'à moins que de s'y trouver forcé par quelque fâcheuse circonstances'.[3]

Copying Hoste's fourth order of sailing, and Morogues's fifth order, Bourdé favoured a sailing order in three columns abreast, which he called the order of convoy. He gave examples of changing from this to the order of battle and order of retreat and back again.[4] A longish chapter dealt with how to restore order in shifts of wind. Only in the last tactical

---

1 Part II, Chapter 8; see *Signals & Instructions,* p143

2 Part III, Chapter 11, Art 7

3 Part IV, Chapter 1, Art 5

4 Part IV, Chapters 2 and 3

chapter in the book (Part IV, chapter 5) does the enemy come into the picture, and here the treatment followed Hoste. In Article 9 of this chapter he stated that a fleet engaging from leeward should stand on as soon as the action began. This would force the weather line to set topsails, and if as a result a ship was disabled in the advance, confusion would ensue, as happened to Byng at Minorca.

Bourdé's last chapter describes the signalling system invented and used by Mahé de La Bourdonnais, a very distinguished officer in the French East India Company's service who, during the War of 1739–48, was governor of Mauritius and Bourbon and commanded the French naval forces in the Indian Ocean. It appears that La Bourdonnais had a description of his system printed in 1746. It was, at last, a true numerical system, with ten pendants, numbered 0 to 9, of plain red, white, blue and yellow and parti-coloured. By this simple arrangement thousands of signals could be made. Each signal needed one hoist only, and could be made from any part of the ship most convenient for sending and reception. The bottom pendant signified the digits, the next above the tens, and so on.

In addition, La Bourdonnais propounded a system of long-distance signals involving a combination of the old and new systems. There were to be eight symbols only, four plain flags and four plain pendants of red, white, blue and yellow. First, a pair of pendants or flags would be hoisted anywhere convenient to indicate whether it was a hundred, ten or digit figure which was to be signalled. Then would come the indicative pendant or flag, which would be valued entirely according to its position. In each case the same position was reserved for each number (1 and 6, ensign staff; 2 and 7, mizzen yard; 3 and 8, mizzen topgallant; 4 and 9, main topgallant; 5 and 0, fore topgallant). The shape and colour, however, were different in each case, according to whether the pendant represented a hundred, ten or digit figure. There were five positions for each pendant or flag, so a particular pair could be used to cover 0 to 9 in each category. Thus, although the main topgallant position was reserved for 4 and 9, the 4 of a four-hundred signal was a white pendant, the 4 of a forty signal was a red pendant and a plain 4 was a yellow flag. Provided the flag or pendant could be properly made out at a distance, the signal could not be misread. In addition to the flag's single possible meaning in the position occupied, there was the additional check provided by the preliminary signal indicating whether the substantive signal was a three, two or single figure number.

La Bourdonnais also advocated numerical night signals by combination of lights and guns to be agreed upon beforehand (he did not lay down a scale of his own). He wisely added that lights should be used entirely in numerical terms, and hoisted in one place. The practice of hoisting them in different positions made it difficult to see them all from any particular bearing.

Publication of French tactical studies in England

The first book on tactics in the English language, *Naval Evolutions: or a System of Sea Discipline,* appeared in 1762.[5] It was published by Christopher O'Bryan, at one time a lieutenant in the Royal Navy, who is said to have entered the Russian Navy in 1739, apparently as a midshipman. He was a son of the Christopher O'Bryan who was made post in 1713, commanded the *Rippon* (60) at Cape Passaro, entered the Russian Navy in 1739 and died in 1742. From this there

5 *Naval Evolutions: or a System of Sea Discipline: Extracted from the Celebrated Treatise of P L'Hoste, Professor of Mathematics, in the Royal Seminary of Toulon: Confirmed by experience; illustrated by Examples from the Most Remarkable Sea-Engagements between England and Holland; Embellished with Eighteen Copper-Plates; and Adapted to the use of the British Navy. To which are added, An Abstract of the Theory of Shipbuilding; An Essay on Naval Discipline, by a Late Experienced Sea-Commander; A General Idea of the Armament of the French Navy; with some Practical Observations; By Christopher O'Bryan, Esq; Lieutenant in His Majesty's Navy,* 1762 (NMM, Tunstall Collection, TUN/105)

A plate from Christopher O'Bryan's Naval Evolution or a System of Sea-Discipline, *London, 1762, following the work of Paul Hoste, showing the ease with which a superior fleet, when to windward, could detach ships to double the enemy rear.* NMM

Firing by Platoon

is reasonable inference that Christopher O'Bryan the elder was the author of the 'Essay on Naval Discipline' which his son included in his text.

Translated extracts from Hoste's book amounted to about a quarter of the younger O'Bryan's work. The experience of the Seven Years War may have tended to discredit Hoste's approach, or at any rate the kind of battle tactics derived from it, but there were, no doubt, British officers who regretted the absence of a tactical manual in English. The extracts from Hoste, covering part of the elementary sections at the start and most of the battle sections at the end, were probably those considered most useful for British officers. Quite apart from the indifferent translation, the presentation was too disjointed to enable the reader to ·appreciate either the strength or the weakness of Hoste's system. The plates were redrawn from the originals, and to the same scale. Hoste's *Théorie de la Construction des Vaisseaux* was also abstracted by O'Bryan.

The 'Essay on Naval Discipline', though dull and prolix, is of considerable value to the historian because it discusses current thought and practice. The author believed that it was more effective for ships to fire their guns 'in platoons' than in broadsides. Half the guns on the lower deck, aft, should be fired first, followed by the aft guns on the upper deck, the forward guns on the lower deck, the forward guns on the upper deck, and finally the guns of the quarterdeck and forecastle. The author also believed that at sea, unlike on land, the line should be formed with the main strength in the centre. Ships in excess of the enemy's line should be in a reserve corps, and the admiral should have a repeating frigate stationed opposite to him. The suggestion was dismissed that the admiral should himself be in a ship of the reserve, though keeping his flag flying in the line. Sea battles were not like land battles, it was argued, and the situation of the general and the admiral were not strictly comparable.

In forming a line of battle, O'Bryan wrote that, except in

the rare instance when a more senior captain commanded a less powerful ship, 'seniority should be regarded, either by giving orders to the oldest captains to lead or be seconds'. Where the fleet was in three divisions with three flags, each flag officer should have the senior captains as seconds 'and each division having a leader upon either tack, according to the seniority of the next oldest captain'. The two junior flag officers could not be seconds to the admiral as that would preclude three divisions. In a fleet of at least fifteen ships there should be three divisions, 'because sometimes a division might be ordered to tack to gain the wind of the enemy, as well as to give the two junior flags command and honour, by their share of division and rank'. Where there were only two flags in a fleet, the senior captain should be given a division, 'that the line may thereby be regular and complete. This [he wrote] is what I have seen and learned in the course of my experience in the navy.' This last seems relevant to Byng's action at Minorca, though admittedly he had had only thirteen of the line, including one weak 50-gun ship.

O'Bryan stressed the importance of reconnaissance. Action, once joined, should be at point-blank range, that is about 300yds, but not closer as otherwise the line would fall into a curve. 'Besides, you cannot as well observe your own line: the enemy's ships may block your view of your own ships, especially if the line be long, which may cause a mistake in the signals.' The notion apparently is wrong that British admirals invariably thirsted for the closest possible action, and that the pell-mell was deliberately sought when fleets were of equal strength.

One of the more interesting sections in the book is that on convoy action, for which O'Bryan employed an example taken from an action over half a century previously. In 1707 Commodore Richard Edwards, escorting a large convoy with three 80s and two 50s, was attacked and heavily defeated by René Duguay-Trouin 'with ten or twelve under his command from 30 to 60 guns'. Edwards, according to O'Bryan, instead of lying-to as he actually did, should have retired in a line-abreast under easy sail; a number of ingenious tactical reasons were given to prove that he would thus have made a better defence. Answering the objection that the enemy might come up and penetrate the line, putting some of the escort between two fires, he argued that ships engaging an enemy ship stationed between them would be in danger of hitting each other's masts and yards.

When Morogues's *Tactique-Navale* first appeared in Paris in 1763 it must have aroused immediate interest in England. With the signature of the definitive Treaty of Paris in that year, Englishmen resumed their visits to France in large numbers. There was an obvious need for a good translation. Instead, in 1767 a partial and wholly unsatisfactory one appeared.[6] Only a small part of the original work was in fact used, the selection being highly arbitrary, with the omission of important sentences. Morogues's entire idea of a system of tactics and signalling, expressed through the agency of a book, with the material organised, explained, numbered and cross-referenced in a particular way, was completely ignored. Instead, a selection of his ideas were jumbled together with other material.

The whole treatment of this translation is reminiscent of Christopher O'Bryan. The publisher, W Johnston of 'Ludgate-Street', was also the publisher of the Hoste translation five years earlier. The translator adds footnotes either

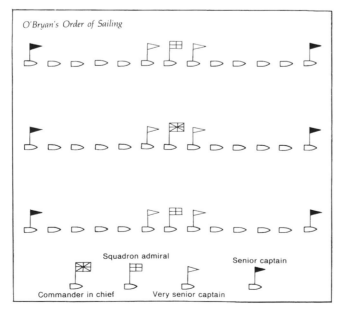

O'Bryan's Order of Sailing

Squadron admiral   Senior captain

Commander in chief   Very senior captain

---

6 *Naval Tactics: or, a Treatise of Evolutions and Signals, with Cuts, Lately Published in France, for the Use of the Cadets, or Guard-Marines of the Academy at Brest; And now Established as a Complete System of the Marine Discipline of that Nation. By Mons De Morogues . . . Translated by a Sea-Officer.* Paris, 1767 *NMM*, Tunstall Collection, TUN/103

laudatory of Hawke, to whom the book is dedicated, or critical of Mathews, who was the standard whipping boy in contemporary tactical discussions. These suggest another influence, that of Lucius O'Bryan, who was brother of Christopher O'Bryan the younger and commanded the *Essex* at Quiberon. Lucius is also said to have served in the Russian Navy in 1739. He died a rear-admiral in 1770 or 1771. He is the most probable author, since the book includes the claim that its author served as a midshipman in the *Orford* under Sir John Norris when still a captain, and was present at the Battle of Malaga in 1704. A comment that the victory at Lagos would have been more complete had the leading British ships reached ahead and cut off as many of the French as possible, by 'bringing to the headmost of the enemy' while leaving the remaining French ships to be dealt with by the rest of the British fleet as they came up, links the two O'Bryan experiences at Passaro and Quiberon. A section called 'Some Familiar Thoughts and Observations on Sea Fights, [and] Naval Discipline' echoes the earlier Essay on Naval Discipline; presumably both were written by the elder O'Bryan. The author stated that he commanded a ship of the line under Byng in the Mediterranean in 1718, which is certainly true of the elder O'Bryan.

Although the translated excerpts from Hoste and from Morogues, respectively, are poor work, the 'Essay' and 'Some Familiar Thoughts . . .' provide valuable information about tactical thought. If we assume that the selection of incidents over half a century old in the Essay make it rather an anachronistic effort, it becomes difficult to account for its reappearance in 1767. On the contrary, it would seem that the 1762 book was well received, and that this was not entirely due to its wretched presentation of Hoste, with whom many British officers must by then have been familiar, especially after the second edition of 1727. Past events, even half a century old, were relevant to the study of tactics because the *matériel* of navies had changed so little. In both works, and especially in the later, there is frequent insistence, particularly in the footnotes, on the need for better tactical discipline and better signalling.

The British 1768 manoeuvres

In the summer of 1768, the Duke of Cumberland was introduced into the sea service under the tutelage of Captain (later Admiral) Samuel Barrington, hoisting his flag in command of five frigates and two sloops. Barrington seized on this unusual peacetime opportunity to carry out extensive exercises in the Channel, issuing tactical memoranda and additional signals to the squadron over the prince's signature.[7] On one occasion, when the squadron was 'forming lines of battle and different manoeuvres, the better to observe them' the prince boarded another frigate which 'kept away to windward to observe signals'.

In memorandum number 3 Barrington provided for attacking the enemy's rear by inverting the order of the line of battle from an original bow and quarter line formation. An unnumbered memorandum has a signal for when the squadron was 'turning to windward on the enemy in a line of battle and the sternmost ship can't weather them on the other tack'; also for the sternmost captain when the squadron was 'to windward of the enemy and he thinks himself far enough astern of the enemy's rear to tack and lead down to it out of the sternmost ship's line of fire'. A similarly worded memorandum, number 6, gives a signal for the headmost ship to

make to indicate that it should be able to weather the enemy in line of battle when turning to windward.[8]

This isolated instance suggests that Barrington must have shared the general concern of men like Howe, Keppel, Rodney and Bryon for tactical and signalling efficiency, although not all of these were specifically concerned either with tactical innovations or a general improvement in signals. The conservative outlook of such commanders, and their commitment to the line of battle, does not imply that they were necessarily unconcerned with ensuring peacetime tactical efficiency.

The obvious need was for a complete revision of the Admiralty Sailing & Fighting Instructions to include the best of what had so far been produced by individual enterprise; this, however, was too revolutionary a project. The original cause of so many errors and limitations was retained simply because it had acquired the authority of tradition and long usage. Besides, re-writing it might have involved the Board in unpleasant controversies. Instead a new edition was printed slightly smaller in size and with fewer pages. A surviving copy of this edition, signed by James Gambier as commodore on the North American Station, 22 August 1771, aboard the *Salisbury* in Boston Harbour, was issued to Captain Benjamin Cadwell of the *Rose*. It included

A plate from the translation by 'A Sea-Officer' of part of Morogues's Naval Tactics, London, 1767, illustrating (fig 11) a chasing fleet edging van and rear squadrons around the chase, and (fig 12) a manoeuvre to double the enemy van NMM

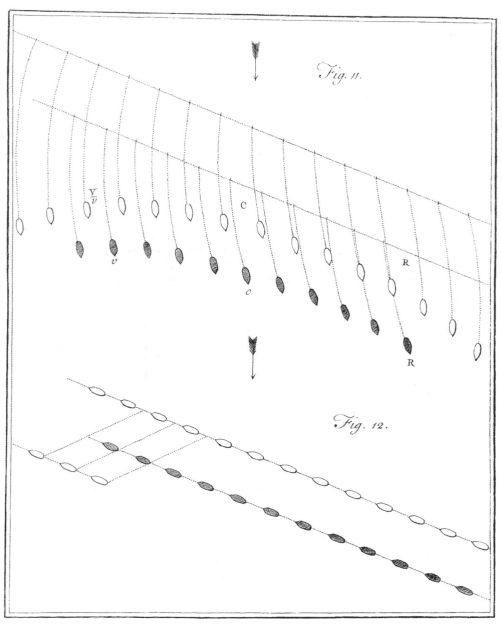

7 *Barrington*, I, pp399–403

8 *Barrington*, I, pp400–1

handwritten additional day, night and fog signals on folio pages, signed by Gambier and sewn in. Sir Charles Knowles states that there was a re-issue of this edition in 1772, but it seems more likely that he was referring to that of 1771.[9]

Lord Howe's undated additions

At some unknown date between March 1762 and June 1776 Lord Howe produced a new printed version of his 1762 additionals, entitled 'Signals & Instructions in Addition to the General Printed Sailing & Fighting Instructions by Day and Night'. In November 1770 he was appointed to command a squadron to be employed in the Mediterranean in consequence of the Falkland Islands dispute with Spain. *Barfleur* was ordered fitted out for him, but he never hoisted his flag. It may well be that the book in question was printed with this command in view, and that it incorporated his most recent thinking. Though almost identical in title with the subsequent Admiralty production of the same period, this is undoubtedly a Howe book, employing Howe's special phrase 'to be observed by the Ships of War'.[10] Taken *en bloc*, the new additional fighting instructions reveal a formidable development of tactical thinking, going far beyond anything reached in the Seven Years War, including Howe's own issue of 1762 of which these are extensions.

The book, which is well indexed by flags, by lights and by guns, contains fifty-three additional day sailing signals and instructions, with one more in manuscript; twenty-one night signals, with five more in manuscript; eight fog signals and forty-nine 'Additional Fighting Instructions Respecting the Duty of Frigates in Action'. In the index the flags and pendants are listed with the places flown, and with blank columns opposite, headed 'Art[icle]' and 'Page' in which the relevant numbers were inserted in manuscript. These page numbers refer only to the instructions themselves and not to an accompanying signal book. Similarly the night sections give the number of lights, how and where disposed, the guns, and the article and page inserted in manuscript. In the case of the fog signals, various details were printed, and the number of guns; the article and page numbers once again were inserted in manuscript. It would appear that Howe, possibly as a new idea, was making a serious attempt at secrecy.

The signals were in traditional form, with twenty-five flags and pendants flown in particular positions, their details written into the text of the respective articles. There is no reference of any kind to an accompanying signal book. The unusually large number of sailing articles is accounted for by the inclusion of a number of orders, derived from different sources, in the form of additional signals issued by particular admirals. Much attention was given to cruising distances and changes of direction, but there is no hint of new cruising formations. The additional fighting instructions seem to represent an early version of the signals and instructions issued by Howe when taking command on the North American station in 1775.

---

9 *NMM*, CAL/130 (formerly Sp/77)

10 *NMM*, Tunstall Collection, TUN/111 (formerly FI/5). This particular copy, unsigned and presumably, therefore, unissued, was part of Lord Sligo's collection of Howe documents. It is a printed folio with one of the usual but undated water marks, and is bound in the usual thick marbled paper covers. The manuscript insertions are in a clerk's hand.

11 *MOD*, Ec/43

12 *MOD*, (formerly *RUSI* 92 and 83[G]. 'not found' at Naval Library, but believed to be NM/92 and 83[G].

13 In the copy issued by Byron, 3 March 1779, *MOD*, NM/92 [?] (formerly *RUSI* 92)

Admiralty signals and instructions

The Falkland Islands crisis may also have been the occasion for the issue of a new official book by the Admiralty, not to supersede the Sailing & Fighting Instructions but to supplement it. The earliest known copy is one issued by Captain the Hon Robert Digby, 7 November 1777, when in command of a squadron cruising off Cape Ortegal.[11] 'Signals & Instructions in Addition to the General Printed Sailing and Fighting Instructions' is arranged similarly to the parent work, but the fact that the word 'Signals' comes first is indicative of a slightly changing view.[12] In the same way, the word 'General' suggests a slightly qualified view of the parent work. Signals and Instructions has forty-four sailing instructions and signals by day, ten by night and eight in fog; and twenty-eight fighting instructions and signals by day, and two by night. There is also the same kind of subject index of signals, with cross references to the instructions, as was in the original Sailing & Fighting Instructions of the late seventeenth century. Nothing else is included. The book was evidently intended to provide a synthesis of all the best and most acceptable developments in tactics since the time of Vernon. Although the parent work was thus no longer the only official book, its ultimate authority was unimpaired; indeed there were specific references to it in the new one. Possible confusions which might arise from this strange dualism does not seem to have worried the Board of Admiralty, since all the signals were different. In fact, by retaining the purely artificial division between sailing and fighting, the Board committed itself to a quadruple series of instructions for the fleet at sea.

The new sailing instructions are not impressive. There were some new chasing signals, but many of the signals were quasi-administrative, the inheritance of the last quarter of a century. No new cruising formations were included.

The new fighting instructions, which were also drafted in the traditional form, included seven articles dealing with the line of bearing, now described as 'bow and quarter line'. The advantage of this as a battle formation was, as Thomas Graves noted on his own copy, that 'the ship ahead will always enfilade the ship that engages her second astern.'[13] Several of the articles merely repeated long-accepted ideas, some as early as Vernon. Article 18, however, provided the instruction and signal which Boscawen had failed to issue at Lagos. It established, and improved upon, Hawke's famous Article 9 of 1747:

If the Commander-in-Chief should chase with the whole squadron, and would have those ships that are nearest attack the enemy, the headmost opposing their sternmost, the next passing on under cover of her fire, and engaging the second from the enemy's rear, and so on in succession as they may happen to get up, without respect to seniority or the prescribed order of battle, he will hoist . . .

This was an improvement on Hawke's additional because it specifically ordered ships coming into action to pass ahead under cover of their own ships already engaged. It actually prescribed what Sir George Byng had managed to do without a signal as long before as 1718.

In other respects the general effect is disappointing to the historian. Hawke's very important instruction of 1759, for pressing home the attack regardless of the line, was omitted. No inspirations seem to have been drawn from Howe's additionals of March 1762. Nothing was added in the way of new cruising formations as a means of increasing the ease and speed of fleet control, apart from the strong emphasis on the bow and quarter line formation. Though reflective of Hawke, the new Admiralty *Signals and Instructions* is on the whole a conservative production. With the sole exception of

Article 18, it did no more than give official sanction to instructions with which the navy had been familiar for twenty years and more, while actually excluding some of the most useful.

The worst structural feature of the compilation is the retention of the original method of indicating signals by embedding them in the text of the instruction. This may have seemed logical because it was a companion volume in the parent work. Nevertheless it was retrograde, not merely on grounds of logic and common sense, but in view of the fact that so many captains, as well as admirals, had become accustomed to issuing their own sets of additional signals, and that, in consequence, a large number of private signal books, constructed as such, were already in use.[14] More serious, though more excusable, was the failure to re-organise the whole signalling system on simpler and more efficient lines. To cope with the new instructions, more flags had

become necessary; this was the only way to make the old system continue to work. To have changed the system would have meant scrapping or completely re-writing the parent work.

Private signal books took note of the new work in various ways. Some had a separate column headed Book, with the reference below, and opposite the signal: 1, for the parent work or 2 for the new one. Others adopted the abbreviation GSA for General and Additional Sailing, GF for General Fighting and AF for Additional Fighting. There is also evidence that occasionally the two books were bound up together, though not necessarily by the Admiralty, and the pages given continuous numbering. This would explain the high figures of some of the page references in the *Terrible's* signal book of 1779.[15]

Pavillon's signalling system

La Bourdonnais's numerical signalling system, if it had been adopted by the French navy, would have been a major tactical development. The Chevalier du Pavillon, however, had the influence necessary to ensure that it was his system which was adopted instead, as it was by a naval council of

14 *MOD,* NM/92[?] (formerly *RUSI* 92) includes a set of the original Sailing & Fighting Instructions, issued by Byron on 26 October 1778, and embellished by the recipient with the flag signals and gun and false fire signals painted in the margins.

15 *NMM,* Tunstall Collection, TUN/1 (formerly S/MS/Am/1)

Chasing signals issued by Admiral Hawke in the Seven Years War. MOD fourteen flag officers assembled at Rochefort on 4 June 1773, and again by a naval council held at Versailles on 17 March 1775 which included d'Estaing and d'Orvilliers.[16] During that summer an 'evolutionary squadron' was formed, in which Pavillon commanded a corvette. During the exercises his signals were given a trial. Next year another evolutionary squadron was formed under Admiral Louis Charles, Comte du Chaffault, for whom Pavillon prepared a complete set of signals. He published his system in 1776, and shortly thereafter was appointed chief of staff to d'Orvilliers, who encouraged him to elaborate his system still further, and to impose it on the Brest fleet for the operations for 1778–9. Admiral Luc-Urbain du Bouëxic, Comte de Guichen, later adopted Pavillon's system, but not de Grasse.

table, filled with numbers which could be read off by reference to a series of flags painted along the top and down the left side. The numbers read downward from left to right. The particular number required was signalled by a hoist of two flags, the top indicating the column in which the number was to be found and the bottom indicating the line. The meaning of the number could then be looked up in the signal book. Pavillon's table contained a hundred numbered squares, and as usual in such a system, the same ten flags were used both for the superiors along the top and the inferiors down the side. These were:

1	Union Jack	2	White and red horizontal
3	Red	4	White
5	White and red cross	6	Saltire white red and blue
7	White bordered red	8	Dutch Jack in oblique stripes
9	Yellow and white	10	Chequered red and white

As in all such tables, the result was confusing. Except along the bottom line, where flag 5 over flag 10 read off at 50, and flag 10 over flag 10 at 100, etc, the table was always ten behind the actual flag numbers. Flag 2 over flag 1 indicated the square numbered 11 (not 21) while flag 9 over flag 5 read off at 85 (not 95).

The hundred signals which could be made using the table provided for the entire handling of the fleet when disposed in three columns, changing the relative order and position of the columns, changing into the various orders of sailing and back again, and forming the line of battle from the various orders of sailing and back again. Also included were signals notifying the general's intention to revert to the traditional signalling system, and to return again to the new, to change to *le grand tableau numéraire*, and to signal numbers for expressing soundings, numbers of ships, etc.

The same ten-squared table, with exactly the same flags, was employed to supply a separate series of a hundred signals for exercises and battle. The sixteen squares occupying the top left-hand corner of the table, which would otherwise bear the numbers 1–4, 11–14, 21–24 and 31–34, were occupied instead by sixteen letters of the alphabet: A, B, C, D, E, G, H, I, K, L, M, O, R, S, T, U. The remainder of the table, that is to say the eighty-four numbered squares, contained signals of a general kind, as well as the usual battle signals, including that for doubling and also concentration on the enemy's centre or van, but not rear. There was no signal for cutting the enemy's line. Numbers 82–98 were entirely for individual ships, designated by their number in the line of battle, to correct their position.

In each of the first sixteen alphabetical squares, one of the standard orders of sailing was also printed, ending as usual

with the order of retreat. To signal these, reference had to be made to a separate *grand tableau numéraire*. Instead of flags, the same letters of the alphabet as were used in the main table were printed along the top and down the side, the numbers in the squares reading from 101 to 356. By means of this table, the fleet could be signalled to change from any one of the standard sixteen orders of sailing into another, the top letter indicating the order the fleet was in, and the bottom one the order into which it was required to move.

On yet another sheet Pavillon produced his original ten-flag table, but with the squares numberd 101 to 200. Here each signal was in two halves, the top, in italics, giving the formation of the fleet at the time, the bottom, in roman print, giving the formation required. These changes of formation were more detailed than those envisaged under *le grand tableau numéraire*, which only dealt with the sixteen standard orders of sailing. The new table of 101–200 signals included changes of formation by the fleet as a whole as well as by squadrons, changes of the relative positions of the squadrons while on different tacks, changes of tack and wearing, and reaching the same positions either from line ahead, reversed line ahead, or from three columns. The signals numbered 101 to 164 were specifically for the fleet in three columns. Signals 165 to 200, although apparently overlapping those given in *le grand tableau numéraire*, in fact covered different movements, always with three columns in mind.

Although it must have been easy enough to change from one table to another by means of a preliminary signal made by a single flag, flamme or cornette, the possibilities of confusion would seem to have been immense, especially as regards reading off the numbers in the tables. What seems extraordinary is that, in spite of so much juggling with numbers and letters, the flags, though each bearing a number, had no significance except in relation to one or other of the tables. How easy it would have been to have expressed 300, 500 or even 900 signals by a truly numerical system.

Although it was an advance on the existing official system, Pavillon's tabular system represents a *cul-de-sac* in the development of signalling. He lifted French signalling out of its Seven Years War defeatist traditionalism, and put it on a new and complicated mathematical basis. Given quickness and accuracy both of reading and sending signals, the system was capable of being used to ring the changes on every possible fleet formation. In this sense it gave scope for tactical sophistications far beyond anything as yet regarded as standard in the British service. It was quite possible to transcribe the contents of Pavillon's complex signal sheets into a manuscript signal book; the problem, however, was that once the French navy turned in the direction of this semi- or even pseudo-reform, it became more difficult to cut through the new mathematical tangles set in train by Morogues, and proceed to the ultimate goal of a true numerical system. In the British service, where neither order, logic, simplicity nor common sense had so far prevailed, the breakthrough was achieved partly by Howe's skill and persistence, and partly by the threatened collapse of all tactical development under the old signalling system.

Howe on the North American Station, 1776–8

When Lord Howe arrived at New York on 12 July 1776 to assume command of the North American Station, the War of American Independence had scarcely begun, the Declaration

16 *Tactique Navale à l'usage de l'armée du Roi commandée par M le Cte d'Orvilliers . . . 1779*, p40 (copy in the library of the Royal Navy Staff College, Greenwich)

of Independence having been signed only twelve days previously. France had not yet intervened and the Royal Navy was solely concerned with operations against the colonists. Nevertheless, Howe at once set about organising his fleet as if for a full-dress naval war with a major power. His instructions and standing orders issued on the day of his arrival included an important battle order (number 17) directing captains to fire as many guns as possible at the first discharge. Apart from the general effect of doing so, this addressed the problems that firing so many together might subsequently be impossible owing to ships drifting out of range or bearing.[17]

On the same day, 12 July, Howe issued his printed Signal Book for the Ships of War.[18] Despite its unassuming appearance it is one of the most important documents in the whole history of naval tactics. It is a real signal book at last, though not an Admiralty one. In form, lay-out and typography, moreover, it has a special character which persisted through all Howe's later work and can instantly be recognised. Its title gradually permeated the navy and was eventually adopted by the Admiralty for their first official printed signal book, issued in 1799.

Howe's signals were only to be used by ships within his own fleet. They were made in the traditional manner, using twenty-three flags and seven triangular flags of simple design flown from particular positions. However, they had several new features. Signals for engaging, forming line, etc, were to remain in force, even though the signal might be taken in after acknowledgment, until annulled or countermanded.

The signals will generally be made without guns, when it may be done with the same effect; unless when it happens that two guns or more are necessary to constitute a part of the signal, or that it does not appear to have been timely observed. And all signal guns will be fired on the same side, to windward in a fog.

Unlike the earlier private signal books, in which all kinds of signals were thrown into juxtaposition on the same page, (largely as a result of successive admirals filling up unappropriated places for flying particular flags), Howe's signals were roughly grouped together. This had already been done by Saunders in his index volume, but not in the signal book itself which he issued.[19]

Such references as were given in the signal book are not to the Admiralty Sailing & Fighting Instructions, but to the page and article of an accompanying printed book of instructions.[20] The title of this book, 'Instructions for the Conduct of the Ships of War, Explanatory of, and Relative to the Signals contained in the Signal-Book Herewith Delivered', its content and wording demonstrate the revolutionary step which Howe had now taken. On the one hand it marked the separation and independence of the new signalling system from all instructional matter and its supremacy over such instructions as remained. On the other, it provided a new,

though subordinate, system of instruction. Unlike the official book, Howe's book consisted partly of standing orders, partly of orders designed to make the admiral's general tactical ideas more easily understood, and partly of explanations of how certain signals were to be executed so as to standardise evolutions. The actual signals were very sparingly described.

These instructions, far from representing a revolt against the line-ahead formation, reveal Howe as a stickler for its effective application, especially when in Article 14 of the fighting instructions he revived the supposedly retrogressive ban on pursuing beaten enemy ships. Nor was anything said about concentration on part of the enemy's line as directed in the earlier undated book, though because of its appearance in the Signal Book this might now have been considered self-explanatory. In Articles 10 and 22 of the 'instructions respecting the Order of Battle', the insistence on flag officers opposing the flag officers in the enemy's line was extremely conservative. So was Article 16, which forbade any captain to quit his station in battle without permission 'though much pressed'. Only in extreme necessity could they do so, but 'Captains deficient in time of action are to be removed from

A page from Admiral Lord Howe's 1776 Signal Book for the Ships of War. NMM

17 *MOD*, NM/88 (formerly *RUSI* 88). Corbett only prints the first six orders in pp87 92

18 *NMM*, HOL/21 (formerly S(H) 21) signed copy issued to Captain Sir George Collier, the *Rainbow*. HOL/22 (formerly S(H)) 22 is an unsigned copy in the original marbled covers. A further copy issued to Thomas Graves is in the *MOD*, NM/56 (formerly *RUSI* 56), and another issued on 13 June to Captain Cornthwaite Ommaney, the *Tartar*, is in the *MOD*, Ec/153. Manuscript copies are in the *NMM*, SIG/B/13 (formerly Sm/111) (without name of owner) and *MOD*, NM/59 (formerly *RUSI* 59), copy made by Thomas Graves.

19 *NMM*, Tunstall Collection, TUN/34 and 35 (formerly S/MS/SY3 & 4)

20 *NMM*, SIG/A/8 (formerly Sp/III), issued and signed 13 June 1776, *Eagle*, at sea, to Captain Cornthwaite Ommaney, the *Tartar*. See also *MOD*, NM/76 (formerly *RUSI* 68), and *MOD* Ec/153.

their ships'. The arrangements Howe ordered for the execution of the line battle, however, were innovative because of their practical seamanship, and because they recognised the need for fleets to operate in a different manner from that of armies.

In his general instruction number 2, Howe abandoned the practice by which the junior ship was required to keep clear of the senior, and ordered the captains of all ships to take whatever action was necessary to avoid collision. In Article 1 of the 'Instructions Respecting the Order of Battle', he ordered ships in line of battle to open out when tacking to give each other sufficient room, and to close up afterwards. In Article 4, he ordered slow ships to fall astern and not disorder the fleet by trying to keep their stations. 'The Captains of such ships will not be thereby left in a situation less at liberty to distinguish themselves; as they will have an opportunity to render essential service, by placing their ships to advantage when arrived up with the enemy, already engaged with the other part of the fleet.' In Article 9 captains were instructed to 'studiously endeavour to obtain by repeated observation and experiment, a perfect knowledge of the proportion of sail required for suiting their sets of sailing respectively to that of the Admiral's ship'. The repeating frigate stationed opposite the Admiral was instructed to copy exactly the sail set by the Admiral so that ships at either end of the line would be able to make their own adjustments.

A year later, on 1 July 1777, Howe issued 'Additional Instructions Respecting the Conduct of the Fleet Preparative to and in Action with Enemy'.[21] In these Additionals Howe revealed more fully the advanced thinking which lay behind the executive form of his Signal Book. Though these were clearly designed to give an even higher degree of precision to the forms of attack included in the main instructions, still greater emphasis was laid on the need for tactical sophistication by individual captains. No fleet at that time could have interpreted these additional instructions intelligently without several weeks, and possibly even several months, of exercising together. Article 1 stated that, whereas 'the chief purpose of a regular disposition' in attack is to avoid exposure to the enemy's fire and the danger of collision with friendly ships, and whereas this may hamper captains 'from taking advantage of the favourable incidents which may occur in the progress of a general action, it is the object of these instructions to facilitate the means of improving such opportunity by an authorised deviation from these restrictive appointments'.

Article 2 ordered that the enemy line was to be engaged throughout its full length, so as to prevent them doubling the British van or rear. If the enemy were superior, captains 'commanding ships of greater force [were] to disregard or pass the weaker and worse sailing ships of the enemy and confine their first endeavours solely to disable the stronger and more active, as their accidental situations in the line may afford opportunity'. Article 7 instructed that, if the enemy were inferior in numbers, the overlapping ships in the British van or rear were to quit the line without awaiting a signal, and 'to distress and annoy any of the nearest ships of the enemy in conjunction with the ships of the fleet particularly opposed to them; or otherwise to assist or relieve any disabled ships thereof, as they can be employed with most advantage'.

21 *MOD*, NM/88 (formerly *RUSI* 88), manuscript copy by Thomas Graves. A comparison of the Instructions of 1776 and Additional Instructions of 1777 with the undated but obviously earlier Signals & Instructions gives the following results, the Instructions of 1776 shown in [] and the Additional Instructions of 1777 shown in (): undated book: Article 7 [7]; 8 [8]; 21 (3); 22–3 (5); 24 (6); 25 [18]; 42 [17]; 46 [19]

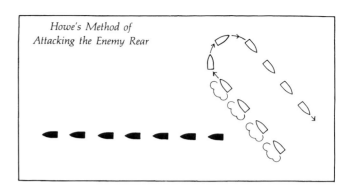

Howe's Method of Attacking the Enemy Rear

Articles 3 to 6 dealt with the situation of the fleet weathering that of the enemy preparatory to a general action. Captains were instructed to expect the admiral to make the signal for the rear to tack first, or for the fleet to tack together, as appeared to be 'most convenient for forming upon the line of bearing best suited to the position of the enemy'. If a number of the enemy's rear ships tacked first, an equal number of British rear ships were to do the same. If the enemy chose to await the British attack, and were on the same tack, 'the Admiral will then most probably choose to continue on upon the same course' until his van was level with theirs before attacking, provided that his ships were at a sufficient distance not to be 'materially injured' by the enemy's fire. However, he might chose to attack the enemy rear. In that case he would signal for the leading ship

to open to a convenient distance, and upon coming up with [the] enemy to give their sternmost ships her fire upon the quarter, then tack or veer and fall into the rear of the [British] line. . . . this method of attack is to be continued in succession by the ships of the fleet, or any particular division, so appointed until further signals, care being taken to leave room as requisite from each other in the execution of this service.

When it was decided to break off this hit-and-run form of attack and bring on a general action, the signal would be made to invert, in succession, the line which had been established primarily 'for the circumstance of coming up at a small distance to windward of the enemy upon the same tack'. The leading ship would then engage the sternmost enemy ship while her next astern would pass her to windward and engage the sternmost but one of the enemy, the whole fleet following the same procedure in turn until the original sternmost British ship engaged the leading ship of the enemy. The object of this manoeuvre was to avoid exposing the leading ships to the fire of the whole fleet of the enemy as they attempted to draw level with the enemy's van, as would be required by the Article 19 of the general printed Fighting Instructions, which stipulated that 'If the admiral and his fleet have the wind of the enemy, and have stretched themselves in a line of battle, the van of the admiral's fleet is to steer with the van of the enemy's and there to engage them.' Much the same tactical idea had earlier been expressed by Barrington in his third memorandum of 1768.

Article 8 provided for the eventuality that 'it may be necessary on some occasions to set the ships of the fleet at liberty to steer for those opposed to them respectively in the enemy's line independent of that necessary regard to the

Chasing Orders, 1777

uniformity in distance, course, and movements to be at other times observed'. When the signal was made, captains were 'to keep those ships to which they are then separately directed upon the same constant bearings, if possible, as they advance towards them', care being taken to avoid collisions 'by a proportionable increase or decrease of sail carried'. On reaching the enemy they were free to engage either to windward or to leeward, taking care not to be drawn away too far; 'being in such circumstances to consider wherein they may render the most effectual service against the enemy and govern themselves accordingly'. Taken together with the signals for attacking, respectively, the enemy van, centre or rear, this instruction seems to mean that captains were to keep steering for the enemy ships they were making for at the moment when the signal was made. If there is some doubt about this interpretation, it arises from Howe's own wording which, though somewhat clearer than his worse utterances, is nevertheless ambiguous.

Article 9 was concerned with preparing ships for action. It ends with the recommendation to captains 'whilst advancing . . . to keep their unoccupied men laid close down upon their decks upon the off side from the enemy ships or', in the case of bombardment of land fortresses, 'works'.

The most striking aspect of these additional instructions is the high confidence reposed in the skill, initiative, good sense and personal restraint of the individual captains. Nothing but a 'band of brothers' could have given full effect to these counsels of perfection. The instructions clearly show Howe's tremendous candour in fully revealing himself to his captains; there is no trace of that arrogance and secrecy so common in admirals.

On 21 August 1778, before leaving the North American station for England, Howe issued 'Signals for the Frigates or other Ships of War appointed to observe the motions of a strange fleet discovered or enemy's fleet, during the night'.[22] This rigmarole, in fact, stands for the germ of a system of night reconnaissance; the end product was to be the system by which Nelson was able to keep in touch with his inshore squadron watching Cadiz during the nights immediately before Trafalgar. The ships, detached on night reconnaissance service, were to carry distinguishing lights, shown so as to be visible to their own fleet but not to the enemy. They were provided with eight signals for indicating the enemy's movements. Howe's Observations stressed the need for showing the lights effectively, especially when the enemy were seen to be drawing away from, or bearing down on, the British fleet. Of the importance of these night reconnaissance instructions there can be no doubt. They were re-issued in 1782 on Howe's assumption of the command of the Channel Fleet, and were a permanent feature of all his future additional instructions.

In the same manuscript there were also fog signals and a hitherto neglected numerical signalling table. The fog signals were made with guns, drums and bells, and there were some experimental gun signals by which one or more guns indicated the page in a signal book, followed after a pause by more guns indicating the number on the page. The layout, however, does not conform with Howe's signal book, and does not seem to have been adopted by him.

The numerical table employed ten flags and a substitute flag to produce 100 signals. The signals could be flown anywhere in the ship. A separate table of flags was also provided.[23] As is inevitable with tabular systems, the flag numbers hoisted bore only an indirect relation to the number they indicated on the table, and this limitation severely restricted the utility of numbered flags. If numbers were to be used, it was obviously easier, quicker and safer to use the ten flags for direct numerical signalling, numbering them 0 to 9, and making hoists of two flags, the top representing the tens and the bottom the digits. Pendants could be used for hundreds, as in the tabular system. This was, of course, the system which had to be adopted in the end. What seems strange is that such a progressive admiral as Howe should have allowed himself to be side-tracked into this complicated *cul-de-sac*.

The accompanying statement, 'by this means the number of any article in the Day Signals may be expressed', implied that the day signals were numbered right through, and that they could now be made by numerical means as well as by the old method of flags flown in particular positions. There is no reason, however, to assume that this numerical table was actually put into use by Howe. The document may have been no more than a project. Sir Charles Knowles claimed that, when he was a lieutenant on the North American Station, he gave Howe a new scheme for signals in 1777, and that he published his own tabular scheme in 1778. It may have been this scheme with which Howe experimented.

By the end of his service on the North American station in 1778, Howe had practically created the tactical and signalling system on which he was to run the Channel Fleet in 1782. So far as we can tell, he had done this without any help from Admiral Kempenfelt, who a few years later was to be deeply involved in the search for tactical reforms.

The National Maritime Museum possesses an undated manuscript volume containing a copy of Howe's signals on the North American Station in 1776, followed by a complete set of numerical signals compiled by the anonymous writer.[24] In his introductory and explanatory remarks this writer states that

The general signals used by the Navy are very deficient in all the requisitions, particularly the letter ones in the confusion of colours in the flags and in the display of the signals; indeed the nature of our signals makes the defect irremediable as the number of the flags used makes the confusion of the colours unavoidable, and the number of orders signified by each flag renders it impossible to show each signal where it would be most visible.

His own proposed scheme was based, he asserts, on French practice and on Howe's signals in America 'with some additions'. These are of interest as an early example of an experiment with a numerical system; the signals themselves exhibit no interesting tactical features.

The younger Knowles's set of signals

In 1777 the younger Sir Charles Knowles published *A Set of Signals for a Fleet on a plan entirely new*.[25] He had entered the navy in 1768 under Barrington in the *Venus*, and served in the West Indies from 1773–6 under Rodney. He was later appointed by Howe to third lieutenant of the *Chatham* on the North American station, and saw much service in the preliminary stages of the war against the American colonists. Knowles's later semi-autobiographical

22 *MOD*, NM/34 (formerly *RUSI* 34), manuscript copy presumably by Thomas Graves

23 The flags were, Union, Red, White, Blue, Half Red and Half White, Half Blue and Half Yellow, Dutch Jack, Red Pendant, Yellow Pendant, and Dutch Pendant. The substitute flag was the St George's Cross or Ensign.

24 *NMM*, SIG/A/8 (formerly Sm/111). The date might well be about 1778–80

25 *A Set of Signals for a Fleet on a plan entirely new* by Charles Henry Knowles, Bart, Lieutenant in the Royal Navy. London: Printed for J Robinson, New Bond Street. MDCCLXXVII. NMM, HOL/25 (formerly S(H) 25)

the conclusion, as Knowles did at the Battle of Grenada, that the French did indeed employ such a tactic.

Knowles was more conservative in Article 18, in which he required ships to keep the line unless specifically ordered to break it. 'NB: the line is not to be broke, if possible, when the fleet are much inferior to the enemy.' A handwritten Article 38, and the directions for the ships of the reserve corps, indicate that Knowles was aware of the value of the line of battle in most circumstances. He wrote that with a very large fleet, forty or fifty sail of the line, the most effective tactic would be action on opposite tacks to the enemy, because the successive broadsides of so many ships would be devastating. The *corps de réserve* should not be used to double the enemy, but rather to reinforce such weak points as might develop in the line of battle, and to provide tow for disabled ships. As an observation in Article 14 Knowles had written:

To double the enemy's line, that is, to send a few unengaged ships on one side, to engage, whilst the rest are fighting on the other, is rendering those ships useless; every [enemy] ship which is between two [of ours], has not only her two broadsides opposed to each of theirs, but has likewise their shot, which cross, in her favour. Therefore, if ships form a double line, it should be on the same side, whether it be to windward or to leeward of the enemy.

In Articles 16 and 17 Knowles expressed the belief that in the open sea it was always possible for a fleet to defeat an attack on its rear by tacking or wearing together, and engaging the enemy on opposite tacks. If the enemy should be caught close to the shore where there was not room to tack, however, it would be profitable to engage them from the rear in succession, inverting the order of battle. If their van and centre tacked to support their rear, 'then the battle will soon be reduced to a scramble, wherein British seamanship and courage will soon show their superiority, as it has ever done on those occasions'.

Knowles stated that when Howe succeeded Keppel at the Admiralty (presumably meaning the first time, in 1782), he wrote offering to show Howe his father's plans and papers, and mentioning his own 'work on naval tactics'.[31] Howe asked to see them all, and was then told by Knowles that the French 'felt the pulse of our Captains and avoided a close action', and that the way to force a close action was suggested by Hoste, 'and is in the French Naval Tactics'.

This Lord Howe executed at the battle of the 1st of June, 1794, when each ship followed her opponent, and thereby did not check the ardour of the men, which was done by the 21st Article of the old Fighting and Sailing [sic] Instructions; the line is not to be broken, unless by signal, until all the enemy's

ships are disabled or run; which Lord Howe changed to 'beaten or give way' though retaining the article.

Knowles wrote that Howe, 'on seeing the French manoeuvre [Article 14 of Knowles's instructions], struck [Article 21] out of his own Fighting & Sailing Instructions, many of which his Lordship altered.' This is quite untrue. In Article 15 of his 1782 instructions, Howe changed the words to 'disabled or broken', and they remained as such until the Admiralty book of 1799 took over Howe's system, when the words were reversed to read 'broken or disabled'. Considering that Knowles's evidence involved an alleged French manoeuvre that did not exist, a fact of which Howe must at least have had some inkling, it can hardly be supposed that Howe acted on those grounds alone.

Knowles was an extraordinary figure, and his claims are bewildering. Yet if he really compiled his Fighting and Sailing Instructions in 1780, and there seems no reason in this case to doubt his written word, these alone entitle him to be regarded as a great tactical innovator. Written in that year they would have done credit to any naval officer, however high in the service. As the work of a lieutenant aged only twenty-five or twenty-six, they were a *tour de force*. In 1782 he compiled an excellent *Ship's Orderly Book*, which he had printed and sent to the Admiralty in 1783. He also wrote day and night signals for a convoy.[32] He was invalided from the navy after the Battle of St Vincent and does not appear to have taken any further part in tactical reform, although his health recovered and, at the age of seventy-six in 1830, he published his famous pamphlet attacking the claims that Clerk of Eldin had inspired Rodney's break of the French line at the Saints.

William Dickson's signals, 1780

Among a batch of papers preserved by Thomas Graves are four sets of signals issued or prepared by William Dickson, Captain of the frigate *Greyhound*, when serving in the West Indies under Sir Peter Parker in 1780.[33] The first is simply a private signal book in the traditional form, but incorporating Parker's additionals. With it, however, is a set of true numerical signals. This employed ten flags, including a green flag, the only known instance of this colour. The location of the flags, at the mizzen, foremast head, and mainmast head, indicated respectively units, tens and hundreds. The flags and most of the arrangements could be changed for security. This system was simple enough 'to be understood by the meanest capacity'. His night signal system was also numerical, but depended upon ten special arrangements of lights to indicate each number, 'by which means 12 Lanthorns only will refer you to the high Article 999'. The fact that these signals were not just a project is proved by their actual issue to Graves, then commander of the sloop *Savage*, by Dickson on 7 March 1780. Further evidence of their actual use is a note stating that 'for all other signals you are to refer to those you have received from Rear Admiral Parker'. Finally there was another numerical code of an even simpler form: ten flags numbered 0 to 9 which could be flown from anywhere in the ship, tens and hundreds being presumably shown by hoists of two and three flags respectively, one above the other. Again, as Dickson carefully noted, the signals could easily be changed by inverting the numbers.

---

31 Admiral Sir Charles Knowles, *Observations on Naval Tactics*, 1830, pp28–29 (*NMM*, Tunstall Collection, TUN/75)

32 All the Knowles instructions, etc, quoted, except the *Observations on Naval Tactics* of 1830, are from the autographed set given by Admiral Sir Charles Knowles to Sir Julian Corbett in 1910 with 'bound together in 1910 with the MS notes from his own copies added' and with explanatory letters from Knowles inserted; *NMM*, Corbett Collection, TUN/41 (formerly S/P/Am/6).

33 *MOD*, NM/92 (formerly *RUSI* 92). Although the document is marked 'a copy' and the signature is not, therefore, in Dickson's own hand, its authenticity is undoubted. See *The Mariner's Mirror*, 40 (1954) no 2, p163, 'Answer', by Commander Hilary Mead.

CHAPTER SIX:

Tactics of the American Revolutionary War: Testing the New Systems

THE Channel and Atlantic operations of the British and French fleets under Admiral Angustus Keppel and the Comte d'Orvilliers, respectively, provide one of the few cases where the instructions, signals and actual battle tactics of rival fleets can all be considered together. The chief tactical event was the Battle of Ushant on 27 July 1778. This battle demonstrated the success of French efforts following the Seven Years' War to reform their tactical system. Its outcome, on the other hand, stimulated important developments in British tactics and signalling. After Keppel's retirement, Admiral Kempenfelt, as captain of the fleet to successive commanders-in-chief, was able to push ahead with tactical experiments, the effects of which gradually reached the more distant stations. The entry of France, and then of Spain, into the American Revolutionary War had led to the Western squadron, or Channel Fleet as it was increasingly being called, taking the primary role in British naval affairs, although very substantial forces continued to be sent both to the North American station and to the West Indies. New tactical ideas and signalling experiments tended to originate in the Channel Fleet and to pass from there to fleets in American waters.

Keppel's signals, 1778

At the time of the Battle of Ushant, Keppel was aged fifty-three and had a high reputation in the service as a fighting captain at Quiberon, a commander of the successful expedi-

Amiral *Louis Guillouet, Comte d'Orvilliers (1709-92).* MM

tions against Senegal and Belleisle, and as assistant to Pocock at the capture of Havana. In 1778 his health was no longer good, however, and he was preoccupied with political concerns. He was not interested in tactical reform, but it must be admitted that he approached his command with great concern for efficiency in his own way. While his fleet was fitting out he drew up his own 'Signals & Instructions in Addition to the General Printed Sailing & Fighting Instructions', and on 28 June 1778, the day after returning from his first cruise, he issued a printed version.[1] This is not to be confused with the Admiralty work of the same title, nor with any of the Admiralty 'additionals' issued during the Seven Years War. Keppel's book was a selection from all these sources, with the articles reworded. There were twenty-four signals for day, sixteen chasing signals, nine for night and five for fog. The eighteen day fighting instructions are reminiscent of Hawke's issue of 1747 and in any case contain nothing new. There are four night fighting instructions.

On 21 August, two days before leaving Cawsand Bay for his third cruise, Keppel issued Captain (later Admiral) Adam Duncan of the *Suffolk* with a copy of the original work, with a handwritten reference to the incorporation of fighting instructions Articles 1 and 2 into a single article.[2] Keppel, apparently, considered it a necessary part of his duty to continue issuing the parent work. Nevertheless, Sir Charles Knowles stated that there was a re-issue of Keppel's own compilation in 1778, presumably for captains of ships newly commissioned for the war. Both books were in fact used during the Battle of Ushant. The schedule of signals which Keppel made from the *Victory* shows that, of the eleven general signals made, six were from the General Sailing & Fighting Instructions, and five from Keppel's own signals.

D'Orvilliers's signals, 1778

Across the Channel the situation was completely different. Unhampered by the British need for warships on the North American station, the Comte d'Orvilliers was free to prepare his fleet at leisure, while the Comte d'Estaing prepared another fleet for intervention in American waters at France's chosen moment. D'Orvilliers was a veteran of sixty-nine, but he had Pavillon as his chief of staff. His signal system for the campaign of 1778 was the most comprehensive and elaborate used until that time. It was not identical to Pavillon's project of 1776, but it is clearly derived from it and shared its strengths and weaknesses.[3]

The system started with a squared five-flag table of letters A to Z, less W. Sixteen of the lettered squares also had

1 *NMM,* DUF/31 (formerly Sm/167), a large folio manuscript dated 29 April, 1778, with a copy of Keppel's signature at the end, and CLE/2/9, printed version, signed and dated by Keppel, the *Victory,* St Helens.

2 *NMM,* DUN/4 (formerly S(D) 4)

3 M d'Antin, *Plans des Mouvemens et Positions, Respectives des Armées de France et d'Angleterre a la Journée du 27 Juillet, 1778*, Brest, 6 October 1778, *NMM,* REC/46 (formerly Sm/7)

The positions of the British and French Fleets at noon on 31 August 1779, as published by d'Antin. NMM

had continued the same as it did, and the signal had not been made to chase, I do not think that any part of the Vice-Admiral of the Blue's division could have come into action at all.'

At about 8am d'Orvilliers signalled the French fleet to wear in succession. This had the effect of bringing them onto the starboard tack heading SW. According to d'Antin's diagrams, they were always well closed up in perfect order, but it seems more likely that by now their line was somewhat extended and not perfectly straight. As this stage, the wind shifted two points to the south against the French fleet, hiding them from the British by 'a dark squall'. The British were coming up in businesslike fashion, evidently intending to attack even though in poor order.

D'Orvilliers sensed a projected attack on his rear, and some of Palliser's Blue division were almost sufficiently advanced to weather him. Sometime between 9.30 and 10am, therefore, he signalled the French fleet to tack together, thus returning to his original course. Next he signalled his new van to run large four points while his new rear was getting into station. He then signalled the whole fleet to come to the wind. As seen from the British fleet, these movements were by no means perfectly executed. No doubt at this stage the French line fell into some disorder.

All this time the British fleet was continuing to work towards the French, but in no very good order; Keppel could

not afford the time to close his fleet in line abreast or bow and quarter line, if he was to be sure of forcing a battle. As Sir Robert Harland remarked (at Palliser's trial),

pursuing the French in the way we did that day, was a bold, and necessary, and allowable stroke of war, where nothing but risk, as it appeared, could stop the French fleet from getting off. The great decisive strokes of the day were to follow, by closing with the enemy and fighting it out.

When the French re-appeared from the squall about half an hour later they were much closer to the British, having fallen somewhat to leeward in the course of tacking. They were now steering north-west again, as when first sighted at daybreak. D'Orvilliers had signalled his van to keep their wind, so as to stop Keppel's Blue squadron from weathering them. Evidently this signal was not meant to apply to the whole van, so that, while the four leading ships turned to larboard in succession, the remainder held their course. Actually d'Orvilliers need not have worried. With his light squadron of three ships already stationed to windward of the van, he was well covered. Nevertheless, a few British ships were slightly to windward of the French van and were forced to give way. Captain Mark Robinson of the *Worcester* stated at Keppel's trial, 'If I had taken any wind I could have weathered more than half the French fleet, the wind having shifted more than two or three points to the westward.'

The British Red squadron was now some two miles from the French van and heading towards them at an oblique angle. It was 10.30am, and Keppel saw that if he was to fight he must do so at once, without attempting to close up his fleet and get it into reasonable order. He made the signal to tack together, which brought him into a rough line ahead formation on the starboard tack parallel to the French and to leeward of them, with Sir Robert Harland's Red squadron leading. The wind now shifted about two points to the west in Keppel's favour, giving him more freedom of manoeuvre. The two fleets were about to pass on opposite courses. The French, despite the partial withdrawal of their van, were in

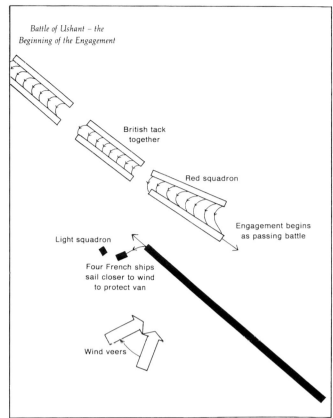

Battle of Ushant – the
Beginning of the Engagement

British tack
together

Red squadron

Engagement begins
as passing battle

Light squadron

Four French ships
sail closer to wind
to protect van

Wind veers

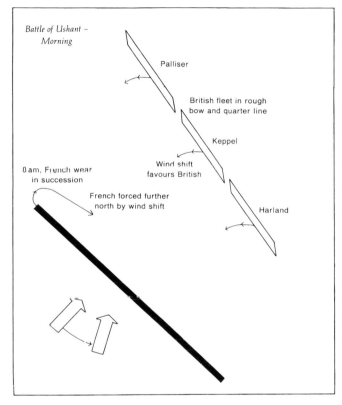

Battle of Ushant –
Morning

Palliser

British fleet in rough
bow and quarter line

Keppel

Wind shift
favours British

8 am, French wear
in succession

French forced further
north by wind shift

Harland

fairly regular order, but the British were strung out in a long irregular line, not even in their correct order of battle. This meant that when the two fleets began to pass each other, the British line had gaps in it of nearly half a mile and that some ships passed much nearer the French than others.

Firing began at about 11 am and at 11.05 Keppel signalled to engage, without having made any preliminary signal for the line of battle ahead after tacking. It says a good deal for the freedom of British tactics and the reliance placed on individual initiative that this omission was easily disposed of at Keppel's trial, Captain the Hon Robert Digby of the *Ramillies* remarking in his evidence that, 'A line of battle in my opinion, always retards. Had the signal been made, the Red division must have shortened sail.' Captain Samuel Goodall of the *Defiance* was equally firm on this point. Asked at Keppel's trial if action would have been possible if Keppel had formed a regular line of battle, he answered, 'No: the admiral had always offered them battle, and it lay in their breast whether they would bear down. Had he kept a line of battle that morning they could not have been brought to an engagement.'

The ensuing battle, with the fleets passing each other at a relative speed of about six to eight knots and with the opposing ships firing at varying ranges, could not lead to decisive results.[5] The battle was a desultory affair. The British ships engaged at widely varying ranges, and were separated from each other by wide gaps, some of the rear ships being up to half a mile apart. On this occasion there seems to be some truth in the standard British assertion that the French deliberately fired to damage masts and rigging. They were to windward, and thus heeling over in somewhat

Battle of Ushant –
Before the Squall

Palliser's squadron
looks likely to
weather French rear

French van runs large,
then comes close to
the wind

French tack together

---

5 John Bazely, Palliser's flag captain in the *Formidable,* stated at Keppel's trial that the maximum speed possible that morning for a ship with all sails set was 'between seven and nine knots.' As, however, the fleets were under the usual battle order of reduced canvas, their combined relative speed is unlikely to have exceeded eight knots.

success. This was an extremely relevant point and has been made subsequently by a whole line of British writers commenting on the American War as a whole. The British tended to employ younger admirals than the French, and no doubt sheer exhaustion at the end of a battle, following hours and hours of manoeuvring, tended to sap the initiative of the veteran French commanders, as in Rodney's case at the end of the Battle of the Saints. D'Antin, in giving the French view of the trials, stressed public uneasiness in England arising from memories of Admiral 'Binke' [Byng]. He naturally refrained from criticising the French command, though this does not necessarily mean that he failed in private to draw some chastening conclusions.

The Franco-Spanish fleet of 1779

The combined Franco-Spanish fleet which sought to gain control of the English Channel in 1779 consisted of twenty-eight French ships of the line and eighteen Spaniards, fairly

---

7 *NMM*, REC/46 (formerly Sm/7), pp117–9

evenly distributed throughout, in three squadrons.[7] A French ship was stationed at the head and tail of each squadron. D'Orvilliers was commander-in-chief, de Guichen commanded the van and Don Gaston Miguel the rear. Five other French and four other Spanish flag officers were also included. Don Luis Córdova, the Spanish commander-in-chief in the *Santissima Trinidad* of 110 guns, with two other Spanish flag officers, was in command of an entirely separate and all-Spanish squadron of observation, of sixteen and sometimes seventeen ships of the line. Separate from this was a composite light squadron of three French and two Spanish ships-of-the-line under a French *lieutenant-général*.

It appears that the French had practised special evolutions before the Spaniards joined them. Thereafter the combined fleet carried out evolutions in company with the squadron of observation and the light squadron, the two subsidiary forces taking up various positions and assuming various orders of sailing. Quite apart from more general questions of command, d'Orvilliers no doubt found it extremely difficult to achieve any tactical efficiency with this huge and strangely organised armada. Was it Spanish pride that produced Córdova's separate Spanish squadron of observation, or was it

The Battle of the Dogger Bank, 5 August 1781, by Thomas Luny (see p 152). NMM

that d'Orvilliers felt he could not effectively absorb any more Spaniards into his own line of battle, and certainly not the mountainous *Santissima Trinidad?*

Presumably experience in the campaign of 1778 lay behind d'Orvilliers' promulgation of a new set of signals in 1779. A sixteen-flag tabular system was provided for 'Signaux Généraux et Particuliers des Armées Navales à l'ancre et dans les Rades', designed to cover all aspects of harbour service; no doubt they saved much time and manpower in using boats to carry orders and messages. Another sixteen-flag table provided for 'Mouvemens Ordres et Signaux Généraux de l'Armée du Roy Commandée par Mons le Cte d'Orvilliers en 1779'. A new feature of both these tables was an additional column of signals written up outside the main tables. A special key flag was provided to indicate reference to this column rather than to the main table. In effect this provided an abbreviated signal system, but in rather a clumsy manner. A new type of table was also introduced, entitled 'Ordres et Signaux Particuliers aux Vaisseaux de Chaque Escadre en ajoutant ou signal la flamme'. It had sixteen flags along the top and eighteen down the side, and covered changes of course and changes of sail and chasing in very great detail,

together with soundings and means of dealing with prizes. Separate pendants were used to indicate which squadron was being addressed.

Two more tables completed the issue for the year. The first was a small table for single-flag general signals, and for single flag plus a pendant signals for fireships. The second was a sixteen-flag numerical table with the squares numbered 1 to 256, used solely, as far as can be seen, for referring to ships' numbers, a list of the fleet with their numbers being written opposite, including twenty Spanish ships. As in the numerical system of 1778, none of the flags themselves was given a number. Nor, of course, did any of the letters of the alphabet appertain to any particular flag. Only thirty different flags altogether, apart from pendants, were employed in the 1778 and 1779 system. Thus many flags appeared in several different tables, while in some tables the order of flags was partially the same as in others.

The mere recital of these two catalogues of signal codes might suggest to the reader that the whole system was utterly illogical and quite unworkable; this conclusion, however, would be wrong. We know from Kempenfelt's observations that he envied the smartness with which the French executed their signals. What, in fact, we have been examining is the internal structure and mathematical relationships of the different parts of the system. For working purposes the system could be condensed, simplified and rewritten in the form of a handy signal book, just as Pavillon had arranged for when drawing up his famous tables of 1776.

A manuscript signal book used in the French fleet in 1779 employed a lettered table with number references to the articles of the instruactions in the squares.[8] The book conformed to what had become the standard French pattern. Sixteen flags were used for the main signals. The first opening of the book showed the first flag in the series painted large at the top of the left and right pages. The sixteen flags were also painted in columns on each page, with the respective article and its number written opposite. At the top of the pages of the next opening appeared the next flag in the series, and so on. A signal could thus easily be read, the top flag indicating the double page on which it appeared, and the bottom flag the actual article or signal written opposite on that page. Making the signal demanded more skill, but as the signals were grouped in logical order, this was not too difficult, and an index with detailed headings could always be made.

In the library of the Royal Navy Staff College, Greenwich, is a vellum bound copy of *Tactique Navale,* as used by the fleet commanded by d'Orvilliers in 1779. It was issued from d'Orvilliers' flagship, the *Bretagne,* on 21 May 1779, signed by Pavillon. At the beginning is the name of the *Bienaimé,* one of the ships in d'Orvilliers's fleet in 1778–9, and with the date 1779 in a different hand. Inside the cover is an inscription stating that the book was captured in the *Ville de Paris* in the Battle of the Saints in 1782. It was evidently out of use by then, since de Grasse had his own signals.[9] It was not a book of instructions; over three-quarters is devoted to seventy-six evolutions in which the existence of the enemy is ignored. There is a sixteen-sided table, employing the usual sixteen flags, and at the beginning of the evolutions is a plate entitled 'Mouvements de Guerre'. Although this book seems disappointing to the historian, it impressed contemporary

8 *NMM*, SIG/C/3 (formerly Sm/9); manuscript signal book, 'Movemens, Ordres et Signaux Généraux de l'armée du Roi, commandée par Monsieur le Comte D'Orvilliers, Lieutenant-Général en 1779, Channel'

9 The book also bears [Prince] William Henry's signature, and was part of his collection. It was transferred from the Royal Library at Windsor to the Royal Navy Staff College in 1928.

opinion enough for a Swedish translation to be printed in Stockholm in 1787.[10]

The Spanish navy entered the war with a system of tactics devised by José de Mazarredo Salazar, *Teniente de Navío* [Commander] *de la Real Armada,* and expounded in *Rudimentos de Táctica Naval para Instruction de los Officiales Subalternos de Marina,* printed at Madrid in 1776 and dedicated to King Charles III.[11] Despite bearing some evidence of the influence of Hoste and Morogues, this is a text book for junior officers rather than a treatise intended for naval officers in general, though it could clearly have been read with profit by all alike. In common with the French writers, Mazarredo said very little about fighting the enemy. Broadly speaking, his tone was sophisticated and undogmatic.

Mazarredo did introduce a new idea, the use of fireships by the windward fleet, if threatened with doubling, as a means of covering its retreat to windward. Salazar also showed himself an innovator in his treatment of breaking the enemy line. He proposed that, when the fleet was to windward, the centre should break through the enemy centre. In the process of breaking through, the enemy's centre ships immediately astern of the break would be forced away to leeward, so disorganising the enemy rear and isolating it. Meanwhile, the enemy van would have no choice but to stand on to avoid being put between two fires, and it would thus become completely separated from the remainder of the fleet. Exactly the same movement might be executed from leeward, though in that case the enemy's rear would be forced to give way to windward, thus exposing itself to the fire of the centre and rear ships of the attacking fleet. The innovation lay in the details of the proposed movements, as shown in the relevant diagrams.

In case of a breakthrough from windward, the ships executing the movement would run to leeward in succession, the leading ship first, and break through in some manner not clearly defined, the implication of the diagram being that the enemy was inferior and that a gap had occurred in their line. Having broken through, the fleet would form a bow and quarter line on the same course as the enemy. This would place them athwart the gap in the enemy line, and on a bearing from each other of five degrees in relation to that line, the leading ship being the most advanced and thus furthest to leeward.

When breaking through from leeward the evolution became a little more complicated, since the ships executing it had first to tack in succession, the rearmost leading. Having broken through, again in a manner unexplained, they were required to tack again so as to get on the same course as the enemy. This would bring them into bow and quarter line athwart the gap on a line of bearing 25 degrees from that of the enemy, the leading ship being in this case furthest to leeward and nearest the rest of the fleet, since it began the movement last. Again there is the implication from the diagram that the enemy fleet was inferior, and that a gap was available to be exploited.

How any evolution such as this could have been carried out by the entire fleet, practising tactics according to the actual battle methods used at the time, it is difficult to imagine, particularly in the Spanish fleet of the time. Nevertheless, Mazarredo was, in this respect at least, an original theorist. His book is extremely well laid out and printed, references being to paragraphs or figures, never to pages. His proposals serve only to emphasise the wide gap which existed between theory and practice.

Mazarredo also drew up a signal book, specifically for Córdova's fleet, which was printed in 1781. It must have been used in the operations against Gibraltar, and it does not seem unreasonable to suppose that Córdova's signalling system was somewhat similar when he first joined d'Orvilliers in 1778. It is true that one of the signal books used in the Spanish fleet during the Seven Years War, a poor affair which only mentioned the enemy in seven articles out of 207, was reprinted in 1765.[12] Nevertheless, the Spanish navy appears to have shared, and more than shared, the theoretical side at least of France's naval renaissance. Mazarredo's signal book of 1781 is in some respects an improvement on Pavillon's. Like the latter, it employed a tabular system, but much less complex. It employed tables 20 by 20, each permitting 400 signals. The flags themselves were not specified on the tables, numbers being used instead, so that the flags employed could easily be changed. The whole business of reference to notional letters of the alphabet, and transposition of signals from one table to another, was avoided.

The signal book was evidently prepared for Franco-Spanish co-operation, as it begins with special signals for indicating Spanish and French squadrons, divisions, frigates, the reserve corps, etc. The 400 signals for use at anchor covered not only every feature of fleet administration, as in the manner of Morogues, but also shore bombardments and landings, no doubt with an eye to the siege of Gibraltar. Twenty special signals allowed for reporting the movements of ships, presumably to be made by private ships. The signals for use under sail by day, made with a combination of

Mazarredo's Tactics

Enemy van stands on

Bow and quarter line

Some ships break enemy line from windward

Enemy ships in disorder and cut off

Enemy fleet disorganised

Ships tack again and form bow and quarter line

Breaking through from leeward

10 *NMM* (formerly Tunstall Collection S/P/SW/1)

11 *Ibid*

12 *NMM*, SIG/C/13 (formerly Sp/177). *Senales que han de observer y practicar los Navios de la presente esquadra de Mando del Capitan General de la Armada, Marques de la Victoria . . . Duda en VIII de Octobre de Año de MDCCLIX, estando embarcado en el navio El Real Fenix,* reprinted 1765.

'cornets',which were swallow-tail flags, other flags, and flags from the table, included a series of battle signals. The series of 400 general signals covered the whole management of the fleet under sail, again in the manner of Morogues. Included, however, was another set of battle signals of the standard kind, but with some interesting extensions and refinements. Number 250 was to advise that the general intended to attack during the night if an opportunity should occur. Number 251 advised that the general did not intend to attack during the night unless a favourable opportunity should occur. Number 252 ordered individual ships to engage when able. Numbers 269-70 ordered the leading ships to steer so as to get to pistol shot of the enemy van, or rear, if possible. Number 271 ordered the fleet to prevent the enemy breaking the line, even at the expense of being repeatedly run aboard. The night signals were made from a table of lanterns and guns. Nineteen signals were provided for use at anchor, and three separate sets of nineteen signals for use under sail.

No-one studying this book could criticise the Spanish either for a lack of useful signals for battle and general purposes, or for over elaboration of signalling technique. Although still tied to the tabular system, their arrangement was brilliantly simple compared with that of the French.

Kempenfelt and the Channel Fleet

Although he was never commander-in-chief, Richard Kempenfelt played a major part in the tactical reform of the Royal Navy. He was the son of a Swedish Jacobite colonel who had accepted service under Queen Anne and had become lieutenant-governor of Jersey. He was highly regarded in the service, especially after being flag captain to Pocock and Admiral Sir Samuel Cornish in the East Indies, but he had only reached the Post List at the age of thirty-nine. He was thus junior in the service to Keppel, Howe, Palliser, Rodney, Byron, Barrington, Rowley, Graves and Hood, though senior to all of them in years. Following the Keppel-Palliser scandal, it had become difficult to persuade any of the leading flag officers to accept command of the Western squadron or Channel Fleet, as it was coming to be called. As a result, the large forces gathered in the Channel in 1779 to resist the combined fleets of France and Spain, comprising thirty-five to forty-five of the line and some thirty frigates, fire-ships, sloops, cutters and brigs in attendance were commanded in turn by Sir Charles Hardy, a resuscitated veteran of sixty-three who died within a year, and by his successor Francis Geary, aged seventy-one, who resigned five months later. His successor, George Darby, an able and vigorous man, nevertheless felt bound to resign when the Whigs returned to power in May 1782.

In May 1779 Kempenfelt was appointed as Captain of the Fleet, or chief of staff, to Keppel's successor, Hardy. He continued as captain of the fleet under Geary even after receiving his flag as a rear-admiral. Under Darby he hoisted his flag as third-in-command of the fleet and a year later held an independent command, during which he achieved a notable victory. After North's resignation he served as second-in-command of the Western squadron under Barrington and then under Howe, meeting his death when the *Royal George* sank in August 1782.

Throughout the spring and summer campaigning season of 1779 the Channel Fleet was on the defensive, despite its size and close proximity to good home supply and repair bases. It was inferior to the combined fleets of France and Spain, which were co-operating for the planned invasion of England. Kempenfelt, however, did not believe that British tactics had to be ruled by circumspection. On 27 July 1779 he advocated a more aggressive approach:

Rear-Admiral Richard Kempenfelt (1718–82), by Cornelis Ketel. NMM

'Tis an inferior against a superior fleet; therefore the greatest skill and address is requisite to counteract the designs of the enemy, to watch and seize the favourable opportunity for action, and to catch the advantage of making the effort at some or other feeble part of the enemy's line; or, if such opportunities don't offer, to hover near the enemy, keeping him at bay, and prevent his attempting to execute anything but at risk and hazard; to command their attention, and oblige them to think of nothing but being on their guard against your attack.[13]

Viewed in broad strategical terms there may be a good case for saying that, despite its relative inferiority to the combined fleets of France and Spain, the British Channel Fleet was larger than it need have been. The invasion threat was not realistically very great, and Franco-Spanish strategical co-operation was not at all good. Meanwhile, in the West Indies and North America, British naval forces were kept inferior to the French so as not to denude the Channel Fleet. It is even arguable that this fleet itself would have been more tactically efficient if it had been kept down to thirty of the line. Whatever tactical device Kempenfelt had in mind for use against a 'feeble part of the enemy's line', however, he was clearly making the condition that, for the weaker fleet to succeed in its harassing and opportunist role, it must be at least tactically equal to the enemy, if not superior. In fact, the reverse was the case.

Judging by his letters to the comptroller of the navy, Sir Charles Middleton, Kempenfelt was not only hamstrung by Sir Charles Hardy's lack of competence due to advancing age, but also by the tactical inefficiency and lack of practice of the fleet as a whole. 'It is with the greatest difficulty,' Kempenfelt wrote, 'I can ever prevail upon him to

13 *Barham*, I, p292

manoeuvre the fleet, he is always [so] impatient and in [such] a hurry to get to the westward, to the northward, or the southward, that he won't lose time to form a line.'[14] Benjamin Thompson, Count von Rumford, who was serving as a volunteer in Hardy's flagship during the summer and autumn of 1779 in order to supervise some experiments in ballistics, wrote a long account to Lord George Germain which gave a damning picture of tactical incompetence, in which Kempenfelt came in for his share of criticism.[15]

You must know we never as yet had attempted but one simple manoeuvre, and that is to draw the ships into a line of battle, one directly ahead of the other, and upon a signal to tack about all together. In this we never have succeeded. It has always been more than two hours before the ships ahead have got into their stations – the line has always been very crooked, and the ships at very unequal distances, and when we have come to put about confusion has commonly ensued, and we have been obliged to end our manoeuvring abruptly by making the signal for the ships to return to their stations in the order of sailing. This has arisen sometimes from one cause, and sometimes from another, but I don't remember a single instance in which we have attempted to manoeuvre when we did not make at least one evidently wrong and contradictory signal.

In one incident reported with chagrin by Thompson, Kempenfelt could find no signal to order a manoeuvre by divisions. When a method was pointed out to him in the book, the wrong signal was flown, apparently one ordering the return of weekly accounts. A similar story appears in Sir John Barrow's *Life of Richard, Earl Howe,* about the period after command of the fleet had passed to Admiral Geary. Exasperated by Kempenfelt's attempts to manoeuvre the fleet, 'Geary at last grew impatient and going up to Kempenfelt, and laying his hand gently on his shoulder, exclaimed with good-natured earnestness, "Now, my dear Kempy, do for God's sake, do, my dear Kempy, oblige me by throwing your signals overboard and make that which we all understand – Bring the Enemy to Close Action!" '[16]

Undaunted, Kempenfelt continued throughout the rest of his life to experiment with tactical reforms.

Kempenfelt's tactical ideas at this time are best illustrated from his letters to Sir Charles Middleton. On 28 April 1779 he wrote:

The course of this [Palliser's] court martial has given me a clearer idea of the action between our fleet and that of France last summer than I had before conceived; and the different state the two fleets were in after the action confirms most strongly what I have always thought; that is, that the disabling of your enemy in his masts and rigging should have no small share of your attention and your fire . . . Owing to the different directions of their fire [our fleet] . . . except the red division [were]so totally shattered in masts, rigging, and sails that for the whole evening of that day they could not all form into a line. The French ships, on the contrary, were so little injured in these particulars that they had the perfect command of their yards and sails . . . There is no strength and force without motion and direction . . . 'Tis plain to me that our fleet, after that action, for all the first part of the afternoon, was at the mercy of the French. Unconnected to succour and support each other, what defence could they have made against the attack of a close, well-formed line of ships? Why the French did not profit from this advantage they had, I can't conceive.[17]

Like Howe, Kempenfelt was in favour of reviving the use of fireships.

I really don't see the necessity of waiting until a ship is disabled before you apply the fireship. I think the best time to use them is in the very beginning of the action; the smoke covers them as well then as afterwards; and is it not best to do

the greatest injury to your enemy as soon as you can to hasten his defeat?[18]

It might have seemed a bad time for experiments, but Kempenfelt felt he had no choice. On 18 January 1780 he wrote to Middleton:

I believe you will, with me, think it something surprising that we, who have been so long a famous maritime power, should not yet have established any regular rules for the orderly and expeditious performance of the several evolutions necessary to be made in a fleet. The French have long since set us the example. They have formed a system of tactics, which are studied in their academies and practised in their squadrons . . .

Indeed, 'tis too obvious to make any arguments necessary to show that fleets as well as armies require rules for the execution of their movements, and that the one stands in need of tactics as well as the other; without which both are unwieldy masses, where force is lost for want of form and order. Superior address in conduct may make up for the want of numbers, but what is to be expected when skill and address are wholly on the side of numbers?

The accounts Kempenfelt had heard of French seamanship at Ushant indicated that what they gained by good tactical control was not lost by any fault in shiphandling. In his opinion,

We should therefore immediately and in earnest set about a reform; endeavours should be used to find out proper persons, and encouragement offered for such to write on naval tactics, as also to translate what the French have published on that subject. They should enter into the plan of education at our marine academies.

But the most effectual way to obtain every wished for reform in your fleet is to find out a man to place at the head of it, who has genius to comprehend every requisite regulation, and has activity and spirit to enforce their execution.[19]

Kempenfelt attempted to create a new tactical system suited to a large fleet opposed by an even larger and better tactically organised enemy, to give this system expression through a combination of signals and instructions, and to introduce these signals and instructions to a body of officers likely to be both suspicious and contemptuous of sophisticated innovations. The best that could be expected of Hardy was that he might somehow be persuaded to sign what was laid before him. In default of personal direction by the commander-in-chief through a series of conferences with flag-officers and captains, Kempenfelt was forced to try to gain his ends by bombarding the fleet with new sets of signals and instructions on a purely *ad hoc*, improvised basis. Never in the history of any navy, at any period before or since, have so many signal books and fighting instructions been issued by one man in the space of only twenty-seven months, during at least ten of which the main body of the fleet was laid up for the winter.

Kempenfelt's tactical reforms

On 31 May 1779, Sir Charles Hardy issued Captain Adam Duncan of the *Monarch* with a large folio copy of the Sailing

---

14 *Barham,* I, pp294

15 W G Perrin, editor, *The Naval Miscellany,* vol III, London, 1928, pp127–31

16 Sir John Barrow, *Life of Richard Earl Howe,* London, 1838 (hereafter cited: Barrow), p141

17 5 September 1779; *Barham,* I, p296

18 *Barham,* I, pp309–13

19 *Barham,* I, pp290–1

& Fighting Instructions, reprinted and bound up with the Admiralty additionals in signal-book form, as previously issued by Keppel.[20] The signals, however, were differently arranged and at the end were a set of printed but unnumbered additional fighting instructions, which were in fact additions to the Admiralty additionals, with signals. They included a signal:

If at any time I perceive the leading ship in the Line of Battle does not steer down enough on the ship she is opposed to, in the Enemy's line, although, I may not be able from the distance, to direct the precise number of points I would have her bear away and would have her bear down on the nearest course she can steer to bring the enemy to action.

Who thought of this? A similar signal in Rodney's book might have been used to resolve Captain Robert Carkett's confusion at the Battle of Martinique (see Chapter Seven below).

So far, Kempenfelt seems to have been feeling his way by using the instructions bequeathed to the fleet by Keppel. A whole collection of instructions and signals, however, issued to Captain Lord Longford of the *Alexander* and bound up together, gives what must be a fairly complete picture of what was happening between May 1779 and October, when the campaign for the year was nearly at an end.[21] Here we find the full extent of Kempenfelt's early reforms revealed for the first time.

Articles 6 to 17 of the day signals, together with an Article 18 in manuscript, are word for word the same as articles selected from Howe's 1782 'Instructions for the Conduct of the Ships of War, Explanatory of, and Relative to the Signals contained in the Signal Book herewith delivered', the companion volume. The general purpose of these articles was, first, to give the admiral better control of the fleet, and second, to standardise methods for carrying out various evolutions. Especially important is Article 9 (p12 Article 4 in the 1782 instructions) for ships unable to keep their station in the line. They were to fall astern so as to avoid disordering the fleet, the gaps being filled by succeeding ships.

The captains of such ships [as have to fall astern] will not thereby be left in a situation less at liberty to distinguish themselves; as they will have opportunity to render essential service, by placing their ships to advantage when arrived up with the enemy, already engaged with the other part of the fleet.

Most of the sailing and fighting instructions and all the instructions for fog were the same as those hitherto thought not to have been issued together until 1781, so that they antedate much of Kempenfelt's work by a year.[22]

The nine night instructions, with headed references in the style of Howe, include Article 9, for captains, in case the enemy was suddenly encountered at night, to have 'discretion to prosecute a closer engagement with the enemy, or to suspend the continuance of their attacks, as they may have reason to think either proceeding most efficacious for the advancement of the public service'. This is entirely in the spirit of Howe's night battle instructions of 1782. It implies a high degree of professional skill and co-operation on the part of captains and a degree of confidence placed in them by the admiral, and is the complete antithesis of everything ordered and aimed at by Rodney. With so many of these 1779 instructions apparently anticipating Howe's work of three years later, it is tempting to make Kempenfelt the inspirer of Howe's innovations; it may, however, equally have been Howe who was supplying Kempenfelt with ideas. This seems in fact more likely, especially in view of a letter written, apparently to Kempenfelt, on 10 September 1779, in which Howe comments,

Admiral Darby's Additional Instructions of 18 October 1779 for the deployment of the fleet en echelon, from Admirals Keppel and Hardy's Signal Book 1778-9. NMM

20 *NMM*, DUN/6 (formerly S(D) 6) On 13 May 1779, Vice-Admiral George Darby, then acting as one of Hardy's divisional commanders, had issued Duncan with a printed copy of the Admiralty Additionals in the standard form with the index (DUN/5 (formerly S(D)), ie, not in signal book form as issued by Keppel a year earlier. To what extent Darby was acting on Hardy's ie, Kempenfelt's) authority is uncertain. *NMM*, COR/61 (formerly Sp/116) is a slightly different printing but undated and unsigned. Darby, Kempenfelt and Duncan, the future Lord Camperdown, had all been members of the court-martial trying Palliser, which had ended only on 5 May.

21 *NMM*, PAK/3 (formerly Sp/135)

22 Barrow, p142

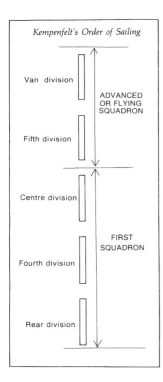

Kempenfelt's Order of Sailing

Van division

ADVANCED
OR FLYING
SQUADRON

Fifth division

Centre division

FIRST
SQUADRON

Fourth division

Rear division

I should be glad to know what part of our signals have been adopted; I should thence be able to form an opinion of the principle upon which that great machine (the Grand Fleet) is to be put in motion. Our signals were adopted rather for a single squadron; and, though most of the articles might be applied to a larger force, (fleets being composed of squadrons collectively arranged), the necessary continuation of the signals being different, the propriety of their use under different circumstances will vary also. In the disposition of them I had those objects in view. But the choice to be made of the expedients which any set of signals has provided for, will constitute, as we know, the ability of the flag-officers.'[23]

The words 'our signals' seem to refer to a joint production.

Bound in at the end of the volume are a number of additional signals and instructions issued in manuscript to Lord Longford by Vice-Admiral George Darby. Fifteen additional fighting signals, issued 1 September 1779, include: number 1, for attacking enemy's rear in succession; number 4, to break through the enemy line and attempt to cut off part of their rear, when fetching up to leeward on the contrary tack; number 5, for attacking the enemy rear and inverting the line; numbers 6, 7 and 8, for doubling; numbers 10, 11 and 12, to signal the admiral's intention to attack the enemy centre, van or rear; number 13, to extend the line to prevent being doubled. Directions for forming the order of retreat follow, and also slightly different directions, dated 18 August and erased. Further signals are dated 27 September. Directions for the order of sailing in three columns in echelon and for the order of sailing in five columns are dated 18 October. At the end are directions for the order of retreat and for forming the line of battle from it, dated 22 September.

With forty of the line under his admiral's orders, Kempenfelt saw that some new squadron organisation was essential, the traditional scheme of the van, centre and rear being inadequate for such a large fleet, and especially so for the kind of offensive-defensive tactics which he had in mind. His idea, as shown in the Order of Sailing, which was the last item in the volume, was to organise the fleet in five divisions under its five flag officers.[24] The centre under Hardy, together with the so-called Fourth division and the rear were then to form the First squadron, which would be the main battle force. The van, together with the Fifth division, would form the Second squadron, which would be regarded as an additional, reserve, detached, advanced or 'flying' squadron. Each squadron was given its own sailing order by divisions. When turning from a close-hauled line ahead to a line with the wind two points abaft the beam, the senior admiral's division of each squadron was to take station to windward of his squadron on either tack. When sailing large, the senior admiral's division was to be to starboard of his other division or divisions. The First squadron, when sailing by divisions in three separate lines ahead, might be signalled to proceed *en echelon*, thus:

Admiral's centre division

Fourth division

Rear division

The leading ship of the Fourth division was to be level with the centre of the Admiral's division and the leading ship of the rear level with the centre of the Fourth division. Divisional commanders might be signalled not to lead their respective lines but to keep station in the middle.

A further version of this cruising formation disposed the Second squadron to starboard of the First squadron, thus:

First squadron	*Second squadron*
Admiral	Van division
Fourth division	Fifth division
Rear[25]	

Following Howe, who followed Admiral Henry Osborn, Kempenfelt introduced the order of retreat for use against a 'much superior' enemy 'or after you have suffered much in action'. This order was also useful for covering a convoy of merchantmen, which could be sent ahead protected by the ships forming the two sides of the angle. The order was to be formed by the fleet being 'upon the sides of an obtuse angle of 135 degrees, formed of the two lines by the wind for the starboard and larboard tacks; the admiral is to windward at the point of the angle and in the centre of his fleet'. To change the order of retreat to the order of battle to starboard, the leading ship on the starboard wing hauled her wind to starboard, followed by the rest of the wing in succession as they reached her wake. Meanwhile, the larboard wing continued together on a course parallel to that originally steered by the starboard wing. On reaching the wake of the starboard wing, they turned to starboard together and the line ahead was formed.

As in Howe's signals, Kempenfelt dealt carefully with alterations of course both together and in succession. A

Kempenfelt's Order of Retreat

Fleet course

Forming line of battle

Ships of larboard wing stand on

Each ship of the starboard wing turns to wind on crossing the wake of the ship ahead

Larboard wing turns together on crossing the wake of the starboard division

---

23 *NMM*, PAK/3 (formerly Sp/135), issued by Darby to Lord Longford at Spithead 18 October 1779 (part printed in *Signals & Instructions*, pp125-31, together with the full list of the line of battle for the fleet).

24 *NMM*, PAK/3 (formerly Sp/135), issued by Vice-Admiral Darby, second in command of the fleet, to Captain Lord Longford, *Alexander*, 18 October 1779 (not printed in *Signals & Instructions*).

25 Kempenfelt to Middleton, 8 April [1780]; *Barham*, I, pp323-4

special feature of the Hardy-Kempenfelt instructions and signals is the use throughout of the word 'attack' in place of 'engage'.

In 1780 Kempenfelt devised yet another tactical organisation for the fleet, although it is unclear whether he persuaded Admiral Sir Francis Geary to adopt it. For a fleet of from fifteen to twenty-six of the line, he was content with the traditional arrangement of three squadrons. If the fleet exceeded twenty-six of the line, he wanted each squadron divided into two divisions, or for a fleet of forty-five of the line upwards, three divisions.[26] This proposed reorganisation is highly relevant to the instructions accompanying Geary's 1780 signal book, which include the order of sailing in three squadrons abreast as the standard formation, with a variation by which each squadron was itself formed into two columns. A large fleet was enabled to sail in six separate lines ahead of a few ships each disposed abeam of each other, or, if required, in pairs *en echelon*.

Kempenfelt's signal books

Meanwhile, Kempenfelt was trying to give better executive expression to his tactical reforms by means of more efficient signalling. Here he was in serious difficulties. It was always possible to make piecemeal issues of additional fighting instructions and orders of sailing, with the necessary signals attached, but it required a revolutionary act to introduce a new signalling system. It was likely to provoke resentment, confusion and disorganisation throughout the service.

Kempenfelt's signal book of 1779 was clearly influenced by Howe's of 1776 but, unlike Howe's, Kempenfelt's book was still tied to the old combined books of instructions. During 1779 Kempenfelt made use of a printed version of the kind of signal book Keppel had introduced, or at least made use of, in 1778. It was a book incorporating the main signals abstracted from the parent work and the Admiralty additionals, with page and article references to these books as bound together with continuous page numbering.[27] The form of the book was entirely traditional, with the flags having signification according to where they were flown, except for a very few special cases of 'where best seen'. All of Kempenfelt's new instructions, including the order of retreat and the fleet order in three divisions disposed abreast in echelon, were incorporated, in each case with explanatory directions and diagrams. Although there were still plenty of spare places in the ship for flying the existing flags, more flags were introduced, presumably to avoid confusion in distinguishing between different flags flown in the same place. Whereas Saunders could work his system of 1760, including all his additional signals and signals for private ships, with twenty-two flags and four pendants, the Channel Fleet signals of 1779 required forty-two flags and thirteen

pendants. Some of the flags were hopelessly confusing.

It is difficult to know the extent to which Kempenfelt really understood Howe's ideas. He wrote to Middleton on 17 October 1779, 'Signals pointed out by numbers, and according to that method observed in those you sent me, I have long been acquainted with. They don't require many flags; however there must be three of each sort, or you can't express a number which consists of three figures of the same rank, as 222 or 333, etc.' This seems scarcely a valid objection to the system as a whole. Either such signal numbers could be left out or a substitute flag used. Kempenfelt continued:

This numeral way of pointing out the signal is very convenient for those to whom they are addressed, for the ready finding out their meaning, and a very great ease to the compiler of the signals. The objections against them are that, for the most part of the signals, you have three flags to show instead of one in the common method; that ships may often be in a position not to see the flag at the mizzen topmast-head; that to ships at a distance it will be more difficult to ascertain the colour of two or three flags than of one; that when signals are simple, with only one position, that, assisted by circumstances, may lead to the knowledge of the signal, though the colour of the flag may not distinctly be distinguished.

Why Kempenfelt cited the difficulty of seeing flags at the mizzen topmast head as an objection to the numerical system, when that was a problem peculiar to the old one, is mysterious. Either the objections were not really his own, or he had failed to understand the numerical system. He concluded:

I think all these objections may be obviated; and in my opinion, the signals by numbers, or those by a superior flag, as used by d'Orvilliers the last two summers, are by much superior to any method we have. I could in a very few days (less than a week) arrange our signals according to both the above methods, could I have those days to myself without interruption.[28]

Were the objections his own or those of others? How could they be obviated? Surely it would have been better to keep the old system for the time being, rather than become committed to an all too obvious second best which could lead nowhere?

Writing to Lord Sandwich, First Lord of the Admiralty, on 9 July 1779, Middleton stated that it was agreed between Kempenfelt and himself 'to introduce the French system [of signals], if practicable, into Sir Charles Hardy's fleet'. 'That gentleman [Kempenfelt] has well considered the subject', he continued, 'but at this juncture it would be dangerous to exchange even a bad set of signals that are generally known for a better which the greatest part of the officers were unacquainted with'.[29] Middleton suggested that Kempenfelt should be brought ashore and that the two of them should, together with 'some other sea officers which I can name . . . simplify this very complex business and . . . reduce it within the practice of the most ordinary understanding. The materials for this purpose are lying by me, but engaged as I am in an office which I own is almost beyond my strength it would be madness to undertake anything new.' Middleton further suggested that Kempenfelt should be permanently withdrawn from Hardy's fleet to act as assistant to the comptroller for the whole administrative work of the Navy Board. Nothing came of either of these suggestions.

On 27 May 1780, Geary, now in command of the Channel Fleet, issued Captain Adam Duncan of the *Monarch* with a larger printed folio, 'Instructions for the Conduct of the Ships of War, explanatory of, and relative to the Signal Book herewith delivered', and followed by 'Signal Book for the Ships of War'.[30] The wording is exactly the same as that used

---

26 See *Signals & Instructions*, pp135-9

27 No printed copy of this signal book has been found. *NMM*, Tunstall Collection, TUN/2 (formerly S/MS/Am/1) is a folio MS copy with a loosely inserted note: 'Robert Callander, Clerk of His Majesty's Ship *Terrible* at Sea, August 12th 1779.' The *Terrible* (74) Captain Sir Richard Bickerton, Bart, was stationed in Sir Charles Hardy's Centre division of the fleet.

28 Kempenfelt to Middleton, 17 October [1779], *Barham*, I, p301

29 *The Private Papers of John, Earl of Sandwich, First Lord of the Admiralty 1771-1782*, G R Barnes and J H Owen, editors, London, 1932-8 (hereafter cited *Sandwich*), III, p42

30 *NMM*, DUN//14 (formerly S(D) 14). *NMM*, SIG/A/14 (formerly Sp/91) is a duplicate copy of the signals unsigned and undated, but with MS insertions, while *NMM*, PAK/4 (formerly Sp/137) contains the signals and instructions (less pp17-8) bound together.

by Howe. Hardy had died on 18 May, almost his last act being to refuse Kempenfelt permission to reorganise the fleet for the coming campaign.[31] Geary had hoisted his flag as commander-in-chief only on 24 May, and was no doubt prepared to sign anything which Kempenfelt put in front of him. It has hitherto been assumed that his particular form of signal book and the accompanying instructions were not issued at least until 1781, but the signed and dated copies issued to Duncan settle this point completely. By abandoning the parent work and the Admiralty additionals, and issuing a true signal book with accompanying instructions, Kempenfelt had become a revolutionary. His adoption of Howe's ideas was a very radical move because the Channel Fleet was the main fleet of Great Britain at the time, and contained five flag officers.

The signal book was on traditional lines but, like Howe's, the signals were numbered serially throughout. In the main they repeated the signals of the previous year, 1779, but also included many fresh ones of an administrative and general nature, up to a total of 397. There were fifty-six spaces for night signals, with some blanks, the signals themselves being shown under the headings number of lights, how disposed, place, guns, and false fires. In addition to the night signals for private ships was a separate section of eleven night signals for 'ships ordered to observe the motions of the enemy', thus

foreshadowing the great system of night reconnaissance used by Nelson with such effect on the eve of Trafalgar.

At the end was a ten-flag table for numerical signalling, arranged exactly like the numerical table possibly attributable to Howe in the manuscript copy of his fog and night signals of 21 August 1778.[32] There was, however, a very important difference in purpose, for whereas the 1778 table was stated to be for the use of ships not possessing the flags used in Howe's signal book, the 1780 table was for general usage. 'The Admiral to intimate his intention of directing the fleet by Numerary Signals will hoist the [Royal] Standard where best observed.' A Union flag hoisted over the Standard was to be the signal for reverting to the normal system of flags in particular positions. Are we to assume that this was Kempenfelt's way of trying to introduce true numerical signalling into the fleet, or did he actually regard the tabular system with all its inconveniences as a desirable half-way house between a true numerical system and the old one? The latter explanation seems unlikely; with so much difficulty being experienced in the recognition of signals, it may be that the numerical table was intended to facilitate distant signalling. Whatever his reasons, why did he introduce into his table a red flag pierced white not included amongst the twenty-seven flags used for normal signalling?

Soon after issuing Geary's signals and instructions of 27 May 1780, Kempenfelt seems to have produced yet another pair, which contained the same ten-flag table for numerical signalling. The printed versions have disappeared, but the Naval Library possesses a manuscript copy of both the signals and instructions, and signed: 'Completed this book, August 22, 1780, Defence at Sea – James Hardy'.[33]

The flags used for the signals were not the same as those in Geary's book, nor the same as those in Darby's book issued for the ensuing season's campaign in February 1781. They are, however, exactly the same as those in an unsigned and undated manuscript private pocket signal book in the Tunstall collection.[34] The dating of this latter book is fairly certain, as the chequered pendants which Kempenfelt told Middleton he had changed in his February 1781 book issued in Darby's name are struck out in pencil and marked 'striped'. On the other hand, the red-white-red and the blue-yellow-blue flags which Kempenfelt implies he had changed at the same time from horizontal to vertical are already in their new form without any sign of erasure or alteration. The explanation for this may well be that, as in the case of so many private signal books transcribed from an official copy which possibly bore heavy corrections itself, the copyist made a mistake which he later corrected.

The instructions which appear only in James Hardy's copy are incomplete, but include the bulk of the fighting instructions in Geary's book of 27 May as well as those issued by Kempenfelt for Darby in the following year. They include several from Howe's fighting instructions of 1776 and his additional fighting instructions of 1777. In fact, the whole trend of Kempenfelt's sailing and fighting instructions during his service in the Channel Fleet from 1779 to 1782 seems to be in the direction of copying Howe.

On 11 February 1781 Vice-Admiral Darby, in command of the Channel Fleet since Geary's retirement in September 1780, issued a 'Signal Book for the Ships of War together with 'Instructions for the Conduct of the Ships of War,

Admiral Lord Howe's 1778 manuscript tabular system for signalling numbers. MOD

Substitute FLAG: White with a Red Cross, in default of that St George's Ensign		Numerary Table.								
	1	2	3	4	5	6	7	8	9	10
	Union	Red	White	Blue	half Red half White	half Blue half Yellow	Dutch	Red pierce	Yellow Pierce bore	Dutch bore
Union	1	11	21	31	41	51	61	71	81	91
Red	2	12	22	32	42	52	62	72	82	92
White	3	13	23	33	43	53	63	73	83	93
Blue	4	14	24	34	44	54	64	74	84	94
half Red half White	5	15	25	35	45	55	65	75	85	95
half Blue half Yellow	6	16	26	36	46	56	66	76	86	96
Dutch	7	17	27	37	47	57	67	77	87	97
Red Pendant	8	18	28	38	48	58	68	78	88	98
Yellow Pendt	9	19	29	39	49	59	69	79	89	99
Dutch Pendt	10	20	30	40	50	60	70	80	90	100

31 Letter to Middleton 8 April 1780; *Barham*, I, p323

32 *MOD*, NM/34 (formerly *RUSI* 34)

33 *MOD*, Ec/50, together in a large folio vellum bound volume

34 *NMM*, Tunstall Collection, TUN/8 (formerly S/MS/Am/8)

explanatory of, and relative to the Signal Book herewith delivered'. These were clearly produced by Kempenfelt, and Thomas Graves significantly wrote the words 'Admiral Kempenfelt' on his copy of the instructions.[35] Nevertheless, in almost every respect they followed Howe's signal book and instructions of 1776. The signal book is a printed folio with many handwritten insertions. There were only twenty-seven flags and seven pendants, compared with the forty-two flags and thirteen pendants used in the summer of 1779. The flags, in which combinations of blue and yellow predominated, were a different selection from any used before. The structure of the book was much the same as that issued over Geary's signature in 1780. There were two flags on each thumbing referring to the left and right hand pages, with a written description of the respective flags at the head of each page. The main signals were numbered consecutively throughout. The fog signals were by guns, drums, bells and muskets. There were fifty-six night signals made with lights in different patterns, generally hoisted 'where best seen', guns and false fires, and there were separate night signals for the private ships. There was a table of compass signals and another for horary signals. The day signals included number 78: 'When fetching up with the enemy to leeward and on the contrary tack – to break through their line and endeavour to cut off part of their van or rear.'

The instructions issued with Darby's signal book were much the same as those issued by Geary in May 1780, and also those of unknown date issued later in the year.[36] As in the case of the signal book, the lay-out and printing were much the same as those of Howe's issue of 1776. Nothing in fact in the whole story of tactics and signalling at this time is more remarkable than the way in which Howe stamped his image on the work of all those who favoured his methods. To compare these with the printed issues of Keppel and Admiral Marriott Arbuthnot, even in terms of typography alone, is to realise that one is in a different world.

The Darby-Kempenfelt signal book included the same ten-flag table for numerical signalling, with exactly the same flags, as appeared in the signals of the previous year.[37] It must be asked again whether Kempenfelt really believed in numerical signalling in its true and unadulterated form. If so, was he prevented from introducing it by the opposition of the traditionalists in the service, of whom there always were a goodly number? The difficulty in answering these questions is not due to lack of evidence but rather to its inconclusive character.

Writing to Middleton on 9 March 1781, a month after the issue of Darby's signals, Kempenfelt said, 'Every article or order in a signal book should be numbered; and then, any ship possessed of eleven flags, different the one from the other, may express signals to any number.' Did he mean that the eleventh flag should be a substitute, the intention being to express signals from 100 upwards with a three-flag hoist and so avoid the use of pendants? Even so it would be impossible to signal three repeats, such as 111, 222, etc. The letter continued:

It has been a common saying that it is an advantage to go by signals that we have been used to, and when a new set comes out, to say we have our trade to learn again. This style was very proper with respect to the different signals used by different admirals formerly, when the signals were jumbled together without form or order, and when a long acquaintance with them was necessary to find out the meaning of any signal that was made in the chaotic state in which they were. But when the signals are formed upon a proper plan they require no study to comprehend them, and when a signal is made you can immediately turn to the article or order alluded to.

Had I remained ashore this winter, which my health very

much required, I think I should have been able to have rendered the signals much more perfect and useful by the helps I have received from others and my own observations. The plan I followed in the signals I made was not that I most approved of. That which I would have adopted – though most evidently the best – I could not get any of the admirals or officers of note to approve and countenance. I therefore followed in a great measure Lord Howe's mode, he being a popular character. The night and fog signals we use are almost entirely his, and both extremely defective. I would have used the French night signals, which are by much superior to anything of the kind that has yet appeared, but I was afraid of prejudice, for not an admiral I showed them to but started objections.

A scheme for fog signals has lately fallen into my hands, which is very simple and clear, which when I have leisure, I shall recommend for use.[38]

This is one of the most famous and most puzzling letters ever written about naval signalling. Did Kempenfelt mean that by introducing his tabular system he was copying Howe, and that by making it thereby an alternative to the normal system he was placating the traditionalists? If so, what evidence have we, apart from the manuscript project of 1778, that Howe ever countenanced the tabular system at all? Certainly it never appears in any of his printed signal books, except for the use of private ships. On the contrary, it seems to have been a Kempenfelt speciality, and copied by him from the French. What then was it that the 'admirals and officers of note' refused to countenance? Was it the true numerical system, with ten flags numbered 0 to 9 with pendants for the hundreds?

Some further light is thrown on Kempenfelt's views by an undated translation which he made himself of the whole of La Bourdonnais' system of numerial signals, numerical distance signals and night signals, as printed in Bourdé de Villehuet's *Manoeuvrier*.[39] This translation was sent to Middleton on about 12 March 1781.[40] No comment was added, but assuming this date to be correct, it was immediately followed by a letter of 14 March 1781 describing the night signals actually in use in the French Navy.[41] These were accurately described as being on

the same principle as their day signals, made with two flags, a superior and inferior. The small table, at the upper part of the page to the left, shows the guns and lights used to express the numbers of the flags in the large table: thus, one gun and one light signifies number 1. The flags in the large table are placed to help the conception by showing the affinity of these with the day signals, the only differences being that lights and guns are substituted to represent flags.

Kempenfelt's comment was: 'This method of exhibiting night signals is certainly very simple, easy to make, easy to be understood, and much less liable to be mistaken than any that have been yet in use amongst us.' Careful reference to the

35 *MOD*, NM/101 (formerly *RUSI* 101) printed folio. Another copy in manuscript is NM/85 (formerly *RUSI* 85), which declares them to be 'signals by Rear-Admiral Richard Kempenfelt'

36 *MOD*, NM/101 (formerly *RUSI* 101) printed folio matching the signal book

37 *MOD*, Ec/50, and *NMM*, Tunstall Collection, TUN/8 (formerly S/MS/Am/8)

38 *Barham*, I, pp430–41

39 First published in 1765. De Sauseuil's English translation did not appear until 1788.

40 *Barham*, I, pp343–8. The editor, Sir John Laughton, seems to have had some reason for tentatively assigning this date to the draft.

41 *Barham*, I, pp348–9

Vice-Admiral Sir Hyde Parker (1714–82) (detail), by George Romney. NMM

signals themselves arouses wonder that a man of Kempenfelt's extraordinary intelligence and technical ability could possibly have penned such a eulogy. Not only are these signals complicated in themselves, but they are based on a dead-end system of flag signals. Admittedly, the existing British night signals were inefficient. Nevertheless, they were amenable to improvement on lines far better than anything so far devised by the French. The only criticism Kempenfelt allowed himself was that the system required too many gun signals. On the credit side, it must be admitted that the French system did not require lights to be shown in particular positions but 'where best seen', and this was certainly an advantage.

Colour and size of signal flags and pendants

Kempenfelt had long been interested in the form of signalling flags. In his correspondence with Middleton he had a good deal to say about the most suitable sizes and proportions, and arrangements of their colours. In a letter of 20 March 1780 he suggested that the 'dimensions which seem to be most generally approved of' for flags was 9yd long by 5½yd deep. The next day he suggested that a depth of 5yd would simplify construction, and on 14 October he suggested that flags for repeating frigates should be 5yd by 2. On 19 February 1781 he changed his mind and suggested flags should by 7yd by 5yd, as 'most convenient . . . for large ships'; 6yd by 4 would be enough for frigates. Signal pendants for frigates he thought should be 6ft at the staff, 2ft at the end of the flag, and 45ft in length. On 2 March he added that pendants for flagships should be 8 or 9ft at the staff, and also 45ft long, but yet again changed his mind about flags, which he now said should all be made to a ratio of three to two. Signal pendants should not be swallow-tailed, and should be in the proportion of three at the staff to one at the end of the fly. Triangular flags should be 14½ft at the staff and 20½ft long.

Crosses on flags should be in width one-fifth of the breadth of the flag. Striped flags should have seven stripes with the darker colour on the outside sections. Chequered flags should have four chequers by three, with two light chequers and one dark next the staff. 'Pierced' or 'bordered' flags should have the pierced part in breadth half the total depth of the flag, and the border all the way round one quarter of the depth. Chequered pendants should have five chequers by two. Pendants striped vertically should have five stripes, the darker colour next the staff. Pendants striped lengthwise should have five stripes with the darker colour on the outside sections.[42]

Kempenfelt believed that every flagship and every repeating frigate should have a set of flags consisting of two of each flag and four of each pendant. This represented a considerable amount of cloth, and the expense evidently worried the Navy Board, which submitted 'to their Lordships whether some general mode should not be adopted to confine this Article within some limited Bounds'.[43]

The Battle of the Dogger Bank

It is evident from Vice-Admiral Sir Hyde Parker's Battle of the Dogger Bank, from the events of Kempenfelt's own engagement with de Guichen in December, and from Howe's relief of Gibraltar in 1782, that the new tactical ideas in the

fleet had very little effect at first on the conduct of battles. Very early on 5 August 1781, Parker, escorting a convoy bound for England out of the Baltic, encountered Rear-Admiral Johan Arnold Zoutman escorting a Dutch convoy intending to enter the Baltic. Parker ordered his convoy to make sail to the west for England, but Zoutman, keeping the Dutch convoy together on his lee side, drew his escorting force together into line of battle. With a north-easterly breeze, Zoutman formed his line under topsails and foresails at a cable's distance a point off the wind on the port tack. Parker was to windward and at once signalled general chase.

Neither squadron could be described as anything but second class. Parker had five of the line, a 50 and a 40. Two of his line of battle ships were good, but the rest of his squadron were old and crank. Zoutman also had seven fighting ships, three of the line, three 54s and a 40, a poor force and no doubt suffering from lack of fighting experience over a period of nearly seventy years. He was supported, however, by his five frigates, which, keeping very slightly to leeward, were ready to fire through the gaps in his line of battle.

Parker's general chase under a press of sail soon had his ill-balanced force strung out in order of speed, which led him a little after 6am to signal for the line abreast. Though the leading ships took in sail to let the rest come up, the squadron as a whole continued to press forward before the wind, Parker being anxious to get to close quarters as quickly as possible. Even so, with his ships in line-abreast at a cable's distance, he presented the ideal target, each ship steering for the enemy and bow to broadside. Nevertheless Zoutman held his fire until 7.55am allowing the British to turn into line ahead parallel to him on the port tack and open fire at what was said to be 'half musket-shot'. If this was an example of old-fashioned chivalry it certainly cost Holland the victory, as otherwise the British ships would have been severely damaged before being able to reply. Following tradition, Parker insisted on bringing his flagship, stationed fourth in the line, against Zoutman, who was stationed fifth. With only seven ships apiece, numbers two and three in the British line were opposed to numbers two, three and four in the Dutch, while the rearmost British ships had nothing to fire at.

Parker realising the unsatisfactory nature of his deployment, soon hauled down the signal for the line and hoisted the signal for close action; all ships then became heavily engaged. At 10.45am he rehoisted the signal for the line, and at 11 he carried his three stern ships with him to leeward,

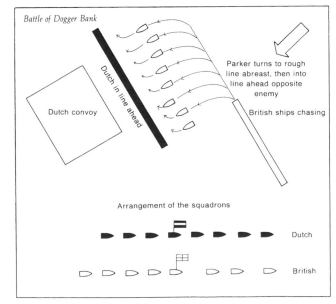

---

42 *Barham*, I, pp313, 315, 334, 337, 338

43 Navy Board Out Letters to Admiralty, 1 February 1781, *NMM*, ADM/B [?]

between his next ahead and the Dutch. At about 11.45 the action drew to a close, with both sides, somewhat unusually, severely damaged in their hulls. Neither was capable of renewing the battle and one of the Dutch ships sank the same night. Both convoys escaped entirely unscathed but, whereas the British reached their destination, the Dutch convoy retreated to the Texel.

Kempenfelt and de Guichen

Kempenfelt was instructed to intercept a French convoy of some 200 ships carrying troops and stores, partly for the West Indies and partly for the East, and escorted by nineteen of the line under de Guichen. He was given only twelve of the line, detached from the Channel Fleet, a force far too weak for such a task. When he sighted the French at daybreak on 12 December 1781, fifteen of their line of battle escort were to leeward, together with part of the convoy, steering west, with the wind at south-east, while the rest of the convoy, with only four of the line in company, were to windward roughly on the same course.

At the start Kempenfelt was to the north and east of the French, but by carrying a press of sail and steering south-west he managed to pass between the windward section of the convoy and the main French fleet. Three out of the four line of battle ships with the windward part of the convoy managed to slip past. The fourth, an 80 and the Marquis de Vendreuil's flagship, escaped only with damage. This very short encounter took place at 12.45pm, after which the British force turned to capturing the transports and merchantmen of the windward group. Two or three were sunk and fourteen captured. Others were struck but could not be secured. All this time, de Guichen was to leeward and though he formed line he was never able to approach nearer to Kempenfelt than five miles. Nor did he make any serious attempt to work to windward on the starboard tack until it was already too late.

Kempenfelt's signals on this occasion show a sense both of urgency and orthodoxy. At 9.42am he signalled for the line-ahead at two cables and to call in all detached ships.[44] Kempenfelt noted,

Had I waited for the slow sailing ships [to take their stations in the line] it would have been dark before we could have got near the enemy, and the convoy would have all escaped. . . . Had I that day known the superior strength of their squadron, and that it was such that I could not in prudence risk an action with, I should have made a general chase of it, and in that case very few of their transports would have escaped; for had any number of our ships been half an hour's sail ahead of us — and many would have been more than that, had the signal for the two-decked ships [to chase] continued out — they would have got in time between the convoy and the ships of war to have intercepted them.[45]

Howe's relief of Gibraltar

Though unimportant in the history of tactics, Howe's relief of Gibraltar in 1782 was a great convoy operation, since the Combined Fleet avoided action even though Howe was embarrassed by the presence of a very large convoy. Three events, however, are worth noting. It was on the way out,

Howe's Order of Sailing

according to Sir Charles Knowles, that Admiral Barrington persuaded Howe to adopt the Order of Sailing in six columns (six ships in each) instead of in three columsn. 'From the sailing in columns and the introduction and use of the signals by numbers', Knowles commented, 'may be dated the era of the improvement in the navy'.[46]

It was also on the voyage out, according to Barrow, that Howe called Barrington and the captains of his squadron on board and asked them to give their opinion in turn, beginning with most junior (as in a court-martial) on the merit of trying to force a night action. Each was in favour of this, supposing it to be the approved line, until Jervis's turn came. A night action, he said, invited confusion and firing into one's own ships, and 'would deprive the British fleet of the advantage of making use of his Lordship's admirable code of day signals, while those for the night were very imperfect'.[47] Barrington agreed with Jervis.

Despite his overwhelming success, Howe was not satisfied with the conduct of all his captains and flag officers. Station-keeping was poor, especially when the ships were caught in the current at the mouth of the Straits, while Rear-Admiral Sir Richard Hughes mistook his station and drew four other ships of his division off their proper course.[48] There was even difficulty in forming line 'advantageously' in the open ocean. Hughes was again at fault, and when he signalled a ship of his division to leave her proper station it was 'properly disregarded by her captain'.[49] Evidently there was still a considerable gap between theory and practice.

Kempenfelt's final signal books

So active was Kempenfelt during the year 1782, until his death in the *Royal George* on 29 August, that it is difficult to place his work in proper perspective. By now he seems to have become convinced that the best immediate course for the Royal Navy was to copy the French system of signals. On 2 February 1782 Sir Richard Bickerton, of the *Gibraltar* at Spithead, issued a printed signal book and instructions to Captain Lord Longford. Any doubts as to whether these were the work of Kempenfelt is dispelled by the evidence of later issues authoritatively attributed to him. Thomas Graves made a careful copy of the instructions, and added to the printed label a note that they had been issued by Kempenfelt.[50] Since the signal book and the book of instruc-

44 'Copy of the minutes of signals taken by Mr Davis Atkins, Acting Lieutenant on board His Majesty's Ship *Edgar* on 12th & 13th Dec 1781, in sight of a French fleet and numerous convoy'; correspondence of Admiral John Elliot, *NMM*, ELL/400 no 58.

45 *Barham*, I, p357

46 *Observations on Naval Tactics*, p25

47 Barrow, pp145-6

48 Not to be confused with Sir Edward Hughes, commander in chief in the East Indies, who fought five battles with Suffren.

49 Barrow, pp154-5

50 *MOD*, NM/53 (formerly *RUSI* 53). Graves's copy is also interesting as it contains a loosely inserted reading list with the following titles: *Manoeuvrier* [Bourné]; Bouguer; Fournier; Hoste; Moroguès; *Naval Evolutions & Naval Tactics* (Steel) *Practical Seamship* (Hutchingson).

tions issued by Bickerton form a pair, Graves's attribution of the latter to Kempenfelt must apply to the signal book as well. A complete copy of both the signal book and the instructions was made by Admiral William Page in 1784 from a printed copy lent him by Captain (later Admiral) Peter Rainier. Page stated that these books were issued by Sir Richard Bickerton to the squadron to take out to India in February 1782, and that they were 'generally said, and no one doubted, to be the compilation and much the invention of the gallant and most able Admiral Richard Kempenfelt'.[51]

Kempenfelt's 1782 book was an exact copy of the signal system used by de Grasse, except that it had only sixteen flags instead of eighteen. A two-flag hoist was employed, and reference was by means of a book with the top flag of the hoist indicating the page, and the second indicating which of the articles on the page was intended. All the flags were the same as those used by Kempenfelt in 1779; none was numbered. To simplify using the book for sending signals, they were grouped according to subject. Howe's and Kempenfelt's new signals were all included, as can be seen in the summary below:

Summary of Kempenfelt and Howe's additional signals, included in Kempenfelt's signal book of 1782

No 218	Attack enemy's rear in succession. (Howe)
No 222	Van to double enemy's van. (1779)
No 223	Van to double enemy's rear. (1779)
No 224	Rear to double enemy's rear. (1779)
No 225	Particular ships to attack enemy ships seeming to be kept in a separate body. (Howe)
Nos 226–30	Engage enemy's centre, van or rear
No 233	Headmost or sternmost ships to signify when they can weather the enemy 'or if to windward of the enemy or on the contrary tack, for the sternmost ship to signify when she is far enough astern of their rear to be able to lead down out of their line of fire'.
No 234	Attack from astern to windward, the headmost ship to engage the sternmost enemy and each ship in succession to pass ahead and to windward under cover of the next ahead until the sternmost ship is engaging the enemy's headmost. (Howe's sixth additional, 1777).
No 235	'When fetching up with the enemy to leeward, and on the contrary tack, to break through their line and endeavour to cut off part of their van or rear.'
No 236	Ships to steer independently of each other to engage respectively ships opposed to them in the enemy line. (Howe)
No 243	'For the van ship to put upon the same tack as the enemy and engage his rear to windward, the rest to do the same as they arrive in his wake.'
No 244	Ditto, and engage his rear to leeward.
No 245	Van ship to steer so as to pass close to the enemy's sternmost ship.
No 246	Ditto, enemy's leading ship.
No 247	'For the van ship to keep no nearer than is requisite, to prevent the enemy's passing to windward of our fleet, and to be at the same time within engaging distance, in case the fleet should come to action on contrary tacks.'

Kempenfelt's instructions provided four orders of sailing, assuming a fleet of fifteen ships in three squadrons. Order number 1 put the second squadron in the van, and the third in the rear. Order number 2 reversed the sequence of squadrons, but kept the squadrons themselves in the normal order. Order number 3 had the squadrons in normal order, but with each squadron in reverse order. Order number 4 was for both the fleet and each component squadron to be in reverse order. When the squadrons were directed to sail in two lines, the first half of each squadron was to form the weather or starboard line, and the second half the lee line. Five cables, or a good half mile, were to be kept between the lines, and two cables between the ships.

Article 9 instructed that

By day, when the fleet is sailing by the wind, a ship of the line from the centre squadron is to keep two points on the Admiral's weather bow, one from the leeward squadron two points on his lee bow, and one from the weather squadron directly to windward, or two points before the Admiral's beam. They are to keep six miles distant from the Admiral, but in thick weather to draw closer.

When sailing large, or before the wind, the ship from the centre squadron is to keep right ahead of the Admiral, that from the starboard squadron two points out to starboard of the fleet, and that from the larboard squadron two points out to port.

In the night each ship is to keep ahead of her squadron about one mile distant.

They are to examine such ships as may pass near them.

The ships are to take this duty in rotation (the three-decked ships excepted) beginning with the senior captain of the squadron.

51 *MOD*, NM/54 (formerly *RUSI* 54). Further printed copies of Kempenfelt's signals and the instructions of 1782 are in the Mead Collection marked 'Kingsmill' *NMM, DUC/49* (formerly Sp/34) is yet another printed copy of the signals, undated. EC/44 and 45 are unsigned and undated copies of the signal book and instructions respectively.

Kempenfelt's Fleet Arrangement

First order	Second order	Third order	Fourth order
Second squadron: 1, 2, 3, 4, 5	Third squadron: 1, 2, 3, 4, 5	Second squadron: 5, 4, 3, 2, 1	Third squadron: 5, 4, 3, 2, 1
First squadron: 1, 2, 3, 4, 5	First squadron: 1, 2, 3, 4, 5	First squadron: 5, 4, 3, 2, 1	First squadron: 5, 4, 3, 2, 1
Third squadron: 1, 2, 3, 4, 5	Second squadron: 1, 2, 3, 4, 5	Third squadron: 5, 4, 3, 2, 1	Second squadron: 5, 4, 3, 2, 1

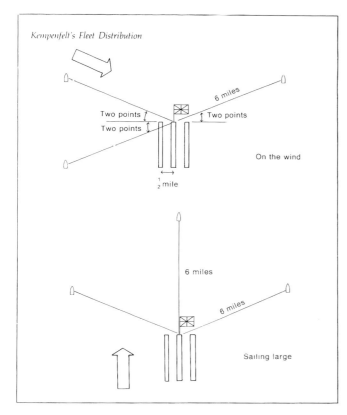

Kempenfelt's Fleet Distribution

The relief is to be in the evening one hour before sunset.

The fighting instructions, firmly based on early Howe, have only three important additions. The meaning of Articles 3 and 12 seems to be that, if a battle looked like being a close thing, the admiral should not necessarily try to weather the enemy. Instead he should be content, at any rate at the start, to engage them from leeward as the two fleets passed on opposite tacks. Article 13 also provided for engagement on opposite tacks:

When weathering the enemy on the contrary tack and signal is made to engage their van, the leading ship is then to bear down to the van ship of the enemy and engage, passing along their line to windward to the sternmost ship of their van squadron; then to haul off close to the wind, the rest of the fleet doing the same in succession.

The object of this last seems.to have been to avoid losing contact with the enemy by passing too far astern of them. By hauling off close to the wind, the weather gage would be retained, and a second attack facilitated.

The difficulty in determining the dates of various changes in Kempenfelt's book is exemplified by a printed signal book signed 'John L Ross, *Ocean* at Sea, 2 June, 1782', and issued to Lord Longford as captain of the *Alexander*.[52] This book was clearly issued by Rear-Admiral Sir John Lockhart Ross Bt, who had commanded the Fifth division of the Channel Fleet under Hardy in 1779 and was now in command of one of its squadrons. Nevertheless the signal book is practically a replica of that issued over Geary's signature on 27 May 1780, and it has the same blank between the ruled lines at the top of the first page.[53] Geary's book, however, shows red-white-red and the blue-yellow-blue flags in horizontal stripes, while the Ross book shows them with vertical stripes. In this respect the Ross book is similar to the signal book produced by Kempenfelt at some time during 1780, but of which only manuscript copies survive.[54]

This, the very last Kempenfelt signal book, completely abandoned the French system, and reverted to the standard system favoured by Howe and used earlier by Kempenfelt himself. It was a printed folio in the style introduced by Howe, thumb-indexed on his standard plan, that is with twenty-seven single flags flown in particular positions.[55] The signals can be described as strictly operational, including signals for use at anchor. There was a minimum of administrative signals, compared with those generally included by the French.

After his successful convoy action against de Guichen, Kempenfelt could expect to command a squadron of the Channel Fleet in the campaigning season of 1782. The position of Lord North's Ministry was becoming insecure, however, and following the loss of Minorca in February, the ministers resigned on 20 March. As a result, Keppel was ennobled and succeeded Lord Sandwich as First Lord. Darby, as a member of the Board, resigned command of the Channel Fleet. He was succeeded by Lord Howe, with Kempenfelt as second-in-command, and the latter transferred his flag to the ill-fated *Royal George*. Howe's orders to Kempenfelt of 30 April stated: 'while you remain on this service you will establish such signals and instructions for the government of the ships under your orders as you may think fit'.[56] Kempenfelt worked through the winter months, and on 4 May, the day after leaving Spithead, he wrote to the Secretary of the Admiralty that he had been 'constantly using all possible despatch for perfecting the necessary signals and instructions for the conduct of the squadron to be in readiness for putting to sea'.[57]

Whether he was referring to his own detached squadron or to the fleet as whole is uncertain. When the fleet was cruising earlier with Barrington, before Howe took up his command, its total force was only twelve of the line and it might well have been deemed a squadron. Anyhow, there is no doubt that Barrington issued Captain George Keppel of the *Fortitude*, one of his twelve ships, with yet another printed signal book drawn up by Kempenfelt.[58] The fact that Barrington's signature is undated raises the possibility that Captain Keppel already had the book, and that it was confirmed by Barrington on taking over the command. In addition, Barrington's name is erased, and 'R Kempenfelt' in his own hand written beneath. Some unimportant instructions issued and signed by Kempenfelt and dated 6 May 1782 are loosely inserted. This shows that, regardless of the date of its original printing and possible issue, it was used by Kempenfelt himself during his last cruise.

There was a flag at the top of each page of the day signals, two flags being painted on each thumbing for the respective double pages when open. Combinations of red and white, and blue and yellow, predominated. The chequered red and white, and blue and yellow, flags were retained. The red-white-red and the blue-yellow-blue flags had vertical stripes. The day signals were numbered right through consecutively from 46 to 412 and there were seventy-nine blanks. Although there was some attempt to keep similar types of signals together in small groups, the battle signals were widely dispersed, presumably to avoid confusion. They

52 *NMM*, PAK/4 (formerly Sp/137)

53 *NMM*, DUN/14 (formerly S(D) 14)

54 *MOD*, Ec/50, and *NMM*, Tunstall Collection, TUN/8 (formerly S/MS/Am/S)

55 *NMM*, SIG/A/14 (formerly Sp/91) is an unsigned, undated and apparently unissued signal book. It has flags exactly as above, though the signals are those of Geary's book of May 1780 (*NMM*, DUN/14 (formerly S(D) 14).

56 Quoted from *Signals & Instructions*, p43

57 Quoted from *Signals & Instructions*, p44

58 *NMM*, Corbett Collection, TUN/42 (formerly S/P/Am/7)

on the page. If there was not enough room, they were printed at the very end of the book regardless of their context.

On 25 August 1779, on arriving at New York, Arbuthnot issued the fleet with a printed book of his own devising, but probably arranged by his secretary, William Green. Although it had the same title as the new 'Admiralty Signals and Instructions in Addition to the General Printed Sailing and Fighting Instructions', it was in fact an entirely different work, printed in official folio form.[5] Ostensibly it was a printed signal book in the old form, each page being divided into three columns for flag, place flown and signification. Under signification, however, the relevant article was printed in full, while the order of the book was governed by the order of the articles and not of the flags. Its main value was for quick reference. Corbett described the contents of the book as 'distinctly reactionary' so far as the fighting instructions are concerned, on the grounds that some of the best articles in the Admiralty Signals and Instructions in Addition were left out.[6] These included Articles 6 to 11 for altering course and for various forms of bow and quarter line, and Article 21 to lead down against opposite opponents and 22 to engage the enemy in reverse order from the rear. On the other hand, Arbuthnot's book included four articles of long standing which had not previously found their way into the Admiralty book.

Sometime in 1781 Arbuthnot issued a printed folio of Additional Signals for the North American Station.[7] This took the form of a three-column page, headed signification, flags and place flown. There were forty-nine articles, including number 24, to engage at night, and number 25, to engage the enemy, if encountered at night, as best able and 'in the closest manner'.

The retirement of Byron from the West Indies command in 1779, and of Arbuthnot from North America in 1781, marks the end of the last stage but one in the attempt to develop tactics and signalling by continuous additions to the permanent Sailing & Fighting Instructions. Rodney issued no signal book at all in 1781. His so-called signals were no more than private manuscript books of the signals made by individual officers to give effect to his general corpus of instructions,

old, new, and personal-additional. These books, of which a number survive, were made, just as in the past, for the use of flag officers and captains, and were, as before, ruled, written and painted by their secretaries and other skilled assistants. At the Battle of the Saints in 1782 Rodney's fleet possessed no official signal book as such. On 18 December 1779, however, before sailing from England for his first West Indian campaign, Rodney had issued those ships accompanying him with his own Signals and Instructions in Addition. On 29 March 1780, two days after reaching Gros Islet Bay,

5 *MOD*, NM/92 (formerly *RUSI* 92), contains one copy of the Admiralty book and two of his own.

6 Corbett was wrong in stating that Article 15 of the Admiralty book was omitted. It appears as Arbuthnot's Article 6. On 17 June 1780, while at Staten Island, Arbuthnot issued three printed additional battle signals for the fleet to form in bow and quarter line, somewhat similar to the instructions previously left out. On 19 August 1780, while watching the French squadron in Newport, he issued a long, detailed and printed but unnumbered 'additional instruction for extending the line of search,' evidently to ensure contact with the French if they tried to slip out. Both sets are in *MOD*, NM/92 (formerly *RUSI* 92).

7 *MOD*, NM/83 (formerly *RUSI* 83).

8 *PRO*, Rodney Papers 15 and 19, and *MOD*, NM/83 (formerly *RUSI* 83), printed in *Signals & Instructions*, pp180–234

The Cancelled instruction

St Lucia, he issued another slightly different set.[8] Both sets had very considerable omissions from, and additions to, the Admiralty version, as well as several minor alterations as regards flags used. Rodney's fifteen additional sailing instructions were largely miscellaneous, including useful and semi-traditional items omitted from the Admiralty book. The most interesting is number 49, for when the admiral 'would have all the ships sheathed with copper chase on any point or quarter of the compass'. Experiments with copper sheathing had been going on for many years and during the course of

the American War a great number both of new and existing ships were so fitted. This was one of the very tangible achievements of the supposedly incompetent Board of Admiralty headed by Lord Sandwich. The French were doing the same thing, and only the Spaniards lagged behind.

Rodney also made some material changes in the fighting instructions section of the Sailing and Fighting Instructions in Addition. He cancelled Articles 3, 9, 10 and 11, dealing with the bow and quarter line. Article 11 was for forming the line in inverted order by 'the second ship passing the

The Moonlight Battle; the Battle of Cape St Vincent, 16 January 1780, by Francis Holman (see pp 164–5). NMM

headmost, and so on in regular succession until the sternmost becomes the leader', with the squadron in bow and quarter line. Graves's comment had been that it would only be satisfactory for ships to form on each other's weather quarter – which it appeared was the intention of the article – if the enemy were coming up with them to leeward. 'But if the enemy is coming up with them to windward they should be formed on the lee quarter of their leader.'[9] No doubt Rodney had himself noted the lack of clarity of this instruction as it stood. He also cancelled Article 26, which directed that, when a fleet were superior to the enemy, surplus ships should quit the line without being signalled, and rake the enemy's van or rear.[10] He probably had two reasons for making this latter deletion. It was against his principles to concede any initiative to his captains, and the instruction, though of some ancestry, was difficult to execute and had never produced much in the way of practical results.

Rodney's four additional fighting instructions contain one quite original article, number 29 or 71 (according to the continuous or non-continuous numbering of various copies), for 'all the three-decked and heavy ships [to] draw out of their places in the line of battle and form in the van [or rear] of the fleet'. Corbett divined in this article both the direct influence of Morogues and the idea later adopted by Nelson at Trafalgar for the three-deckers to lead. Morogues, however, never went beyond stressing the tactical value of heavy ships compared to two-deckers, a mere emphasis on the obvious. Nor did he provide any signal for their special employment. Rodney never made use of his own signal in any of his full or partial engagements, perhaps because he never had as many as five 'heavy ships' of 90 guns or over at his disposal except at the Battle of the Saints. At the Saints he adopted the traditional plan of having one for each of the three flagships and placing the remainder one ahead and one astern of his own flagship in the traditional position of the commander-in-chief's heavyweight seconds, thus giving additional strength to the centre. If flag officers were not to be deprived of their traditional claim on heavy ships, and there were only a tiny sprinkling of such ships, how could such ships be moved ahead or astern of the fleet without depriving the flag officers of control of their respective squadrons? Possibly, like Morogues, Rodney was thinking in terms of a possible future fleet including enough three-decked private ships to allow a much freer disposal of force. Even Howe's exceptionally powerful fleet of thirty-five of the line at the First of June included only seven three-deckers,

of which six were flagships whose admirals carried divisional responsibilities. Though apparently standing for everything retrogressive in tactics and signalling, Rodney voiced the demands of all his great successors when he wrote: 'On, my Lord, the three-decked ships are what must maintain the sovereignty of the sea; nothing can withstand them . . . I had rather have ten three-decked ships than eighteen two-deckers, and would answer with my head for their defeat.'[11]

Rodney's articles, variously numbered in different versions as 73 and 74, 31 and 32 with sub-division 33, or 26 and 27 with sub-division 28, which began, 'When the Commander-in-Chief means to make an attack upon the enemy's centre', or van, or rear, have become famous through an abortive application at the Battle of Martinique. This was clearly a signal of general intention and Rodney deliberately refused to commit himself on the means of execution. An attack on the enemy's centre might require a very different approach from an attack on the enemy's rear, quite apart from the state of the wind and weather and the relative disposition of the two fleets, especially if the fleets were on opposite courses. To have issued more elaborate instructions and signals would have been contrary to his general attitude, since any attempted sophistication of tactical approach would have necessitated at least giving his squadron commanders and possibly even his captains that kind of individual discretion which Howe favoured, but which he himself strongly opposed.

In November 1781, in temporary command of the British fleet in the West Indies, Hood issued a set of signals rather less extensive than those used by Rodney either in 1779–80 or in 1782, and with only thirty-seven flags in place of forty.[12] On 23 January 1782 he issued instructions with signals for his abortive attack on de Grasse's fleet at anchor in Basseterre Road, St Kitts. His plan was to attack in succession from windward, each ship engaging the first three ships only of the enemy; then to veer short round and fall onto the rear in succession.[13]

On 19 February 1782 Rodney arrived at Barbados to resume command of the West Indies station and the same day issued a corpus of printed instructions even more elaborate than Arbuthnot's.[14] Typical is his alteration of Article 27 of the 1703 Sailing & Fighting Instructions making it absolutely clear that 'no ship, or any particular part of the Fleet whatsoever, shall presume to tack, or wear, whilst in action, unless by signals, or order from the Admiral so to do'.[15] Article 31, for forming the line, which had read, . . . 'and if he would have those lead who are to lead with their starboard tacks aboard' or 'those lead who are to lead with their larboard tack aboard', were altered by Rodney in handwriting to: 'and would have him lead who commands in the second post', thus shifting the emphasis of the signal from ships to flag officers. Corbett links this change with the third charge against Keppel of having failed to make the signal in its old form.[16]

Article 32, the last, was given a long addition in handwriting to the effect that signals made to the fleet when in line of battle and duly repeated and acknowledged, should remain in force even though hauled down, until actually annulled 'by the annulling flag, or until it be naturally superseded by the purport of some subsequent signal'. Nevertheless, it was added, signals for the line-ahead or abreast or engaging would be kept flying as long as the commander-in-chief judged proper for the fleet to continue in line or engaging. Corbett saw in this an attempt to prevent a repetition of the misunderstanding about Rodney's intention to attack the enemy's rear at the Battle of Martinique.[17]

Rodney also issued in 1782 a new printed version of the Admiralty Sailing and Fighting Instructions in Addition.[18] The first twenty-three of the thirty-one printed fighting

---

9 *MOD*, NM/83 (formerly *RUSI* 83); *Signals & Instructions*, p220

10 *Ibid*, pp227–8

11 Rodney to Sandwich, 25 April 1782, *Sandwich*, IV, pp264–5

12 Manuscript copy of Hood's signals issued to Captain Cornwallis of the *Canada*; *NMM*, Tunstall Collection, TUN/10 (formerly S/MS/Am/10). Loosely inserted is a sheet of additional signals for landing and re-embarking the marines, seamen and the 69th Regiment used in the abortive attempt to relieve the British garrison of St Kitts.

13 *MOD*, NM/24 (formerly *RUSI* 24), Hood's signals issued to Captain Graves of the *Bedford*, with various additions, 12 January 1782.

14 *MOD*, NM/94 (formerly *RUSI* 94)

15 *Signals & Instructions*, pp274–5; from the copy in the Library of the United States Navy Department, Washington

16 *Signals & Instructions*, p279

17 *Signals & Instructions*, p280. Rodney made some useful additions to the original night sailing signals of the parent work: 5 and 6 for chasing NE, NW, SE, and SW and giving over the chase; 7 'If in pursuit of a number of the Enemy's ships, and the headmost brings to any of them, and passes on without taking possession of them;' signal follows; on which chasing ships further astern will keep a careful look-out and take possession of them. He also issued eight extra fog signals. See the copy in the US Navy Department Library but not quoted by Corbett.

instructions were the same as the first twenty-eight of the original Admiralty issue, less the five which Rodney had cancelled in 1779. Articles 24 to 28 contained his handwritten additions of 1779, for the heavy ships to draw out of the line, for close order in battle, and for preparing to attack the enemy van, centre or rear. Article 29 provided the signals for the second or third in command to lead the fleet with his divisions. Article 30 gave signals ordering the van or rear to close up on the centre, and forbidding individual ships to chase a beaten portion of the enemy fleet. Article 31 directed in detail how, should some ships have become separated from the fleet, the admiral would form a temporary line of battle ahead or abreast by signals to individual ships as to their temporary station. This article was originally devised for Rodney by Captain Walter Young.

Rodney's signals in 1782, judged by numerous examples, were arranged as before in the traditional manner with forty flags. Many of these were underemployed, so that he had a large number of positions in reserve for extra signals, though it is doubtful if this entirely offset the risk of confusing some of the flags.[19]

On 20 March 1782, only a few days before the Saints, Rodney issued an order of sailing by which the commander-in-chief in the *Formidable* was to lead the centre squadron and the fleet, with Hood in the *Barfleur*, followed by the van 'two miles broad on the weather quarter of the *Formidable*, that is about 3 points abaft her weather beam'. Rear-Admiral Francis Samuel Drake in the *Princessa* was to lead the rear 'two miles from the *Formidable* narrow on the lee quarter, that is to say to leeward and astern withall'.[20] Each squadron was to sail in arrow-head formation with the flagship leading, the remaining ships being disposed outwards from their own flagship. Thus in a squadron numbered 1 to 11 in order of battle in line ahead with the flagship number 6, the order of sailing would be as shown in the diagram on this page.

On 10 April 1782, the day after the abortive action with de Grasse, Rodney issued an additional Article 32 to his printed additional fighting instructions:

If at any time from baffling winds or other cause I think it expedient for my division to lead the fleet in line of battle for attacking the enemy . . . And if I would have the van [or rear] division become the centre . . . The Leading ship of my division is then to lead the fleet, and the leading ship of the division next to follow is, as above alluded to, of course to be next to the sternmost of mine and so on.[21]

On 25 April, thirteen days after the Battle of the Saints, the British fleet met off Haiti and then finally separated, Hood taking the main body to watch the French at Cap François while Rodney took his prizes and the badly damaged ships to Port Royal, Jamaica. On the very same day Hood gave out a list of additional signals. These included the following instruction: 'When fetching up with the enemy to leeward and on the contrary tack', if the admiral wished the fleet 'to break through their line and endeavour to cut off part of their van or rear', he would fly a chequered yellow and blue flag at the foretopmast head. In the main body of the signals, for which in the particular manuscript book cited there is neither date nor ascription, is the instruction: 'When standing towards the enemy in a line of battle on a contrary tack or being actually in battle', a red and white quartered flag would be flown at the fore, or the same with a red pendant over it at the mizzen topgallant, mastheads 'for any particular ship to lead through their line and all the rest to follow in due and close succession to support her for that purpose'. The same signals appear in two manuscript signal books, one entitled Hood's Signals of 12 January 1782. The additions are in both books dated 27 April.[22]

The signals did not provide for a general manoeuvre of breaking the enemy's line. They were solely concerned with the tactical situation presented by the Battle of the Saints, and the implication is that the admiral might signal the leading ship or any other ship to execute the manoeuvre if opportunity offered. The fleets would have to be passing close to each other and the enemy would have to be sufficiently close-hauled to allow for the necessary deviation off course to windward by the British. Neither the manuscript signal book actually used by Commodore Edmund Affleck at the Saints, nor any of the so-called signals 'by Rodney' contain either of these signals, which indicates that Rodney, alone amongst the leading admirals, possessed no signal for breaking the enemy's line at the time the battle of the Saints was fought.[23] When Admiral James Pigott arrived to supersede Rodney he was persuaded to take over Hood's two signals for breaking the line, and issue them to the whole fleet. The process by which Hood guided, or tried to guide, Rodney's successor can be read in his own letters to Jackson.[24]

Some attempt was made to improve the signalling system. In one of Hood's signal books of 1782 an additional left-hand column was ruled, and headed on every page 'N Signal' indicating an intention to number the signals right through in the manner of Howe's and Kempenfelt's earlier books, while leaving the traditional arrangement of flag, where flown and signification intact. This intention was never fulfilled but, in the same year, a West Indies signal book was produced, with Hood's and Pigott's additional signals included;[25] all the signals were numbered in the main text, though not in the index, with painted flags at the end. The signal for breaking the enemy line, with a flag chequered yellow and blue, was number 11, while the signal for any particular ship to do so, with a flag quartered white and red, was number 243, again ascribed to Pigott. The signal referring to additional fighting instruction number 34, for divisions in parallel lines in echelon, was given the signal book number 265, very near the end.[26]

Admiral Robert Digby on the North American station issued a signal book on 5 October 1782.[27] It was the work of John McArthur, later secretary to Lord Hood and joint biographer of Nelson, and it was used experimentally by a small cruiser squadron for three months.[28] It is a well arranged printed book, thumb-indexed, with the flags and

Rodney's Order of Sailing, 20 March 1782

18 *MOD*, NM/94 (formerly *RUSI* 94): but see also NM/43 (formerly *RUSI* 43) in MS.

19 *MOD*, NM/28, 31 [wrongly dated in the catalogue as 1792S] 43 and unidentified file (formerly *RUSI* 28, 31, 43 and 94); Tunstall and Corbett Collections, *NMM*, TUN/3 and 6, (formerly S/MS/Am/3 and 6)

20 *MOD*, NM/92 (formerly *RUSI* 92) and *NMM*, Tunstall Collection, TUN/3 (formerly S/MS/Am/3)

21 *MOD*, NM/94 (formerly *RUSI* 94) and US Navy Department Library, Washington; printed in *Signals & Instructions* pp300-1

22 *MOD*, NM/24 (formerly *RUSI* 24), and *NMM*, Corbett Collection, TUN/6 (formerly S/MS/Am/6). The signals in the second book appear twice over, the book being drawn in the usual tumb-indexed form but with a second part in which the signals are fully repeated, including coloured flags, thumb-indexed according to subject matter. In the first part the signal for a particular ship to perform the manoeuvre is attributed to Pigot.

23 *NMM*, Tunstall Collection, TUN/3 (formerly S/MS/Am/3); and *MOD*, NM/28, 31, and 43

24 David Hannay, editor, *Letters of Sir Samuel Hood (Viscount Hood) in 1781-2-3*, London, 1895 (hereafter cited *Hood*), pp138-61

25 *MOD*, NM/18a (formerly *RUSI* 18a), labelled 1790 but actually 1782

26 Instruction number 33 in *NMM*, Corbett Collection, TUN/6 (formerly S/MS/Am/6)

27 *MOD*, NM/76 (formerly *RUSI* 76) and *NMM*, KEI/S/9 (formerly Sp/9)

28 *Signals & Instructions*, p58, where, however, the description of the book towards the bottom of the page is somewhat misleading.

Captain (later Vice-Admiral) the Honourable John Byron (1723–86), by Sir Joshua Reynolds. NMM

positions on the old system and the signification columns partly left blank and partly completed in manuscript. Some of the twenty-four flags, however, are of unusual design, for example, a blue flag with the sinister chief quarter white; a flag divided diagonally blue and yellow, and the reverse; a white flag with a red lozenge and a white flag with a thin perpendicular pale blue stripe down the middle. Though the signals are not numbered in the main text, there is a separate index of the signals with numbers for use with a numerical table of ten flags selected from the main series of twenty-four. There is also a separate numerical table for private ships, which includes only the established flags, since only the admiral or senior officers would have had the experimental set. There is nothing very new about the main signals, except one for the van of the fleet to pass through the enemy line.

Amiral Jean-Baptiste Charles Henri Hector Théodat, Comte d'Estaing (1729–94), bust by Houdon. MM

Byron and d'Estaing

The Battle of Grenada (1779) is yet another example of a straggling British attack against a well-formed French line, though unlike the battles of Toulon and Minorca, the British mistake was due less to pedantic adherence to the line than to complete neglect of it. The British commander-in-chief, Vice-Admiral the Hon John Byron, aged nearly fifty-six, had a great reputation as a navigator and explorer, but Sir John Laughton was probably correct in asserting that 'It is very doubtful whether he ever saw a fleet extended in line of battle before he saw the French fleet on the morning of 6th July, 1779.'[29]

His opponent, the Comte d'Estaing, was aged fifty. He had been an infantry officer until the end of the Seven Years War, when he was promoted to *lieutenant-général*, and soon after the peace to governor-general of San Domingo. Then, by one of those acts of military metamorphosis of which only the French service was capable, he became simultaneously *Chef d'Escadre des Armées Navales*, and *vice-amiral*.[30]

The Battle of Grenada resulted from d'Estaing's capture of the island from the British on 4 July 1779, while Byron was clearing a home-bound convoy. By nightfall on 5 July d'Estaing had re-embarked most of his troops in transports in St George's Bay, covered by his battle fleet of twenty-four of the line. He knew that Byron was coming south to attack him and he had his frigates on the lookout. That evening they reported Byron's approach. As expected, he was steering south along the west side of the island with an off-shore breeze, E by N. D'Estaing kept his fleet at anchor until the last moment, chiefly because he had so little breeze at Georgetown that he was in danger of being carried out westward by the lee current if he put to sea before the dawn breeze took effect. At 4am, however, his fleet put to sea with orders to form line ahead, *par rang de vitesse* (in order of sailing speed), on the starboard tack, that is, as quickly as possible and without regard for stations in the official line of battle. Even so, he had some difficulty in getting his ships formed.

Byron meanwhile was coming along the western side of the island, helped by the freshening dawn breeze. He had with him eighteen of the line and astern a convoy of troops ready for the recapture of the island, guarded by three of the line under Rear-Admiral Joshua Rowley. Though Byron was well aware of the enemy's presence and indeed hoped to catch them soon after daybreak while still engaged in attacking the fortress, his fleet was still unformed, despite the fact that he had been prevented from sailing fast by the presence of his convoy. At 5.15am on sighting the French fleet to the south, apparently huddled in confusion and numbering only fifteen of the line, he signalled general chase, hoping to score a surprise, and at the same time he signalled Rowley to leave the convoy and rejoin the fleet with his three ships.[31]

This general chase was Byron's undoing, since it tended to disperse his ships instead of collecting them together. Both Anson and Hawke had been careful to form in line ahead before making their attacks against weaker French forces in their respective battles of Finisterre. Moreover d'Estaing was not to be caught so easily. By now the dawn breeze had reached him and his line was beginning to form on the

29 Source unknown

30 This happened as a result of the reorganisation of the French Navy by which new regiments were created to act as marine gunners and marine infantry. See Calmon-Maison, *L'Amiral D'Estaing*, 1910, Chapter ix.

31 Byron's own account is printed in *The Mariner's Mirror*, 30 (1944), pp88–90

starboard tack on a roughly north-west course. Byron still did not realise the truth, and thought that he faced a distinctly inferior force. At 6.30am he signalled for 'the five or seven ships nearest the enemy to draw into a line-ahead in order to engage the enemy in their rear, endeavouring at the same time to get up with their van'. This was Hawke's old signal of 1747, intended for use against a fleeing enemy, and it was quite inapplicable here as the enemy rear was further away than their van by at least two miles. Hence not only was the element of surprise – in so far as it ever existed – completely sacrificed, but the five leading ships had to sustain the fire of a number of French ships to starboard before they could wear round and engage the French rear on a parallel course.

Had Byron attacked the enemy van he might have been more successful, as he would then have engaged the nearest French ships instead of the furthest, and would thus have been able to bring his whole fleet more quickly into action in support of his leaders. It is true that he had no signal for attacking the van, but it would seem there was nothing to have prevented him from signalling 'general chase to the south-west', that is directly towards the leading French ships, followed by 'engage' and then 'engage closer'. Considering the kind of flag officers and captains he had, this should have produced the necessary pell-mell attack on the French van. Alternatively he could have signalled for the leading ships to lead more to starboard, followed by the signal to engage. More conservatively, he could have signalled his leading ships to shorten sail, and then have used Howe's additional signal of 1762, 'for the squadron when in chase to engage the enemy in line of battle according to the order in which they may successively arrive up with them'.

Byron's fundamental mistake seems to have been his inability to keep his fleet formed in reasonable order during the night. What was its formation at sunset on the night of 5 July? What orders did he send or what signals did he make immediately before sunset with regard to his night sailing formation? He was steering a steady course and he had the northern tip of the island of Grenada ahead of him to act as a check on his position. With a steady breeze and the need for not outstripping his convoy, it seems incredible that had his ships been in a fixed formation, they should not have been able to maintain it better during the short summer night.

It is much to Byron's credit that, when he realised that it was he who was outnumbered and not d'Estaing, he signalled

for closer action and did everything possible to force it. It is also to his credit that he was well supported by the whole fleet. Indeed the attack of the three leading ships, the *Sultan*, Alan Gardner, *Prince of Wales*, Vice-Admiral Barrington, and the *Boyne*, Herbert Sawyer, was so impetuous that they drove two of the French rear ships out of the line with severe damage and casualties. Having already worn round, Sawyer and Barrington were trying their best to get up with the enemy's van, Gardner's *Sultan* being entangled to leeward with the French ships.

At 8.30am Byron signalled the remainder of the fleet to wear in the wake of the leaders. The bulk of the British fleet was still straggling south, close inshore, and now began wearing round in succession, making what was practically a hairpin bend so as to follow the leading ships and get on a starboard course parallel with the enemy. The British fleet was thus steering in the opposite direction from the French, in order to lead round onto the same course and then try to draw level with the French van. Having lost the choice of heading the French at the start, they were now compelled to sail twice the distance to achieve the same result.

Three British ships were so out of position to leeward in the run south that they found themselves fired on by the whole French fleet as it passed on the opposite tack. After an hour the indefatigable Barrington and Sawyer came up from the south and passed between these ships and the French, still trying to get up with the French van. Meanwhile Byron had sent a schooner to order two ships to leave the line and pursue the two rear French ships which had been damaged at the start of the action. This shrewd attempt, however, proved abortive, and the schooner now passed on to the north to order Admiral Rowley's *Suffolk* to attack the leading French ship as it came north. Rowley obeyed this order with great bravery and skill and to the relief of the three British ships which had been too far to leeward, and which now lay crippled as successive French ships fired on them in passing.

Meanwhile, Captain Robert Fanshawe of the *Monmouth*, another of the ships recalled from guarding the convoy, acting without orders, cut right across the loop now formed by the British line and engaged the ships of the French van, suffering very severely in the process. At 10am Byron, realising that his whole attack was falling into disorder, cancelled general chase and signalled both for the line of battle ahead at two cables and for close action. As each ship tried to reach its proper station in the line there was much delay, some ships having to tack and then wear round again.

D'Estaing was now in a winning position. Except for the two crippled ships in rear, his line was intact. The British fleet, with three ships badly damaged and several hardly able to keep station, was still not properly formed in line. He thus had twenty-two effective ships against what now amounted to sixteen or seventeen. His van overlapped the British, and on his starboard bow was the British convoy. He could either cripple the British still further and then go for the convoy, or else make as if to attack the convoy and so force the British to fight him at a tactical disadvantage. As he drew further ahead, Byron kept close to the wind so as to cover his convoy and avoid being doubled. At 2.30pm he hauled down the signal for closer action.

D'Estaing was now clear of the British. His line was well formed and he could choose his method of attack. Instead, however, of attacking Byron or heading for the convoy, he put about – some of his ships tacking, some wearing. He then steered south-east back towards St George's Bay. Soon after, Byron signalled to tack together, putting the two fleets roughly on the same course but some distance apart. He could do nothing to save the three ships so severely damaged earlier through being too far to leeward; they were still further to leeward now and nearly in line with d'Estaing's approach. The French ships, however, held their course and contented themselves with firing at the British stragglers as they passed. The battle was over and no British ships were lost.

D'Estaing was strongly criticised by Suffren, captain of his van ship, and by other officers, for not pursuing his advantage. His aim seems to have been to prevent Byron slipping inshore of him, seizing the harbour of Georgetown, capturing the French troops lying there in transports, bringing in his own convoy, and landing the troops to recapture the fortress. Yet had he beaten Byron decisively there could hardly have been a Battle of the Saints. The decisive factors in his failure were probably three: his genuine preoccupation with the safety of Grenada, the 'custom of the service' with regard to maintaining the safety of the fleet, and, perhaps, a cautious respect for the fighting qualities of the individual British ships.

The Moonlight Battle

On the British side, George Brydges Rodney stands out not only as the most controversial naval commander of the American War but also as the most controversial figure in the whole history of British tactics and signalling. In many ways he was unsuited to be a commander-in-chief, being violently contemptuous of those he disliked, and there were many. He had been a difficult man to serve under, even at the age of forty-four. Now, at sixty-two and broken in health, he was full of energy one moment, vacillating the next, highly suspicious of his subordinates, and inclined to treat them vindictively. Launched into the tropical West Indies, he proved himself both a tyrant and a martinet. Nevertheless, he achieved a very great victory. He was a tactical reformer and was popularly credited with 'breaking the enemy's line', although in fact he had actually refused to adopt the new signal specially devised for such a manoeuvre. When in good health and not exhausted he was still capable of driving a great fleet into action, assisted by his faithful assistants Sir Charles Douglas and Captain Walter Young, and his private physician Sir Gilbert Blane. He may well have studied Morogues, for he accepted the notion that a ship-to-ship action between equal forces offered little hope of success. To win, it was necessary to concentrate a superior number of ships against an inferior, leaving the remaining enemy ships disengaged.

On his way out to the West Indies via Gibraltar, Rodney scored an important tactical success. In the early afternoon of 16 January 1779, with his powerful fleet formed in good order, he was off Cape St Vincent when a Spanish fleet of eleven of the line and two frigates was sighted in some disorder. Rodney was ill with gout and his flag captain, Walter Young, directed the battle. Young wanted a general chase, to cut off the Spaniards from Cadiz.

This the admiral opposed and ordered the signal for a line of battle ahead at two cables distance. This I opposed in turn, conscious that a great deal of time and distance would be lost, and proposed the signal for a line of battle abreast at a cable's distance. At the same time (as he was confined to his bed) I kept out a stiff sail on the *Sandwich* that no time might be lost; and, as soon as I discovered the enemy making sail to get off, the signal for a general chase was made. When our ships got up the signal [was made] to engage to leeward and in succession, and the signal for close action at the same time.

The admiral's ill state of health and his natural irresolution, occasioned our shortening sail frequently, which prevented me from bringing the *Sandwich* so early into battle, as I could have wished; and it was with difficulty that I succeeded at last, as he attempted several times to have the ships called off from chase; but on representing to him that they were too far ahead to either see or hear our signals, and that it was our business to be at the head or near to them, that we went on and stopped one of the enemy who was prepared to evade us.[32]

Admiral George Brydges, Lord Rodney (1719–92), first Baron Rodney, by Sir Joshua Reynolds. NMM

---

32 Young to Middleton, 24 July 1780, *Barham*, I, pp65–6. The British ships gained on the Spanish either because they were coppered or because they set more sail, or both. The signal to engage to leeward in succession is Howe's signal for inverting the line of 1776 and earlier, number 18/61, *NMM*, Tunstall Collection, TUN/111 (formerly F1/5). The leading British ship was to engage the rearmost enemy while the second nearest ship, regardless of the order of battle, was to engage the next but one astern of the enemy, and so on. Text in *MOD*, NM/92 (formerly *RUSI* 92), and printed in *Signals & Instructions*, p223.

The battle lasted from 4pm until after midnight and seven of the eleven Spanish ships were sunk, taken or wrecked. Though Captain Young claimed to have directed the battle, Sir Gilbert Blane, in attendance on Rodney, suggests that it was a more equal partnership. In any case, Rodney took the entire responsibility for what was done and also the praise. Even if Young did not entirely direct the battle, the action itself was extraordinary, and praise was justified. A night action on a lee shore in a gale was a serious risk, as was discovered next day when the British fleet were in some danger of being driven ashore.

Young, in the same letter in which he described the battle, deprecated the use of general chase against an equal or superior enemy, particularly if formed in line of battle or in close order. In such cases he favoured a close line abreast, if the British fleet was to windward, as the best formation and one from which 'any evolution is instantly put in practice'. He considered one cable distance sufficient for this formation, as distances were invariably overjudged.

The Battle of Martinique

On the morning of 17 April 1780 the British fleet of twenty of the line and a 50 was to windward of the French fleet of twenty-three of the line, at four to five leagues distance. The wind was east and de Guichen was forming the French line on the larboard tack, that is, heading south. At 4.30am Rodney signalled for the line ahead on the starboard tack so as to close him, and at 5.45am for the line ahead at two cables. At 6.45 Rodney made the signal for Article 74 of his new additional fighting instructions, which read, 'When the commander-in-chief means to make an attack upon the enemy's rear he will hoist a flag half-blue half-yellow at the main topgallant mast head with a white pendant under it'.[33] There were separate signals for attacking the van and centre. In his own words, 'I gave notice, by public signal, that my intention was to attack the enemy's rear with my whole force; which signal was answered by every ship in the fleet.' But this is not the same thing, for whereas the instruction merely indicates the point of attack, Rodney's statement implies an actual concentration of force. Yet nothing was said about this in the instruction, nor provision made for how it was to be done (for instance, certain ships astern might be detached to double on the enemy's rear and so put them between two fires). It was no more than a signal of intention, requiring further signals to give it effect. In the manuscript signal book made for Commodore Edmund Affleck, the signal is condensed to read 'to attack the enemy's rear', as if implying an immediate executive purpose which, if baulked by the enemy, might automatically cease to operate. Here, therefore, was a further source of possible confusion. Rodney is known to have been exercising his fleet in tactics a fortnight before the battle, but there is no evidence that he had gone through his additional fighting instructions with his flag officers and captains, article by article, to see that they were properly understood. Sir Gilbert Blane, writing twenty-nine years later, said that all the captains had been informed of his plan of attack either orally or in writing; if this is true, Rodney's explanation of this particular point must have proved inadequate.

For the time being, however, the signal clearly held, whatever its interpretation. Rodney continued manoeuvring to give it immediate effect, his further signals being as follows: '7.00am line ahead at one cable distance' (this presumably was a closing-up movement on the centre at the

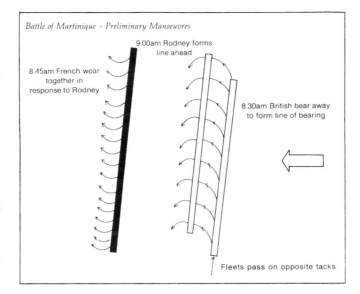

Battle of Martinique – Preliminary Manoeuvres

9.00am Rodney forms line ahead

8.45am French wear together in response to Rodney

8.30am British bear away to form line of bearing

Fleets pass on opposite tacks

moment when the bulk of his fleet was level with the French centre and rear); and, '8.30am, line of bearing, the ships to bear N by W and S by E of each other, at two cables' (the wind being then on the starboard beam). The extension of two cables distance was presumably to avoid bunching and possible collision should that course need to be altered again before reaching the enemy.

De Guichen seems to have divined the plan and sought to thwart it; at 8.45am he wore his fleet to the north, thus offering a van-to-van attack, which of course was not to Rodney's purpose. At 9am Rodney replied by signalling his own fleet to haul to the wind and to form line ahead at two cables on the port tack. This brought the two fleets on opposite courses once more, and at 9.42 Rodney's signal was repeated.

The fleets were now some three miles apart and the situation was again becoming favourable for Rodney's plan, the two leading ships in the British van having reached slightly beyond a point level with the rearmost ship of the French, with the fleets roughly parallel. So at 10.10am Rodney signalled for the fleet to wear and come to the wind on the starboard tack, thus coming once again onto the same course as the French. At 10.18 the signal was repeated, and at 10.19 the signal made for the rear ship to wear. At 10.36 Rodney signalled for the line-ahead at two cables. By this

Battle of Martinique – Coming to Action

10.10am Rodney wears to come onto same course as the French

11.50am Rodney signals for each ship to steer for its opponent – but some ships misunderstand

33 *Signals & Instructions*, p230

down on a lasking course to engage, instead of making a slanting approach each directly astern of the ship ahead. In other words, Graves's signal was intended to produce the same result as a 'sailing' signal for forming a line of bearing. In view, however, of how Arbuthnot's signal actually reads – 'If at any time I would have the leading ship in the line alter course to starboard, I will hoist a flag half red, half white at the main topmast head' – the captains cannot be blamed for acting as they did. The similarity of his dilemma with that encountered by Byng at the Battle of Minorca is quite remarkable.

Seeing his whole plan of attack going awry, Graves repeated his signal twice, and also signalled various individual

ships to get into station and the rear to make more sail. At 3.34pm he made signal number 5 from Arbuthnot's additional fighting instructions of 1779, for the fleet then in line of battle to alter course to starboard and at 3.46 for the line at one cable.[61] 'The enemy's ships advancing very slow and evening approaching, the Admiral (judging this to be the moment of attack) made the signal for the ships to bear down and engage their opponents – filled the main topsail and bore down to the enemy.[62] This was followed by the signal for close action, and at 4.11pm the signal for the line was hauled down, 'that it might not interfere with the signal to engage close'. Nevertheless, the *Montagu*, twelfth in the line, luffed and opened fire at too great a range, thus forcing the ships of the centre astern of her also to luff so as to avoid overrunning and getting between her and the enemy. At 4.22 Graves 'hoisted the signal for the line-ahead and hauled down the signal for close action, the ships not being sufficiently extended'. But five minutes later the line ahead signal was hauled down again and the signal for close action rehoisted.

61 *Signals & Instructions*, p238

62 Graves's account, *Sandwich*, IV, p185. These words are from the official narrative endorsed in Graves's private letter to Sandwich; hence 'the Admiral'.

Graves tried to set an example by steering straight for the enemy, but soon found himself obstructed.

Meanwhile, the van of the fleet had begun to engage the four leading ships of the French van, which were very much exposed to windward. The French ships succeeded in beating the leading British ship out of the line, but were themselves heavily punished. De Grasse signalled to them to bear away to leeward but this proved expensive as it meant exposing their sterns to the British broadsides; the fourth French ship was badly hit in the process and her captain killed, and another ship was in danger of capture. Altogether only about eleven of the twenty-four French ships became engaged, the remainder being too far to leeward. On the British side only ten ships out of nineteen were seriously in action, and of these, ten suffered considerable damage to masts and rigging. They included the *Terrible*, now so crank that she leaked from the shock of discharge of her own guns and had to be burnt six days later.

All this time the British rear remained practically inactive,

Hood being under the impression that the signal for the line was still flying and that it had never been hauled down. There is much confusion of testimony here but, at any rate, the signal for close action was flying at 5.20pm, and the signal for the line was not, so that at 5.30 the British rear bore down – much too late. At 6.23 Graves finally hauled down the signal for close action and hoisted the signal for the line ahead at one cable. The French could be seen bearing away to leeward. It was nearly sunset, and at 6.30 all firing ceased.

The manoeuvres of the next nine days had little tactical significance and, when Graves sailed for New York on 17 September, Cornwallis's surrender became a certainty.

Reactions to defeat

Both Graves and Hood were greatly dissatisfied with their tactical fiasco, Hood especially so as he must have felt that he was partly to blame. Although he made no comment whatever on the failure of the approach signals to produce the

The repulse of the French in Frigate Bay, St Kitts, 26 January 1782 (dated 1783) (see pp 177–8). NMM

enemy rear ships; if towards the centre, the cut would still force the rear of the windward fleet down to leeward. Clerk admitted that it would be dangerous for a cut to be made by this method towards the enemy van.

In a remarkable foreshadowing of the Battle of Cape St Vincent, Clerk postulated a situation in which a large fleet was advancing in an irregular line abreast with the wind abeam. At daybreak and in foggy conditions, the fleet might be firing signal guns for getting into their proper order. A much smaller fleet, warned by the gunfire, might appear suddenly ahead on the opposite course and about level with the centre of the larger fleet. The commander of the smaller fleet could at once form a line ahead and cut right through the larger fleet. He could then tack together or in succession, concentrating on that part of the larger fleet to windward of him and leaving the leeward part unable to intervene. The parallel with the Battle of St Vincent is extraordinary, though St Vincent himself said that the claim that Clerk had 'any merit' in that action itself was 'totally out of the question'.

If an admiral was determined to make an attack from windward, Clerk's proposed method was for concentration on the three rear ships of the enemy line.[102] The particular ships detailed for the attack would have to be the best sailers in the fleet; the remainder coming up well to windward of the attacking group, in order to give support, should be in loose order rather than in line-ahead. Clerk argued that, once the three rearmost enemy ships were engaged, and their speed checked, their loss would be inevitable. Whatever the rest of the enemy fleet did in an effort to save them would fail. If the fleet tacked in succession or together, the supporting ships of the attacking fleet to windward would be there to deliver raking broadsides. If they wore in succession or together, the attackers could also wear and drive the leeward fleet into flight. The three victims would remain in the clutches of the attackers. Indeed, by this time the fourth and fifth rearmost ships might well have been engaged and surrounded by the attackers coming down *en masse* from windward. Clerk considered various methods of cutting off the enemy's rear ships, but not the two specially advocated by Howe: attacking in succession, each ship then taking station in the rear; and attacking by inverting the line, the first ship attacking the

enemy's rearmost, and each succeeding ship passing the ship ahead of it on the disengaged side.

It was possible that the fleet whose rear ships were in danger might react more quickly by wearing together before the attack was launched. In that case, the same ships, now the headmost of the enemy fleet in reverse order, could be heavily attacked. The leeward fleet might indeed extricate itself, though the windward fleet had still a chance to harry their new rear. Clerk considered the possible changes of wind, as well as partial breezes, and concluded that, once the three rearmost ships had been heavily engaged, even the most favourable wind would not enable the rest of their fleet to extricate them without becoming involved in a pell-mell.

An easy objection to all these schemes for cutting the

---

102 p123 (second edition, 1804)

Other ships come up when possible

Attack on the Three Rearmost

Three fastest ships engage three enemy rearmost

Enemy tacks in succession to relieve three rearmost

After tacking, has lee gage and is in danger of being raked

Enemy tacks together

Enemy in disorder, and in severe danger of being raked

enemy's line from leeward is that Clerk never allowed sufficiently for vigorous counter action by that part of the windward fleet which had not been cut off. In general, he was far too much inclined to dismiss them as out of the battle, demoralised, and making off after realising the difficulty of helping their friends. To this, however, he had an even easier reply, although he never made it specifically. If the ships of the windward fleet did succeed in intervening, the result would be a real pell-mell which is exactly what the attackers, assuming they were the British, most desired.

Clerk undertook a lengthy analysis of the amount of fire given and received in action fought by fleets passing each other on parallel and opposite courses. At very slow speeds, no individual ship could possibly fire or receive more than one broadside at or from any other individual ship. At higher speeds their guns would bear on each other for less than a minute and it might not be possible to reload before the next

ship in the opposing line came directly abeam.

Clerk's bold statement on cutting the enemy's line as a deliberate tactical procedure must not be confused with anything that took place at the Battle of the Saints, on which occasion the three movements were all 'accidental and unpremediated'. In Part IV, which was first published in 1804, Clerk reviewed the Battle of the Saints at great length. Rodney's accidental cutting of the French line was shown to have reaped all the advantages that Clerk had prophesied. He did not, however, suggest at any point, in either his narrative or observations, that either Rodney's or Affleck's or Gardner's cutting through the French line was in any way inspired by anything that he himself had written. In everything he wrote, Clerk was extremely modest, especially in view of his apparently heterodox teaching. It was his enthusiastic and less cautious followers who made exaggerated claims on his behalf.

The Battle of the Saints, 12 April 1782 (detail), by Nicholas Pocock (see pp 179–84). NMM

CHAPTER EIGHT

Tactics of the French Revolutionary War

THE stimulus of defeat, and an empire lost, ensured that tactical reform continued in the British fleet following the American War. Howe was at the Admiralty from 1783 to 1788, with a single break of eight months. His administration was by circumstances one of demobilisation and retrenchment, but tactics and signalling continued to develop, stimulated also by the feeling that unrest in France and the alarming progress of the Revolution were inimical to any real peace settlement. Reform was inevitable with such men at the head of the service as Howe himself, Hood, Alan Gardner, and Jervis.

On 18 May 1785 Vice-Admiral Mark Milbanke, then port admiral at Plymouth, issued a set of numerical signals with 145 signals listed.[1] On 27 June 1786, Commodore Alan Gardner, commanding on the Jamaica station, issued another. The signals Gardner devised were not, it must be admitted, of a very progressive kind. There were no battle signals at all, except that to engage. No signals for fleet formations were included. The day signals for private ships were on the old system, as also were the signals for calling officers. Fog signals were by a table, employing guns first to indicate the column, and then after a pause of two minutes, the horizontal line. Despite its limitations, however, this set of signals was a pioneer code by a man whose reputation had so far rested on his fighting qualities alone. On 16 October 1787, at Spithead, Lord Hood issued a set of ninety signals worked on a true numerical basis, but accompanied by a squared table apparently used to express the same numbers.[2]

Admiral Philip Patton's system, 1787

Not all British tactical progress was in the right direction. Philip Patton, a highly capable and experienced officer who had been constantly at sea since 1755, also produced a new system. He began his work well before 1784 (when his first draft appeared[3]) and his book was finally printed at his own expense in 1787. By that time, events had already passed him by; Howe had begun to move towards a true numerical system. Patton got no further than attempting a compromise between the old and new systems, presenting a scheme for operating both at once. In his letters to Sir Charles Middleton (now Lord Barham) advocating his own scheme of signals, and castigating those of Howe and his ex-flag captain, John Leveson Gower, Patton seems hardly to have understood current trends in tactical thinking, quite apart from his admission that he found it difficult to obtain access to their latest codes, or to those of Hood. Even in that age of free discussion, security was beginning to be an issue.

Middleton, whatever he might have said to Kempenfelt by way of encouragement, was a complete reactionary who knew nothing of tactics, and had never seen a battle in his life. In a splendidly blimpish effusion, with a veneer of apparent common sense, he asserted that seamanship was enough to win battles.

> The more I consider the situation of the fleet, the more I am convinced of the necessity of confining your movements to the old Fighting Instructions, and which are certainly better adapted to fleets who fight on an equality than the more modern craze. Simple movements with the assistance of good leading ships are within the reach of the meanest capacity: but when additional signals are given out without system or order, as I am afraid is the case with all those I have yet seen, it must only bewilder the judgment and increase the confusion . . . A facility for forming and tacking the line seems to comprehend the whole of what is necessary at present. . . All I fear at present is the confusion of our signals; but however bad they may prove, they will do no harm if left asleep, and however ignorant our commanders may be at setting out, their seamanship will give them great advantages in forming and keeping the line as soon as they have been practised so as to understand it.

Perhaps Jervis was not being too severe when he wrote with characteristic violence of 'that damned fellow Sir Charles Middleton . . . the utmost extent of his abilities having gone no further than . . . compiling with the aid of that dull dog Patton the most voluminous stupid code of signals that has been exhibited by the signalmongers of the present age'.

As time went on, Patton became more and more worried about his failure to obtain official recognition. He abandoned the idea of sending a printed copy of his book 'to every admiral upon the list', because of the expense, but seems to have distributed widely both the book and an eight-page pamphlet on naval signalling.[4] He toyed with the idea of seeking the patronage of his young ex-shipmate the Duke of Clarence, but foresaw difficulties.[5] Nothing came of a grandiose plan of his that the Admiralty should appoint a jury of flag officers to select the best signal system amongst the many projected.

Patton's book is a large and cumbersome folio weighing over 6lbs. It contains 352 numbered pages, many left blank, together with many unnumbered blank pages. It was fit for use only on a cabin table.

1 *NMM*, HOL/39 (formerly S(H) 39). The book is in manuscript and is unfinished.

2 *NMM*, HOL/39 (formerly S(H) 39); see also *NMM*, ADM/BP/7, Navy Board to Admiralty, 1 October 1787. Hood evidently requested signal colours with a printed set of ten flags, quite unlike those used for his signals of 16 October.

3 *Barham*, II, pp368-85

4 *A System of Signals combining the method commonly used in the British Navy of making the signal from fixed places of the Masts or Rigging with a Numerary Method By which the Flags are showed where they may be best seen. So as to gain the Advantages of both these Methods: at the same time, Securing to the Commander in Chief, a Certain proof, that his signals have been seen and understood, in every Ship under His Command, NMM*, Corbett and Tunstall Collections, TUN/38 and 43 (formerly S/P/Am/2 and 8) [Tunstall notes other copies in the Mead Collection and the National Maritime Museum, which have not been identified]; and *A Short Sketch of the use and importance of Naval Signals* (*NMM*, Tunstall Collection S/P/Am/3). There is a clear reference to the *Sketch* in *Barnham*, II, p375.

5 Prince William Henry, Duke of Clarence, was given or obtained a copy; *NMM*, Corbett Collection, TUN/43 (formerly S/P/Am/8), signed 'William Henry' and corrected in manuscript by the Prince's hand.

The system of day signals it contains depended on the use of ten flags, numbered 0 to 9, and four pendants, which allowed for fourteen groups of signals, up to a total of 925. Flags hoisted at the foretopmast head expressed units, at the main topmast, tens; and at the mizzen topmast, hundreds. Signalling by the old method was facilitated by having the two methods printed on the left and right openings of each double page. By using the three topmast heads and the mizzen peak, together with one or two of the four pendants hoisted either above or below, or both above and below, the flag in question, a large number of options was possible with a single flag. In the case of the groups signified by the pendants themselves, another, and even a third pendant, was hoisted below the indicative one. In all but three of the groups, Patton confined himself to forty-two signals, and in all but two groups there were seven signals on each page. With maddening eccentricity, however, he omitted numbers from his series, without any discernible reason. An outstanding virtue of the system, however, is that every ship in the fleet was envisaged as possessing all ten flags and pendants, which would permit two-way signalling, and repeating of signals.

Each group of signals was preceded by a set of general orders, which provided a useful and necessary codification of service practice, the majority of which presumably were already established. Group 4, for evolutions in the order of sailing and for bringing to, tacking and wearing, included the standard movements, and most of Kempenfelt's signals and instructions for the period 1779–81. Group 7, line of battle and evolutions, was thoroughly in tune with the Howe-Kempenfelt developments. The general form and tone of the signals were to give the admiral flexibility of control, and to ensure proper station keeping and tacking and wearing procedure by individual ships. There was no hint of evolutions for their own sake.

The night signalling system allowed for seventy signals, divided into six classes of ten signals, except for Class 3, which had twenty signals and was restricted to private ships signalling the admiral. The admiral could make his signals in one of three ways. Lights could be employed to indicate by their number the class of the signal, and by their arrangement the number of the signal within the class. The admiral would probably also use gun signals to indicate the number of the signal. A third possibility was that the admiral might not use lights at all, but would signal entirely by guns, one to six to indicate the class, followed after a pause by one to ten to indicate the number.

Every ship was required to acknowledge the admiral's signals by showing at the mizzen peak the correct number of lights for the class, and by firing musket shots to indicate the number of the signal in the class. If the admiral had made his signal entirely by guns it was to be answered similarly with musket shots. To make their own signals, private ships employed three lights in different arrangements to indicate the twenty signals of Class 3. They could also signal by means of guns for the first ten, and false fires for the second. The admiral acknowledged the signal by showing a single light at the ensign staff. There were only ten fog signals, made using guns and acknowledged by muskets.

By combining the old and new systems, Patton evidently pleased nobody, quite apart from the fact that the tide was now flowing strongly with the new. His night signals were confusing in themselves, quite apart from the mad-hatter explanations he provided. The tactical signals did little more than reproduce the progressive ideas of 1779–81.

'An Essay on Signals', 1788

By contrast with Patton's work, *An Essay on Signals by an Officer of the British Navy, 1788*, is a neatly bound volume measuring only 7 by 4½ in.[6] The author proposed a true numerical code using ten flags, three triangular flags, and a cornet. He provided 1051 signals, arranged in alphabetical order according to the first word of the signal. The normal procedure was to fly the flags one above another at the foremast, but for long-distant signalling the hundreds could be flown at the mizzenmast head, the tens at the main, and the units at the fore. A flag at the mizzen peak denoted thousands. Triangular flags were used for substitutes. The yellow cornet indicated that the number of the signal was to be read as a numeral. A really distinctive feature of the book was the device employed for night signalling. This was a square signal lantern with numbers printed on the four rectangular crown-glass windows. Gun signals were used for fog signalling. Up to six guns fired at different rates, from quick firing to six-minute intervals, made possible thirty-two different signals.

In the main the book is simply a signal book, without an instructional element. One of the best features is the very large number of interrogative signals, requiring some kind of numerical answer, which enabled the admiral to gather information from the fleet about numbers of sick and casualties, latitude and longitude, soundings, how many days supply of various stores on board, etc.

Unfortunately the author spoiled his effort by devising an additional set of thirty-two quite unsuitable flags to be used as single-flag battle signals. Supplementary flags for a single-flag signalling had already been introduced by Kempenfelt, but the object here was to provide special and distinctive flags to be used in battle only, and preferably with the admiral in a frigate. Their complex design and colouring made them hopelessly indistinct. Some were not merely confusing but bizarre: the sun rising from a strip of blue sea against a red background; red with a blue border and with a red cross on a yellow fly; yellow with a red horizontal stripe through the middle thickened near the fly and with a blue oval over all; red and with the Man in the Moon in yellow and white. Nevertheless, the book is a notable contribution to signalling literature, even if it never received encouragement and came too late to have practical value.

The National Maritime Museum possesses an ingenious experimental numerical signal system in manuscript, in which all the signals could be made by a hoist of two of the ten numerical flags, which were so designed that they could be flown upside down to double the number of options.[7] In 1788 Rear-Admiral the Hon John Leveson Gower, who had been Howe's first captain at the relief of Gibraltar in 1782, and had come to command of the Channel Squadron, issued another set of numerical signals. However, his system was not at all progressive as it depended upon an eight-sided table.[8]

Lord Howe and Sir Roger Curtis, 1788–9

The close association between Howe and Sir Roger Curtis dated from the American War, when Curtis was appointed to be Howe's second captain on the North American Station. Subsequently Curtis had played a distinguished part in the defence of Gibraltar, and he was destined to be Howe's First

---

6 *NMM*, Corbett Collection, TUN/37 (formerly S/P/Am/1) with the book plate of Prince William Henry, Duke of Clarence and William IV. The plates are marked 'Published 25 Aug 1787 [sic] by S Hooper.'

7 Source unknown

8 *MOD*, Ec/5

Vice-Admiral Sir Roger Curtis (1746–1816), by Mather Brown. NMM

Captain at the Glorious First of June. There is no evidence to suggest that, apart from being a good chief of staff, he had any wide experience of tactics or had made any serious study of signalling.

Howe resigned from the Admiralty for the second time in July 1788, and sometime during that year Curtis devised a new set of signals and instructions, of which a copy in the form of a general index to the signals, initiated by Curtis and dated 1789, still survives.[9] This book is frequently quoted in a set of handwritten discussion notes dealing with draft signals and instructions, written jointly by Howe and Curtis in September 1789, which throws much light on the Howe-Curtis partnership.[10] There are also frequent references to Howe's signals and instructions of 1782. In many cases, criticisms and objections appear in the same hand as that of the originating item, suggesting that these had been copied by the author of the draft from a paper already received from the other partner.

It is impossible to imagine Howe at any moment in his career, and certainly not at this particular moment, proposing anything as retrogressive as Curtis's 'Admiral's Signals', which employed a table of sixteen flags, along with ten

supplementaries, presumably to signal numbers. As usual, the numbers assigned to the tabular flags had only an indirect relationship to the number of the signal. The fact that they had any kind of relationship at all makes them all the more confusing. For example, flag hoist 1 over 1 equals signal number one, but flag hoist 5 over 9 equals signal number 73. The same flags could also be used to provide single-flag signals of importance, such as number 3, 'to alter course together', and number 7, 'to engage the enemy'.

The notes provide a wealth of information about the progress of Howe's and Curtis's thoughts about tactics. For example, Curtis proposed a signal, no 70, which would require the fleet 'to cannonade the enemy's ships within reach, aiming at their masts – the guns to be pointed by the most expert gunners'. Howe queried it, wondering whether 'under *any circumstances* this appointment in a signal book would be advisable? And, if not productive of an *evident* effect, it might not tend to relax the Ardor of the men?'

Curtis dealt very fully with night signals, using Howe's method. Howe had always stuck to his original system of 1776, classifying signals according to the number of lights shown, followed by 'how disposed', (ie, in a square or triangle, etc.), followed by where shown in the ship, and finally by guns and false fires if any. In fact, however, Curtis provided no indication of where the signals were to be shown. There are numerous references in the notes to Howe's signal book of 1782 and Curtis's of 1788.

These documents show that Curtis had little appreciation of what Howe was aiming at. Far from feeding him with new ideas, he was merely offering outmoded, pedestrian devices. In October 1789 Howe wrote to an unknown correspondent that 'the looseness of our present system of tactics in the navy, if any system may be properly said to exist, is such that I cannot say I have quite made up my mind upon the plan that I would recommend for publication'.[11] The following year, however, Howe was given command of the Channel Fleet formed as a result of the Nootka Sound dispute with Spain. In responding to the occasion, Howe grasped the nettle and issued to the fleet a true numercial signals system.

Howe's numerical signals, 1790

The Channel Fleet was an impressive force of thirty-one ships-of-the-line divided into three squadrons, each of two divisions. Admiral Samuel Barrington, with Jervis, commanded the van, Howe with both Lord Hood and Sir William Hotham commanded the centre, and Sir Alexander Hood with Sir Richard Bickerton commanded the rear. The order of sailing was still in three parallel columns in the same order as in line of battle, but with Howe supernumerary and leading the starboard or weathermost column.[12] The fleet's tactical efficiency failed, again, to reach Howe's exacting standards.

The *Signal Book for the Ships of War*, issued to the Channel Fleet, marks the final development of Howe's signalling and tactical system.[13] It was the long-delayed masterpiece for which the more progressive British sea officers had been waiting. It was re-issued by Howe himself as well as by Hood, Jervis and others, and remained the standard form for the navy until the amalgamation of his signals and instructions in the first official Admiralty signal book, which was issued in 1799, and remained in effect until 1816.

Twelve special-purpose flags were supplied, three of which could be inverted, and these could be employed for a number of miscellaneous single-flag signals. Otherwise, all inhibitions were cast away, and a true numerical system established. Five columns were provided in the signal book for numeral signals. In the first was entered, in writing, the

9 *NMM*, Corbett Collection, TUN/7 (formerly S/MS/Am/7), a marbled covered notebook 7¼in × 4¼in.

10 Seven double foolscap sheets and four smaller ones, some headed 'Sr R C' and one dated 'Sept 1789'; *NMM*, Tunstall Collection, TUN/72 (formerly HC/MS, once in Lord Sligo's collection of Howe Papers)

11 Barrow, pp142-3

12 MS line of battle and order of sailing, signed by Howe, 10 July 1790; *NMM*, Tunstall Collection, TUN/19 (formerly S/MS/R/4); *NMM*, DUC/1.2; and the former *RUSI* 146, which may be in the *MOD*, but has not been identified. Tunstall identified *NMM*, 37 MS 1792, as another undated copy, but it cannot be traced. *NMM*, DUC/2 (formerly Sm/91) has order of sailing signed by Howe 19 October 1790; squadrons to be ½ mile apart, divisions 'nearly at the same distance;' ships three cables apart in fair weather.

13 *NMM*, MKH/A/2/n/4, signed by Howe, *Queen Charlotte* at Spithead, 25 July 1790, and issued to Captain Keith Stewart, *Formidable*

number of the signal, the second gave the 'purporte', the arrangement of the signals being determined by the alphabetical order of the words in this column; in the third column was entered a brief 'signification'; in the fourth and fifth columns the page in the instructions, and the relevant article number were noted. All signal numbers and many signals were inserted in handwriting, especially the battle signals, possibly for security reasons. There were 160 numbered signal spaces in all, with many blanks.

Howe's introduction of numerical signals caused the Board of Admiralty, on 26 June, to request the Navy Board to supply flags and other necessary signal colours, provided they should judge them 'fit to be adopted'. It is interesting that the practicality of the design was referred to the expert judgment of the Navy Board. On 28 June the Navy Board gave their approval, and on 2 July they supplied the Admiralty with a complete list.[14]

Important and otherwise new signals in Howe's 1790 'Signal Book for Ships of War'

Article 19 (MS) 'Ships of the fleet are at liberty to fire upon the enemy in passing them, though not prepared to bring them to a general action immediately.'

Article 20 (MS) 'Particular ships or divisions of the fleet to attack or harass the rear of the enemy or such part thereof as the detached ships may be of competent force to engage for giving opportunity to bring on a genereal action.'

Article 22 (MS) 'To quit and withdraw men from ships captured after having destroyed or disabled them so that the enemy cannot carry them off if time permits, and to join the fleet.'

Article 40 (MS) For passing through the enemy's line in either direction [as in Signal 75 of 1782].[15]

Article 41 (Printed) 'The ships to take suitable stations for their mutual support, and engage the enemy, as arriving up with them in succession.'

Article 42 (MS) 'Each ship of the fleet to steer for independently of each other the ships opposed in situation to them in the enemy's line.'

Admiralty Flags of 1790 designed by Lord Howe
1 Red
2 White with narrow red border
3 Blue, white, blue vertical
4 Yellow, with narrow black edging top and bottom
5 Quarterly white and red
6 White and blue diagonally divided
7 Blue with a yellow saltire
8 Yellow and blue vertical
9 Blue, white and red, horizontal
10 Blue with a white centre
11 White
12 White with a blue cross
13 Red with a white cross
14 Chequered yellow and blue
15 Union
16 Red and white horizontally divided
17 Blue and yellow vertically divided
18 Blue and white striped horizontally

19 Blue cornette
20 White and black divided horizontally
21 Yellow
Flag sizes: large ships 12ft broad by 14ft long; frigates 10ft broad by 12ft long.
Pendants sizes: large ships 8ft at staff, 36ft long and 2⅔ft at end of fly; frigates 6ft at staff, 27ft long and 2ft at end of fly.[16]

Howe was in a dilemma when it came to signals from private ships. Despite all his urbanity and his willingness to seek the opinion of his subordinates on grave tactical matters, he was still an autocrat, for whom it was unthinkable that private ships should have the full set of ten flags, plus the substitute flag and the hundred pendants. As a result, sound principles adhered to with such praiseworthy persistence were cast aside and the private ships given an eight-flag table providing sixty-eight numbers, together with substitute and annulling flags, a pendant signifying numbers only, and an interrogatory pendant as well. To make themselves understood the private ships had a parallel set of twelve flags and two pendants. Of these, only one was contained in the

Admiral Lord Howe's signature, dated 25 July 1790, at end of his Signal Book for the Ships of War, *issued to Captain Keith Stewart. NMM*

Nº of Lights.	How disposed	Place	Guns	False Fires.
5	— equal Height —	Where most easily seen	— / — if	requisite —
5	Ditto	Ditto	2 quick	
5		{ Where most visible to the Admiral, and as little as possible to the strange Ships.		

14 *NMM*, ADM/B.P./10 (1790)

15 Sir Julian Corbett was under the impression that this form of the signal was 'for an entirely new manoeuvre' (*Signals & Instructions*, p255) and based a good deal of his argument about Howe's tactical ideas on this erroneous supposition.

16 *NMM*, ADM/BP/10 (1790)

admiral's numerical series, the rest being drawn from the special-purpose flags. At the head of the table it was stated that it might 'be used by the admiral occasionally'. The night signals – both for the admiral and for the private ships – were much the same as those in the signal book of 1782. In view of these complications, it seems unlikely that Howe was dissuaded from allowing private ships the numerical signals because the Navy Board refused to issue more flags. The choice seems to have been his own.

For many years there was some doubt as to the actual authorship of the accompanying 'Instructions . . . Explanatory of, and Relative to . . . the Signal-Book herewith delivered', issued by Howe in 1790. Luckily there is a copy in the Mead Collection, signed 'Howe, 9 October 1790, *Queen Charlotte* at Spithead'.[17] This was after the fleet had finished its cruise but was prepared if necessary to put to sea again. In layout, general content and actual text, it is practically a reprint of Howe's instructions of 1782, and thus of 1776 as well. So, once more, there is evidence that the system of tactics devised for the Royal Navy, under which all the great victories of the Revolutionary and Napoleonic Wars were won, was already promulgated before the American War of Independence had properly begun.

The Instructions are divided up in exactly the same way as before: 'General Instructions'; 'Instructions . . . Preparatory to, and in Action with the Enemy'; fog; sailing by night; battle by night. There are twenty-four battle instructions in place of the previous twenty-two. Numbers 23 and 24 are for exercising particular squadrons or the fleet – additions which would have been useful in 1790 no doubt. There is also number 25 added in writing, to the effect that, if a detachment of the fleet was ordered to attack part of the enemy fleet but was so far ahead that it risked being 'overpressed by a superior force', it should confine itself to 'retarding the progress of the enemy' until the admiral arrived with the rest of the fleet, and brought on a general action. There seems to be a hint here of the 'advance squadron'. The fog, night sailing and night battle instructions are the same as before.

Howe also reissued his printed 'Additional Instructions Respecting the Conduct of the Fleet Preparatory to and in Action with the Enemy', dating from 1 July 1777, and almost identical.[18] It seems curious that by this stage he should not have thought of incorporating his additionals into his main book of instructions. No doubt keeping them separate was the way of least resistance, the Royal Navy having been long accustomed to additionals of every kind. In view of the amount of time he had been devoting to tactics and signalling during the previous year, it is hardly likely that his action, or rather inaction, was due to laziness. A more likely reason is that he was already in the process of producing a complete volume, entitled 'Additional Instructions for the Ships of War', which appeared in 1793. In the meantime, he was experimenting with the separate sections of this volume by issuing them piecemeal.

On 10 July 1790, while the fleet was still preparing at Spithead, Howe issued a 'Separate [printed] Instruction respecting the signal with a white flag with a blue cross, page 9 of the Signal Book.'[19] This is a preparative signal (Article 1) which ushers in a highly ambitious series of instructions for night battle. Article 2 states:

If the preparation so required refers to a fleet or squadron of the enemy present, either greater in force and number of ships than the British fleet, or which the Admiral may for other reasons mean to defer bringing to action until at, or after, the close of day, when he judges it may be done with greater advantage in two or more divisions acting independent of each other; and it then happens that the circumstances of the case do not admit of a more leisurely communication of his plan; he will signify the form (if to be different from any before delivered) in which he would have the ships drawn up for conducting the attack after dark; by showing the flags expressive of the different divisions, in succession at proper intervals and putting abroad with each divisional flag the signal pendants distinguishing the several ships of which he would have (in such succession of place) each of the two or more divisions to consist. The commanders and captains are thereupon to take their stations accordingly; and to keep their ships in readiness for proceeding to engage the enemy in the order so pointed out when the proper signal is made for engaging the enemy at or after the close of day. But the divisions of the fleet, for the purpose of this signal, will be to remain as in the established form of battle expressed if no such change is to be made in the disposition thereof as in this article provided.

This article is quoted in full not so much as a further example of Howe's prose style, which unfortunately did not improve with age, but as evidence of the lengths to which he was prepared to go in order to ensure a satisfactory form of attack under hazardous conditions. Article 3 seems to suggest a ship-to-ship engagement, though Sir Julian Corbett saw in it Howe's intention, like Nelson's at Trafalgar, to contain the enemy with his own centre division.

On the same day, 10 July, Howe issued two further documents. Of the first, 'Instruction respecting the formation of the Order of Battle', the main purpose of the most important article, number 1, was to insure correct leading by the head or tail ships of the respective squadrons when the fleet tacked or changed formation.[20] 'Of the Order of Sailing', consisted of ten articles and was his first full-dress attempt to regularise the cruising formations adopted by the fleet at sea.[21] It was the original version of what appeared in his Additional Instructions three years later. Article 1 reads: 'When the fleet is formed in Order of Sailing, in three columns or parallel lines, it is meant that the Commander in the Second Post should keep with his Squadron or Column, to Starboard when Sailing from the Wind, and to Windward

17 Mead Collection, *NMM*, TUN/127 (formerly Mead/Ins/4) Another copy in the Mead Collection, *NMM*, TUN/128 (formerly Mead/Ins/5), is inscribed 'Rear-Admiral Kingsmill' as the cover label.

18 *NMM*, Sp/80, signed Howe 10 July 1790; see also *Signals & Instructions*, pp319–26, reprinted from the Bridport Papers, *BM*, Add Ms 35194. Compare pp108–16, Howe's additionals of 1777. In the 1790 reissue the last four paragraphs of Article 9 are in manuscript.

19 Issued to Vice-Admiral Sir Alexander Hood (Lord Bridport) and printed, in *Signals & Instructions*, pp327–31

20 *NMM*, DUC/1/3; issued to Captain John Duckworth, *Bombay Castle*. The same instructions, issued to Sir Alexander Hood, are printed in *Signals & Instructions*, pp331–3

21 *NMM*, DUC/1/4; issued to Captain Duckworth

Howe's Order of Sailing

of the Centre Column by the Wind, on both Tacks.' Article 2 established that the columns were to be one and a half to two miles apart, with the ships at about three cables distance, and that the fleet would steer no nearer than a point from the wind to enable the most leewardly ships to keep their stations. Article 3 directed the fleet to keep station on the leading ship of the weather column. The admiral would himself usually take station about three cables ahead of the leading ship of the weather or starboard column. In Article 6 captains were warned not to fall out of station even if the ship immediately ahead of them did so. Article 8 ordered that when the fleet sailed in two columns the same regulations were to apply as when sailing in three. In Article 9 frigates, fireships and smaller vessels were ordered to keep to windward when the fleet was sailing by the wind, and to take station outward when the fleet was sailing large. And in Article 10 it was ordered that leading ships which became the sternmost, and vice versa, as a result of tacking, wearing or altering course, were 'to continue in this performance of the services incident to such changes in their situation, until replaced by some subsequent evolution in appointment'.

On 24 August 1790, a week after sailing from Torbay, Howe issued a 'Memorandum for the more particular explanation of the Admiral's intentions respecting the order of sailing'.[22] It was an exhortatory document on station keeping, drawn partly from Articles 6 and 7 of his Order of Sailing. In particular, it was noted that ships were often seen to fall to leeward for want of sail aft. The summer's cruise proved once again the tactical inefficiency of the British fleet, at least by Howe's own standards.

The fleet returned to Spithead in the middle of September and, in anticipation of an autumn cruise, Howe issued on 19 October a new order of sailing and a new order of battle.[23] The order of sailing was for thirty-seven ships divided into three squadrons each of two divisions. This time the admiral was usually to lead the starboard, or weathermost, column. In the order of battle, Barrington commanded the van and Sir Alexander Hood the rear, and the admiral in the *Queen Charlotte* was supernumerary to the centre. On the same day Howe issued two further instructions, based no doubt on the experiences of the recent cruise. The first, entitled 'Instruction respecting the disposition of the fleet in order of Sailing by divisions, denoted by the signal flag, half black, half white', was very long and confused, no clear distinction being drawn between divisions and squadrons, if indeed a distinction was intended.[24] The second instruction, headed 'Observation', sought to define the respective functions and responsibilities of squadron and division flag officers.[25]

Lord Hood and John McArthur, 1791–3

Howe was ordered to strike his flag in December 1790, once the dispute with Spain was over. Such was his zeal that, although he was already aged sixty-four and far senior to all

flag officers still able or willing to continue facing the 'rigours of the service', he described this event as 'my professional annihilation'. The initiative for tactical reform passed to Lord Hood, who for the next two years had command of the Channel Fleet, which continued on a war footing because of the 'Russian Armament' scare. A plan of Hood's fleet while lying at Spithead in two lines, shows, in the southern line, three divisions each of six ships commanded by Leveson Gower, Admiral Samuel Granston Goodall and Cosby; and in the northern line also three divisions, commanded by Hotham, Vice-Admiral Jonathan Faulknor and Vice-Admiral Sir Richard King. Hood in the *Victory* is sixth in the centre division of the northern line.[26]

Hood's position as tactician and signaller is difficult to determine. He was certainly a bitter critic of Rodney for his self-seeking at St Eustatius, and for not pursuing the French after the Saints. Yet he seems to have accepted Rodney's general ideas of tactics and fleet discipline, though he was quick to insert a signal for breaking the line once he had the West Indies fleet under his own command. Like Howe, he was a stickler for fleet discipline and station keeping. It is doubtful, however, if he was really interested in advanced tactical ideas.

In signalling, he was much influenced by his new secretary, John McArthur, who since 1782 had been working on the signal system he had presented to Admiral Robert Digby for use on the North American station.[27] The code which he eventually produced and submitted to the Earl of Chatham, then First Lord of the Admiralty, in 1790, was a much more ambitious but far less satisfactory affair.[28] It was in fact an adaptation of the French scheme, earlier adopted by Kempenfelt. Twelve flags were used and though no actual list is given, they were referred to as parti-coloured and quartered, white apparently being always one of the colours. The flags 'in their natural order' have the red, blue and yellow colours uppermost or next the mast. Each flag, however, could be inverted or reversed, which seems to imply reversing the flags divided vertically as they cannot be changed by being turned upside-down. The treatment of the quartered flags is uncertain.

Four tables of signals were provided. In Table 1 each of the twelve flags is used singly, after which the twelve flags are used in pairs, number 1 at the top, and each of the twelve under it in turn, followed by number 2 with the rest in turn under it as before. This produces 144 signals which, with the twelve used singly, make 156 in all. Table 2 has the top flag each time in its 'natural' form and the bottom flag inverted or reversed. No single flags are used, so that the total is 144. Table 3 has the flags singly, reversed, making twelve signals, and they remain reversed at the top when the flags are used in pairs, thus providing 156 signals. Table 4 has all the flags reversed and no singles; total 144. The grand total is thus 600.

Highly elaborate security measures were offered for changing the flags monthly, weekly or even daily, by combining the Dominical Letters and Golden Numbers with Lunar and Solar Cycles, 'changes might be made every lunar month for thousands of years'. Since the success of the scheme depended on the accurate use of the twelve flags and avoiding mistakes in reversing and inverting, changes in the flag values would have been most inadvisable.

When we come to the signals themselves, however, the confusion inherent in the scheme becomes fully apparent. Instead of the four tables being used to group similar types of signals, each table includes a good deal of material common to the rest; signals for attacking and engaging appear in all four tables with the reference phrase, 'Fighting Evaluations' at the bottom of the page. Confusion also exists in 'Line of Battle',

22 *NMM*, DUC/1/6; issued to Captain Duckworth

23 19 October, 1790; *NMM*, DUC/1/7 & 8; issued to Captain Duckworth

24 *Signals & Instructions*, pp333 5; issued to Sir Alexander Hood

25 *Signals & Instructions*, pp335; issued to Sir Alexander Hood. It seems possible that this Observation was only issued to flag officers and not to Captains.

26 Cecil King, article in *The Mariner's Mirror*, 24 (1938), pp176 83

27 *MOD*, NM/76 (formerly *RUSI* 76), and *NMM*, KEI/S/9 (formerly Sp/9)

28 Manuscript presentation copy to Lord Chatham, *BM*, Add Ms 38812; and manuscript presentation copy to the Duke of Clarence, *MOD*, Ec/143

which often includes what are clearly battle signals. There is little of tactical value that is new. Number 89 'For the van division to force through the enemy, break their line and double upon them', is a repeat of his signal of 1782. Private ships' signals are included in the main body.

It had apparently been Hood's intention to give McArthur's signals a trial during 1791, but the Admiralty refused to accept them. Whatever the real reasons may have been, the decision seems sound. McArthur therefore undertook to rearrange Howe's signals of 1790 for immediate use by Hood, which, in theory, proved acceptable to all parties. The signals were re-arranged alphabetically and were more fully indexed, but the result was too complicated, and the book seemed likely to defeat its most obvious purpose – simplicity in handling.[29] In the inflated and egotistical style which has damaged his reputation, McArthur wrote in his *Memorial*,

> Finding that some scruples of delicacy intervened in the adoption of any new plan of signals which would supersede that of Earl Howe's numerary code, the memorialist, therefore, made a new arrangement of his Lordship's day signals by simplifying the form of the Indexes . . . and engrafting many new ideas and instructions of his own. In this new arrangement plates or engraved plans are originally introduced by the memorialist explanatory of the principal evolutions, together with instructions illustrative of them; and the whole, with a paper containing the memorialist's observations and reasons for so making material alterations, were transmitted by the late Admiral Sir Hyde Parker to Lord Howe, who was pleased in the most flattering manner to approve of the alterations made *in toto*.[30]

McArthur also compiled new night signals from some 'undigested manuscripts' left by Kempenfelt. 'These day and night signals were approved by the Lords Commissioners of the Admiralty and were ordered to be printed under the inspection and superintendence of the memorialist, and were accordingly issued for the first time to the ships belonging to Lord Hood's squadron.' If his claim to have introduced the printing of diagrammatic plates of the standard sailing and tactical evolutions is correct, his contribution to the literary presentation of tactics is important.[31]

The 'Instructions and Standing Orders with Separate and Additional Instructions . . .', which Hood issued to the Channel Fleet in 1791, were firmly based on Howe. All the items in Howe's Additional Instructions appear, except the 'Instructions respecting the Disposition of the Fleet in Order of Sailing by Divisions . . .' and the unimportant rendezvous item. The Order of Sailing, however, is rearranged and cut down to eleven articles (the last in handwriting) with an added Memorandum, incorporating some of the material

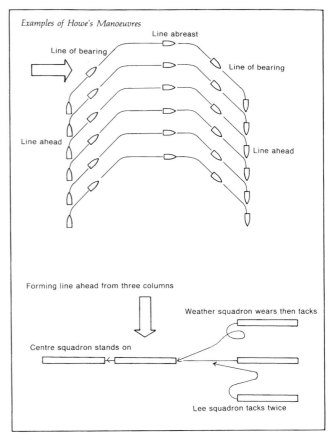

Examples of Howe's Manoeuvres

29 Manuscript of 'Lord Howe's Day and Night Signals classed under General Heads, Alphabetically sub-divided and adapted for a Commander-in-Chief's ship. The whole arranged for Lord Hood in the Russian Armament, summer 1791. By John McArthur, Secretary', (formerly *RUSI* 133, present location may be in *MOD*, but the piece has not been found).

30 John McArthur's *Memorial* to the Admiralty, 9 November 1807

31 *PRO*, Admiralty Papers, Secretary's Department, Promiscuous M.4 (notes by Sir Julian Corbett). Some confusion about the use of signals at this time is due to the reliance placed by Corbett on the manuscript private signal book made by Lieutenant John Walsh of the *Marlborough* (formerly *RUSI* 33 but present location unknown). The scheme of painting of this book is that each flag is at the head of a separate page to represent the top flag of a two-flag hoist. The ten flags are then painted below to indicate the bottom flag of the hoist. Pendants are added for the hundreds. The flags are those of Howe in 1790.

32 *NMM*, Mead Collection, TUN/128 (formerly Mead/Ins/5), issued '*Victory* at Spithead, 13th June 1791 to Captain Robert Kingsmill, the *Duke*', signed 'Hood' and countersigned by McArthur; *NMM*, MKH/4/2/n/5 is another copy issued on the same day.

from the missing articles. In addition, Howe's Standing Orders of 1776 are reprinted as the first item in the book.[32] There are the same twenty-two battle instructions with the extra two already printed by Howe in 1790 and the twenty-fifth added in handwriting. In view of the great tactical ferment stirred up by the American War it seems extraordinary that Howe's early tactical work should have remained pre-eminent. In the field of signals it was, perhaps, less surprising; he himself had made the crucial change in 1790 which took the British service ahead of the French, and the lay-out and general arrangements of the signal book remained much the same as for 1776.

The following year, 1792, Hood had command of an evolution squadron which he exercised by signal during July and August, sometimes keeping his flagship out of the line. Starting with fifteen ships disposed in two and then in three columns, he practised them in forming line ahead on the lee column by the weather column proceeding to leeward in line of bearing, thus taking station ahead. From line ahead, with the wind west, they altered course together four points to starboard, thus coming into line of bearing, and a further alteration together of four points to starboard brought them into line abreast; and, with a further and similar alteration, they were in line of bearing again on the starboard tack. With the wind again west and the fleet in line ahead, they tacked together into line of bearing, altered course together four points to larboard, and then hauled to the wind together on the starboard tack. With the fleet in three columns in line of bearing, they formed line-ahead, the leeward column taking the lead. With the wind WNW, and the fleet in order of sailing in three columns slightly *en echelon* on the larboard tack, line ahead was formed to leeward by the two weathermost columns wearing to starboard, and then coming to the wind while the lee column passed ahead of them to lead the fleet. A whole day was spent in wearing in succession, the sternmost ship first, an evolution likely to leave large gaps in the newly formed line unless carefully executed. It was necessary to begin the evolution with the rear ships if the line was well closed up, as otherwise there was a risk of collision.

From the order of sailing in three columns, line-ahead was formed on the centre column, the weathermost column wearing and then tacking and the lee column tacking and re-tacking so as to get into their respective stations astern; this took forty-three minutes. With the fleet in line-ahead the line was inverted from van to rear by the leading ship shortening sail and the second ship making all sail and passing to windward, then herself shortening sail when at a proper distance ahead, the remainder running ahead similarly in succession. This took an hour and forty-five minutes. With the fleet again in three columns, line-ahead was formed on the weathermost column by the other two tacking and then retacking on reaching their position astern. With the fleet sailing in two columns on the starboard tack the larboard (and leeward) column tacked in succession, rear first, and then retacked, so as to form a larboard line of bearing. The columns would then be at an obtuse angle and ready to form the order of retreat, which apparently took only twelve minutes.[33]

None of those evolutions was very difficult for ships with good officers who were trained to work as a fleet. They were also severely practical as a means of enabling the admiral to form his fleet as advantageously as possible for an attack, either to get the wind of the enemy or to station his stronger or weaker ships at one or other end of the line. The timing of performance is of little value as a gauge of competence without an exact knowledge of the prevailing weather conditions and tidal currents. Nevertheless, they provide a valuable and possibly unique record of the kind of training attempted in time of peace.

On 7 May 1793, a fortnight before sailing for the Mediterranean, Hood, then at Spithead, issued his *Signal Book for the Ships of War*, with day and fog (but not night) signals, countersigned by McArthur.[34] This book is labelled on the cover 'New Arrangement', indicating that it was a variant of Howe's 1790 book, now regarded as standard for the whole navy. The main difference is that it is thumb-indexed for the signal numbers, which dispensed with the general-purpose page headings. More important is the set of eight diagrammatic plates illustrating the meaning of particular signals, just as claimed by McArthur in his *Memorial* (see above). It seems certain, therefore, that McArthur was actually responsible for this useful innovation, which ensured that captains had a standard reference for the practical interpretation of particular signals. Where necessary, the diagrams include a scale of cable lengths. Of course, the idea was taken from the French, but its value to contemporary British naval thought is proved by the incorporation of this set of plates in the Admiralty book of 1799.

On the same day, Hood issued his accompanying book of instructions.[35] Its aim was to consolidate Howe's Instructions and Additional Instructions, together with new material, in a single volume.[36] The arrangement must be McArthur's: it starts with a good index to all the articles and contains Howe's 'General Instructions', 'Separate Instruction . . . close of day' (Additional), 'Formation of the Order of Battle' (Additional) and fog instructions; it ends with a complete reprint of Howe's original Standing Order, issued on the North American Station in 1776. There are, however, three items which characterise the book as a Hood-McArthur production. The 'Order of Sailing' of seventeen articles is not the same as Howe's, though it incorporates a good deal of his material. To some extent the articles overlap the General Instructions but at least the wording is more precise than some of Howe's vaguer drafting. Article 8, which warned that when tacking in succession captains were not to be

misled if the captain ahead failed to put his ship in stays, is virtually the same as Article 9, 'ships not to be thrown out of their station by the inattention of their seconds ahead and astern' (marginal summary). Article 17 covers much the same ground as Howe's Additional 'Instruction respecting the disposition of the Fleet in Order of Sailing by Divisions', which is omitted as a separate item.

A clear innovation is the 'Instructions for the Conduct of the Fleet in the Execution of the Principal Movements of the Evolutions'. This Howe-like heading covers detailed references to the diagrammatic plates in the Signal Book, and provides a direct link between the two volumes, in addition to signal references which could be inserted against each article in the Instructions.

Another innovation is the form given to the thirty-seven 'Instructions . . . in Action, with the Enemy'. Howe's original twenty-six instructions and nine additional instructions were amalgamated with a certain amount of rearrangement and rewriting, especially towards the end.

Hood also issued on the same day *Night Signals & Instructions*,[37] which is by far the most interesting of the three volumes. It combines Howe's night signals of 1790 with a shortened form of his 1790 night instructions, extracted from his main instructions, and prints them in a single volume. Howe's principle of signals in one volume and instructions in another, with an additional volume, is abandoned in favour of a single volume incorporating all signals and instructions releating to the night. The fact that in 1799 the Admiralty not only applied this principle to the night signals and instructions but to the day and fog signals as well, shows how far-reaching was the influence of the Hood-McArthur partnership.

On Christmas Day 1793, when the allied naval forces were evacuating Toulon, Hood reissued his three volumes without alteration.[38] These seem to have continued in use after his recall,[39] and were certainly used by his successor's successor, Sir John Jervis.[40]

McArthur's contribution to all this has never been explained, but it is reasonable to assume that he influenced Hood very considerably. As admiral's secretary, he countersigned all copies of signals and instructions issued to the fleet over Hood's signature, so that these matters must have been

33 Formerly *RUSI* 154a; present location may be the *MOD*, but it has not been identified. *Signals & Instructions*, p68n, gives details including the times for certain evolutions, as taken from McArthur's plans.

34 *NMM*, SIG/B/64 (formerly Sp/17), issued to Captain Keith Elphinstone [the future Lord Keith], *Robust*. Admiral Holland, apart from confusing his footnote references, seems to have thought this to be a Howe book; see *Development of Signalling*, p13

35 *NMM*, KEI/S/10(c) (formerly Sp/19), also issued to Captain Elphinstone.

36 Although the first dated copy of Howe's Additionals is 5 July 1793, (*NMM*, Tunstall Collection, TUN/109 (formerly FI/3)), they must previously have appeared as individual items.

37 *NMM*, SIG/B/63 (formerly Sp/18), also issued to Captain Elphinstone. At the time it was first discovered, Corbett regarded this copy as unique (see manuscript notes inserted in cover pocket). There is now a duplicate copy, *NMM*, KEI/S/10(b) (formerly Sp/18 (1)).

38 *NMM*, KEI/S/11(a) (formerly Sp/20), a day and fog signals labelled 'New Arrangement'; KEI/S/11(b) (formerly Sp/21), night signals and instructions; KEI/S/11(c) (formerly Sp/22), instructions identical to KEI/S/10(c) (formerly Sp/19) of 7 May 1793, but a different printing.

39 *NMM*, DUN/20 (formerly S(D) 20), Night Signals & Instructions issued by Sir William Hotham to Captain Bazely, *Blenheim*, 18 March 1795.

40 *NMM*, Mead Collection, SIG/B/68, Signal Book for Ships of War, Day and Fog, 1795, issued to Robert Calder by Admiral John Jervis, Gibraltar, 19 November 1795, copy; and *NMM*, SIG/B/68 (formerly Sp/133).

textbook. From a contemporary British point of view it would have been considered too long and repetitive, even by the leisurely and discursive standards of Howe. It began with a definition of tactics, an explanation of sea terms, a description of ship and fleet organisation and definitions of the various orders of sailing, particularly in three columns. To the usual list of advantages and disadvantages associated with the windward and leeward positions, a new one was added. Ships in the windward position, if they had the misfortune of having a mast go by the board, would be in danger of their guns setting fire to the sails, which would have fallen to leeward. A section on 'Principes pour la Chasse' was very much in the style of Hoste. The second paragraph of a section on 'Mouvemens de Guerre' was devoted to avoiding combat. On the other hand, far more space was given to boarding, described as 'une action de hardiesse et de vigueur', than might have been expected. Cutting the enemy line was described as 'très delicate et hardie, et ne peut être entreprise que par un Général consommé dans le métier, et qui a sous ses ordres une Armée très-exercée et leste dans ses mouvemens'. The signal technique was a straightforward variant of the tabular system, using a numerical table which could represent a double set of any nineteen flags.

In order to judge the practical value of this book, it is necessary to understand its background. In comparison with Morogues's book, which must have been rejected from the start as quite unworkable, Tactique et Signaux is a monument both of practicability and common sense. All ordinary necessities were provided for in a simple and efficient manner. Signals could be made by reference to the numerical table and could be read in the same way, though the table is printed in comparatively small type, – the squares containing three figures measuring only 7mm by 7mm. Real doubts, however, arise from the feeling, admittedly prompted by hindsight, that many of the tactical signals were never likely to be used, and perhaps were never even expected to be used. In the hands of a vigorous admiral, determined to defeat the enemy, they could easily have proved their superiority over those in the contemporary British signal book; but the situation, as we know, was very different. The signals inherited the elaboration bequeathed to the French navy by Hoste. The subtleties of attack postulated as 'très delicate' as well as 'hardie', provided for over-direction by the general, and encouraged him in this course. If French tactics were a reflection of the book, the book also reflects French tactical thinking, the kind of thinking which had often led to success but never to decisive victory.

When the Revolutionary war began, the French had one great asset, a true fighting spirit, as shown by their unflinching acceptance of Howe's attack on the First of June. This spirit was frittered away as time went on, by admirals who failed to draw the conclusion – obvious enough to the soldiers – that the quadrilles and pirouettes of the eighteenth century were no longer what was required. Of course the sea service was not able to benefit from the concept of *levée en masse*. Nevertheless, the Revolution offered a real chance to revise the whole system of signals in the light of pressing naval requirements. By tying themselves to the tactics of the *ancien régime* without possessing the technical skill to exploit them properly, the French found themselves at a serious disadvantage when facing the British, now tutored by Howe's reforms. By continuing to overload many of the signals with so much instructional matter, they failed to differentiate sharply enough between orders and explanations. By failing to adopt the true numerical system of signals, even after careful consideration, the French navy placed its commanders in a situation of serious tactical disadvantage.

It may perhaps be argued that these decisions were taken by the officers of the *ancien régime* in 1786 and 1787, and that the navy of the Republic was almost bound to accept them, in view of the executive and administrative confusion then prevailing. This is true but there was plenty of opportunity to make changes after that, especially during the peace of Amiens. The truth seems to be that the French navy retained all that was least valuable in the tactical and signalling system of the *ancien régime*, without acquiring any of the advantages which might have flowed from the acceptance of the revolutionary spirit and revolutionary techniques.

An attempt was made in 1798 to give the French navy a proper signal book when *Signaux Generaux, de Jour, de Nuit et de Brume, à la Voile et à l'ancre, à l'usage des Armées Navales de la République Française* was published at Lorient.[53] The book, however, was constructed on the plan first used in the American War; it was a complete throwback. The signals were the same as those given in *Tactique et Signaux* (1787), though the numbering and arrangement was slightly different. The usefulness of the book was, however, increased by omitting all preliminary matter to the actual signals.

In 1799 a new plan was adopted of printing *Tactique et Signaux* in two separate sections. The main part of the work appeared as before under the title of 'Signaux Généraux', but omitting all the tactical diagrams used to illustrate Articles 1–15 and 28–65.[54] The introduction was slightly altered, and there were new preliminary instructions for the reporting frigates. The numerical table numbers and flags were increased from nineteen to twenty, thus enabling the 'Ordres Particuliers' to be increased from fifty-seven to seventy-seven, though the new ones are unimportant. The single-flag signals were slightly different, and again overloaded with observations. The main day signals under sail were differently numbered and arranged. Articles dealing with the light squadron were extracted from the main series and put in a supplementary section at the end of the book; these were increased from six to thirty-eight by transferring articles previously shown in the chapters dealing with the duties of frigates.

The missing diagrams and general tactical introduction were printed in the second section of the book, which was entitled *Tactique à l'usage des Armées Navales de la République Française*.[55] The diagrams illustrate the attack from windward, doubling the enemy's rear from windward and leeward, doubling the enemy's van from leeward and cutting the enemy line from leeward. In each case the doubling or cutting ships were to resume their original course as soon as they had executed the main manoeuvre. This meant the same course as the enemy, since none of the manoeuvres was contemplated with the fleets on opposite courses.

A new printing of the Signaux was issued in 1801, in which the material remained practically unchanged. The Tactique was reprinted in 1807 without any substantial alteration. What seems somewhat surprising is that the words in the title 'à l'usage des Armées Navales de le République Française' were still retained after the formation of the Empire.[56]

53 *MOD*, NM/60 (formerly *RUSI* 60). It is signed 'Jean Bart,' presumably the great-grandson of the famous admiral.

54 *NMM*, SIG/C/6 (formerly Sp/90). This copy is inscribed 'Vice-Admiral Latouche à St Domingo' and thus seems to have been used by La Touche de Tréville when commanding in the West Indies in 1803 4.

55 *NMM*, SIG/C/6 (formerly Sp/90). The title page is dated 1805, but since the *Signaux* were published separately as early as 1799 the *Tactique* was probably published as a separate work at the same time.

56 *NMM*, SIG/C/7 and 8 (formerly Sp/41 and Sp/41 {bis}). The second copy, printed at Lorient, is inscribed 'Donné à la Maj. . .[?] le 26 Avril 1807.'

The Vicomte de Grenier, 1787

A striking contrast to the sterility of the official thinking on tactics is a book published in Paris in 1787 by the Vicomte de Grenier. *L'Art de la Guerre sur Mer, ou Tactique Navale* is a short work of about fifty pages in which Grenier set out an entirely new and revolutionary system of naval tactics.[57] Based on his own war experience, he asserted that two tactical objectives were fundamental: first, to concentrate the whole force against part of the enemy's, the remainder of the enemy being meanwhile rendered 'nulle', and defeated later; second, never to expose any part of the fleet to the enemy unless it were 'flanquée', so that an enemy attack could effectively be defeated. 'Never', wrote Sir Julian Corbett, 'had the fundamental intention of naval tactics been stated with so much penetration, simplicity and completeness.'[58] Grenier condemned Hoste, Morogues and Pavillon in revolutionary style, as men concerned only with setting ships in order of battle, 'et non celle d'attaquer avec advantage un ennemi et de s'en déffendre le mieux possible'. Their so-called tactics amounted, in Grenier's view, to no more than changing from the order of sailing in three columns into line ahead and back again, while the only fighting formation they offered was 'une seule ligne au plus près'.

Grenier believed that experience had shown the close-hauled line ahead to have several disadvantages for battle. A change of wind could easily break the line, as could poor station-keeping, or an enemy attack which disabled ships. During the necessary evolution between the order of sailing in three columns to the order of battle, the fleet was exceptionally vulnerable. Once in line the enemy was free to concentrate against one part of it. Hawke attacked Conflans's rear at Quiberon 'sans aucun ordre'. At Ushant and Grenada similar attacks were frustrated by the skill of d'Orvilliers and La Motte-Piquet and their captains, combined with the inefficiency of the British, rather than by any virtue arising out of the French line ahead formation. A further disadvantage of the single line of battle was that a commander could not assist part of his line which happened to be disabled without exposing another part to danger. In any case, the commander, being in the middle of his line, could not communicate efficiently with either end of it. Under these circumstances, battles were generally indecisive, with the commanders remaining in sight of each other until night, when they would slip away unobserved.

The idea of attacking from windward, Grenier wrote, was nothing but a legacy from galley warfare, when the wind aided the rowers in driving home their main form of attack, the ram. For broadside sailing ships, the windward position only helped the weaker fleet refuse action and retire; it was also easier in the windward position to ensure that smoke did not accumulate between decks. The leeward position was better because it enabled the lower tier of guns to be fired; this was especially the case when ships had been on a long cruise and lacked stability, because of the consumption of large quantities of victuals and ammunition. The leeward position also facilitated protecting disabled ships, which could drop away yet further to leeward. Grenier's idea was that every effort should be made to keep part of the enemy fleet to windward, and then to concentrate to attack it.

Grenier replaced the single close-hauled line of battle with a 'nouvel ordre de bataille', which consisted of the fleet in three divisions disposed along three sides of 'un losange régulier'. The order of sailing of the three separate divisions could be varied considerably, but must always result in one of two forms. Either one of the divisions would be in line ahead with the other two in line of bearing, 'en échiquier', or two divisions would be in line ahead, and the other on line of

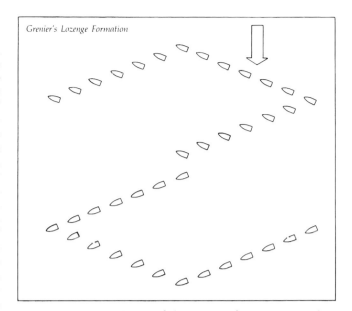

Grenier's Lozenge Formation

bearing. An essential part of the lozenge formation was that the furthest ship to windward and the furthest ship to leeward must be on exactly the same relative bearing to the wind. All movement in the face of the enemy must be calculated from the position of the leewardmost ship.

In defence, the lozenge formation did not leave any ships exposed to attack unsupported, and any attempt by the enemy to double either end of the division nearest to it or to cross its 'T', either ahead or astern, could be met by one of the other two divisions. By an ingenious system of battle tactics the lozenge could always be handled so as to put the enemy at a serious disadvantage. A and C divisions could always come to the help of B by tacking, or come up to take station on either side of B. A and C could also be used to cut the enemy line in two places. Alternatively A and C could pass together either ahead or astern of B so as to concentrate against the enemy, or to protect B against a concentration.

The real difficulty about accepting Grenier's lozenge formation as the best solution for the tactics of his time is that he never really developed it as a form of attack. Despite his criticism of Morogues, he was concerned with attack only in a theoretical manner. His concept of concentration against part of the enemy fleet was clear and effective, but how did the lozenge help to bring it about?

Grenier's willingness to accept the leeward position gives away his own essential defensiveness. He really regarded the lozenge as a formation which might give an inferior fleet a chance to counterattack, should the superior fleet show signs of weakness; whatever happened, the lozenge could never be doubled. Admittedly Grenier was correct in claiming that the lozenge formation would facilitate control. Signals would be seen more easily and quickly, and the admiral would be in a better position to gauge the situation faced by his more distant ships. Nevertheless, Grenier's failure to give details of other than defensive benefits of his formation, except for some very general remarks about the chase, suggests that his lozenge should be regarded as no more than one of a number of methods for avoiding the rigidity of the line.

The virtue of Grenier's work is primarily that it was genuinely concerned with naval warfare, not naval parades. His influence on the Dutch must have been considerable; at the time of the battle of Camperdown in 1797 the lozenge

57 Source unknown

58 Corbett, *Fighting Instructions*, p286. Corbett, however, misquoted Grenier's second dictum.

freedom of action as was consistent with efficient action by the fleet as a whole. At the same time he did everything possible to ensure efficient station-keeping and the prompt and efficient execution of all the complicated evolutions which his three- and six-column cruising systems entailed.

A good example of his careful balance of initiative and command control is an addition to his additional instructions which stated that when the signal number 166 was made for 'the ships severally to take such stations as most convenient at the time, without regard to any established order of sailing,' flag officers and other divisional commanders were neverthe-

less to keep their appointed stations with reference to the commander-in-chief, and to make such signals as were necessary to their own squadrons and divisions.[63]

---

63 Issued to Sir Alexander Hood and printed in *Signals & Instructions*, p340. There is no date, but Corbett assigns it to 1794. The reference to p12, Article 17 of the Additional Instructions is puzzling, as Article 17, already on that page, relates to the stationing of fireships to windward of the line. An additional reference is made to signal no 166: 'the ships severally to take such stations as most convenient at the time, without regard to any established order of sailing.'

His tactical object was to devise some means of forcing a supposedly unwilling enemy to action. The traditional device, ordering ships in excess of the number in the enemy line to harry the enemy van or rear, was inefficient. First, it could only be undertaken when the enemy ships were near enough to be counted. Second, the choice of ships to perform the harrying depended entirely upon which tack the fleet was sailing on at the time. D'Orvilliers had had a light squadron at the Battle of Ushant, and had formed part of Cordova's Spanish fleet into a squadron of observation, but it is doubtful whether the tactical function of these groups was the same as that which Howe now required for his fleet.

Howe revealed his plan on 20 April 1794 in a letter to Sir Alexander Hood, with a copy of the instructions he was issuing to Rear-Admiral Pasley, who was to command the advance squadron. This squadron was to consist of four 74s, three, including Pasley's, from the first division of the van and the fourth the last ship in the whole line of battle. When the signal was given, Pasley was to take station to windward of the weather column of the fleet and to intercept any enemy ships crossing ahead of the fleet, or as directed by the rear-admiral. In the presence of any considerable force of the

The Battle of the First of June, 1794, by Robert Dodd (dated 1795). NMM

Rear-Admiral (later Admiral of the Fleet) John Jervis (1735–1823), first Earl St Vincent, by Sir William Beechey. NMM

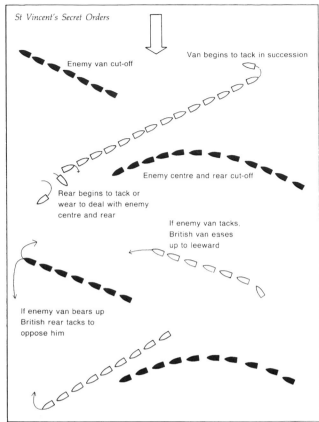

When in the following year he succeeded Hotham in command of the Mediterranean Fleet, he took over Hood's system employed on that station, presumably because the fleet was familiar with it.[86] He made a most important addition, however, by inserting as signal number 50 a tactical order: 'When approaching the Enemy's fleet on opposite tacks and the Admiral finds he cannot weather it, he may judge it expedient to make this signal . . . Vide separate Secret Instruction.' The Secret Instruction itself was an entirely separate document issued ten weeks later, and it opened up a dazzling series of tactical options.[87]

The purport of the Secret Instruction was that, in the event of the enemy succeeding in gaining the weather gage, the British fleet was to decrease sail, form into a strong body, and to force through the enemy line from leeward. It was then to tack and engage those enemy ships thus cut off from their centre and rear, and to prevent them reuniting. Meanwhile, the British centre and rear would engage the enemy centre and rear, doing everything possible to prevent them from reuniting with their van. Jervis had an answer for any manoeuvre the enemy attempted in an effort to reunite as a single line. Largely these British movements would have to depend upon the initiative of individual captains.

The Secret Instruction alone is sufficient to label Jervis as an incisive tactical thinker. Some observers might, however,

criticise it as far too complicated and theoretical. Jervis would undoubtedly have accepted this criticism, but would have added, no doubt, that he proposed to raise the fighting efficiency of his fleet to such a pitch that the instruction would no longer seem impracticable. There is no suggestion anywhere in the instruction of containing part of the enemy fleet with a possibly inferior number of ships in order to effect a superior concentration on the remainder. Jervis's idea was, evidently, that by breaking through the enemy's line, he would separate and disorganise the enemy ships, and that in the general pell-mell British fighting efficiency would prevail.

Jervis's objectives were illustrated by a set of seven extremely interesting tactical diagrams, apparently prepared for him by Captain Robert Calder, his Captain of the Fleet.[88] Number 1 shows the British fleet of seventeen ships in line ahead, having broken through the enemy line. Seven enemy ships (presumably the enemy van) are shown retreating to windward, and eleven isolated ships to leeward, the fleets having apparently met on opposite courses. The leading British ship is seen tacking, and the rearmost as having tacked or worn – in each case, presumably, to head off the isolated enemy van ships to windward from helping the remainder of their fleet to leeward. This looks very like what actually happened at the Battle of St Vincent, but with the starboard and larboard positions reversed, the rearmost ship imitating Nelson's action. The date of this diagram is uncertain. It is given in a different form from the rest, and is the only one which even hints at breaking the enemy line.

Number 2 shows the British fleet in line ahead about one and a half miles to windward of the French fleet, also in line ahead and parallel with the British. When the signal is made for the British fleet to bear down and attack, the leading ship will have to sail nearly a mile and a half on a diagonal course before she can open fire on her opposite number. This will take about twenty minutes, during which the leading French ship will have sailed about one mile. The ships in both fleets

86 *NMM*, SIG/B/68 (formerly Sp/133), day signals issued from the *Lively* 19 November 1795 to his Captain of the Fleet, Captain Calder, *Victory*, 'or of any other ship or vessel in which I may hoist my flag.' The book is thumb-indexed along the top and bound in blue morocco. The Mead Collection contains an exactly similar copy issued to Captain Thomas Wall, *Defence*, as well as a copy of the Night Signals, in Hood's manner, both similarly dated.

87 *NMM*, DUC/1/18. This copy, issued to Captain Duckworth on joining the Mediterranean Fleet in the middle of 1798, is dated 31 January 1796, showing that it was first issued ten weeks after the signal book.

88 *The Naval Miscellany*, II, pp301–7

are shown at two cables distance. The implication is that the French will continue on their original course at 3 knots and the British go down to attack at about 4½ knots.

Number 3 is signed Robert Calder, *Victory*, 19 August 1796. It shows a fleet of thirty ships in line abreast, in wedge formation and in the order of retreat and in line of bearing on either tack 'comfortable to the signal 32'. That signal was 'for the ships of the fleet to be kept on the larboard line of bearing from each other, though then sailing on the starboard tack or the contrary on the starboard line of bearing, though then sailing on the larboard tack'. Jervis issued a general order, '*Victory*: at sea. 9th May 1796', relative to signal 32. In case of a shift of wind forward, 'the leading ship becomes the axis' (instead of the flagship) and the other ships of her squadron or division were to make sail and hug the wind in succession as much as possible, the sternmost ships most, so as to retain their relative positions. If the wind came aft, the sternmost ship was to become the axis 'and those ahead of her are to make sail and cling to the wind, the headmost ship most.' [89]

Number 4, signed 'Robt Calder', as also were numbers 5–7, is explained as showing the British fleet attacking in line of bearing from windward and the French fleet turning away four points to leeward and in succession. 'The English fleet must then force sail and bear up also in succession, one point: the English fleet will then be eight and the French six points from the wind. Each ship to bear up instantly her opponent

does, by which they will preserve their bow and quarter line', and so be able to engage ship for ship 'as before directed'. The British fleet, having borne up four points in order to begin their approach, would need only to bear away one point more and force sail to reach round the French line as it turned away, and so force an action. There was no suggestion of breaking the enemy line.

Number 5 shows the situation arising should the French turn away four points together so as to avoid action. The British fleet, already approaching in line of bearing, 'must then alter their course together, one point to starboard, steering SSE; and force sail'. Number 6 shows a further attempt by the French to avoid action by wearing together. The British fleet, still shown as following their original mode of attack, 'must then alter their course together to starboard six points, steering SW by S and increase sail'.

Number 7 shows an extension of the French avoiding action taken in number 4. This time, instead of merely turning away from the British in succession, the French wear in succession, thus forming a new line on the opposite tack. The British answer to this was highly sophisticated. They must alter their course together ten points to starboard, when the fourth or fifth of the enemy's van [is] in the act of wearing and steer W by S, shorten sail and keep the main topsail shivering, by which manoeuvre the rear and centre of the French fleet will be greatly annoyed in wearing and their van unable, from their position, to fire upon our ships, until they have opened astern of their rear ships. The English fleet will be enabled to close up with them in the same manner on their starboard quarter in lieu of the larboard, as if they had not wore the lines being reversed.'

89 General Order by Sir John Jervis, 9 May 1796, *The Naval Miscellany*, II, pp224–5

A tactical diagram by Admiral John Jervis's flag captain, Robert Calder, showing the means by which, should the French Fleet wear together, they might still be brought to action, about 19 August 1796. NMM

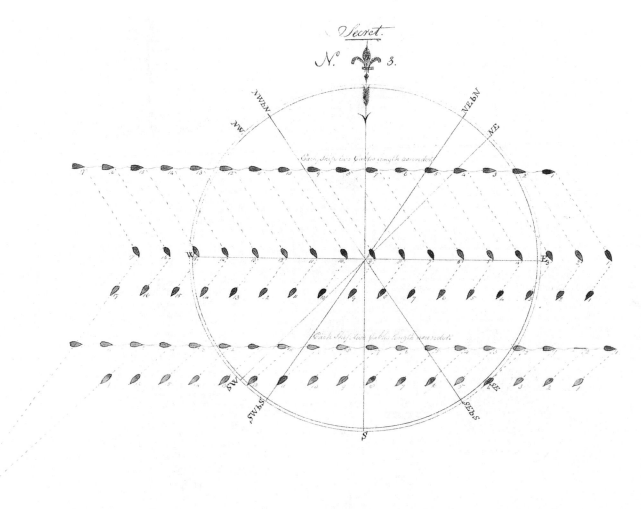

This means that the French centre and rear would have to receive the fire of the whole British fleet while passing on the opposite tack preparatory to wearing round in succession following their van. Thereafter the British would be able to engage their new line now formed on the starboard tack.

On 3 November 1796 Calder issued Jervis's instructions for the look-out frigates to keep two points on the admiral's bow. By day they were to be one to three leagues distant, and by night one to three miles, according to the weather. They were to make sail ahead of the fleet at 4am, and shorten sail at 4pm, so as to attain their proper distances. 'All night signals should be avoided as much as possible by these frigates, as they always cause confusion, more or less, in a fleet.' Instead, one of the frigates was to report back. On 28 March 1797, immediately after the battle of St Vincent, Jervis issued an order of sailing for the fleet in two parallel columns one mile apart, with the ships at intervals of two and a half cables.[90]

The leading ship of the larboard column was to be level with the second ship of the starboard column. The admiral was to be two points on the weather bow of the starboard column and three points forward from the beam of the leading ship of the larboard column. He issued an identical order of sailing on 12 April 1798.[91]

Jervis's thinking about cutting the enemy fleet from leeward, shown in Calder's first diagram, evidently should not be taken as indicative of any tendency towards radical ideas generally in tactics. His night signals and instructions of 1798 followed very closely on Hood's New Arrangement of 1793.[92] Two days after Nelson had won the Battle of the Nile, 3 August 1798, Jervis, now Earl St Vincent, issued a copy of his 1796 instructions.[93] These were exactly the same as those he issued in 1795, themselves a copy of those issued by Admiral Alan Gardner in 1793.

On 17 December 1798 St Vincent issued further signals and instructions, to come into force on 1 January 1799.[94] The day and fog signal book resembled the Hood-McArthur New Arrangement of 1793, except that the signal numbers were printed throughout. There were 181 instead of 162 signal spaces, the squared table was for nine instead of eight signals, and there was a table of horary signals.

The Battle of Cape St Vincent

At dawn on 14 February 1797, Jervis had his fleet well closed up in two parallel columns. The wind was WSW, and

Admiral John Jervis's Order of Sailing in Two Squadrons, issued to Sir Charles Thompson, Vice-Admiral of the Blue, 19 August 1796. NMM

90 *MOD*, NM/35 (formerly *RUSI* 35)

91 Copy sold at Hodgson's Rooms, 3 April 1908

92 *MOD*, Ec/164, bound in blue morocco. *NMM*, SIG/B/72 (formerly Sp/134), unsigned, is similarly bound and was possibly St Vincent's personal copy. *NMM*, Tunstall Collection, TUN/22 (formerly S/MS/R/ 7) is an undated manuscript copy of the whole of St Vincent's Night Signals, with explanations, directions and instructions, in pocket book form.

93 *NMM*, SIG/B/69 (formerly Sp/30) and *MOD*, Ec/165

94 *NMM*, SIG/B/70 (formerly Sp/95), issued to Captain Henry Savage, *Warrior*. It much resembles *NMM*, Sp/17, but appears to be a fuller and later printing of KEI/S/11(a) (formerly Sp/20). The Mead Collection has a copy similarly worded about coming into force in the New Year.

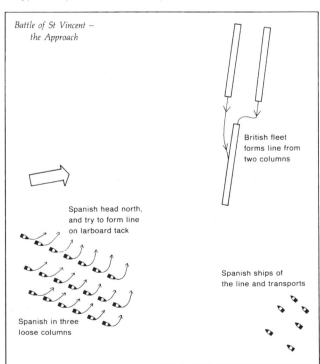

Battle of St Vincent —
the Approach

British fleet forms line from two columns

Spanish head north, and try to form line on larboard tack

Spanish ships of the line and transports

Spanish in three loose columns

HMS Victory *raking the Spanish at the Battle of Cape St Vincent, 14 February 1791, by Robert Cleveley.* NMM

he was steering a little west of south. He was warned of the Spanish fleet's presence and had cleared for action twenty-four hours previously. His look-out frigates soon reported the Spanish fleet in what turned out to be complete disorder. It was a hazy dawn, but gradually he made out to starboard (and therefore to windward), twenty or more Spanish ships-of-the-line and frigates, while very slightly to larboard, and to leeward, a smaller group of apparently eight or nine of the line. Jervis had fifteen of the line with three flag officers and a commodore, Nelson, under him. Realising his tactical advantage, he held on his course until 10.57am, when he signalled (number 31) 'To form the line of battle ahead and astern as most convenient', thus allowing his fastest sailers to draw to the head of the line. No signal was made for carrying a press of sail, since the instructions enjoined all ships to carry sail equivalent to that of the admiral. After altering course only one point to larboard and then back one to starboard, Jervis at 11.28am signalled (number 40) 'The admiral means to pass through the enemy's line'; two minutes later he signalled 'to haul the wind on the starboard tack'. So well was

he placed at dawn that his line of advance had scarcely changed in the course of nearly three hours. Considering night movements and the hazy weather, this was largely luck. Nevertheless he risked much: had the windward section of the Spanish fleet shown anything like an indication to form a real line and engage, their superior firepower could, on paper, have enabled them to immobilise his leading ships at long range and to have overwhelmed the remainder by sheer weight of metal.

The situation in the Spanish fleet, however, was chaotic. The fleet was short of men, short of supplies, short of training and in every way deficient, though, as before, the flag officers and captains possessed excellent signal books and detailed tactical manuals. At dawn on 14 February the Spanish Admiral, Don José de Cordóva, had his fleet in the order of sailing in three columns, steering ESE with column commanders leading their respective columns. This part of his fleet consisted of twenty of the line. To the east, and hence to leeward, were the smaller group of ships already sighted by the British and judged to be a detached division. In

Battle of St Vincent – the Engagement

Spanish attempt to pass British rear

Excellent supports Nelson

Nelson leaves line to attack Spanish van

Moreno's ships tack and retreat

Culloden supports Nelson

British fleet tacks in succession

Spanish transports, etc

fact they consisted of four large merchantmen loaded with mercury and protected by two of the line and some frigates. Had Jervis known this he might well have captured the whole convoy and so seriously have damaged the Spanish economy.

When various Spanish ships began to report enemy ships astern of them, Cordóva took a quick and, as it turned out, disastrous decision. Realising that he was heading away from the enemy, and wishing to keep the weather gage, he signalled the fleet to form a line of battle 'as convenient' on the larboard tack. For a British fleet of the same size, well trained under Jervis, to have attempted such an evolution, suddenly and in hazy weather, would probably have led at least to temporary confusion, but it is not likely that Jervis, Howe or Hood would ever have attempted it, since it involved reversing the order of sailing in columns and at the same time inviting a general scramble to form a single line on the opposite tack. In addition, it gave no opportunity to the admirals leading the three columns to get to the heads, or even the centres of their columns in the reversed order. In fact all the Spanish ships seem to have been left astern, including, of course, Cordóva. Three of his ships – Admiral Juan de Moreno's flagship, together with another 112 and a 74 – failed to wear smartly enough and fell to leeward. Cordóva signalled to them to wear again and take station in the rear so as to avoid the advancing British.

At 11.31am, only five minutes after signalling his intention to pass through the enemy line, Jervis's leading ship, the *Culloden*, opened fire on one of Moreno's group. All three tacked independently and then withdrew to leeward towards their convoy. As the British line continued on its course, firing to starboard, the weather part of the Spanish fleet began to pass them on the opposite tack, pursuing a slightly diverging and north-westerly course.

At 12.08pm Jervis signalled the British fleet to tack in succession. Having forced the enemy fleet to separate, his first object was achieved, and he was now free to pursue the

retreating Spanish force to windward. Meanwhile his rear ships could continue to keep off the convoy group to leeward by firing their larboard broadsides as they passed. At 12.15 he repeated the signal to pass through the enemy line, presumably to reassure his rear ships that they were doing the right thing in leaving Moreno to leeward as they came up, and then tacking in succession. To have tacked earlier might have prevented this essential separation and have allowed the Spanish rear and the convoy group to rake the British ships as they came round in stays.

Moreno and the two ships with him did in fact attempt decisive intervention from leeward and succeeded in disabling the *Colossus,* fifth ship in the line, so that she missed stays and fell away to leeward. This caused a considerable delay and though Moreno's ships were eventually checked, their intervention certainly hampered the British pursuit of the Spanish weather division.

Cordóva realised that, despite his initial setback, he could still make use of his vastly superior strength. He decided to double on the British rear by getting his van ships to wear round on the opposite tack. By this manoeuvre he might capture the rearmost British ships and frigates and save the mercury convoy from capture, as well as protecting Moreno's three ships. Unfortunately, he could obtain no effective response from his flag officers and captains, despite signalling and hailing. Nevertheless, he succeeded in initiating a movement to close the British rear with a number of ships, his own included, though no proper line was formed.

Jervis saw that the Spaniards were slipping away from him, partly because of his delay in making the signal to tack and partly because of the delay caused by the disabling of the *Colossus.* Although the *Culloden, Blenheim, Prince George* and *Orion* were well round on the larboard tack and heavily engaged, the leading Spanish ships had reached far ahead of them, while others were bunching up further to windward. Nor was it any longer possible, in view of the delay caused by the *Colossus,* for the British line to continue its hairpin bend movement in smooth, uninterrupted fashion. Nor was there any further need to worry about Moreno who was now in retreat, with one of his ships working right round the British rear to leeward in order to rejoin Cordóva. The British rear ships, therefore, had an opportunity to turn directly to starboard and attack the main body of the Spanish fleet by the quickest route. At 12.51pm Jervis made the famous signal, number 41, 'The ships to take suitable station for their mutual support and to engage the enemy as arriving up with them in succession'.

With the line of battle thus abandoned, flag officers and captains were free to act to the best advantage. This did not mean that they were free to do just as they pleased, but rather to engage the enemy in a new line or part of a line as well and as quickly as they could.

Nelson may or may not have seen the signal, but in any case he had already anticipated it. Being the sternmost ship but two, he could see the need for cutting across the Spanish line of advance or, more accurately, retreat.

… perceiving the Spanish Fleet to bear up before the wind, evidently with an intention of forming their line, going large – joining their separated division – or flying from us, I ordered the *Captain* to be ware, and passing between the *Diadem* [next astern] and *Excellent* [rearmost ship], at ten minutes past one o'clock, I was close in action with the Van, and, of course, the leewardmost of the Spanish Fleet.

In this way he engaged the *Santissima Trinidad,* supported by two other 112-gun flagships as well as an 80. The *Excellent* (Collingwood), however, executed the same manoeuvre and was later followed by the *Diadem.* In the meantime, the *Culloden*, still leading the original line, was coming up together with the *Blenheim* and other leading ships. Thus an improvised line or quasi-line was formed along the leeward flank of the Spanish fleet. Further to the rear the remainder of the British fleet, having tacked, were now formed in a line somewhat to windward of the Spanish and were thus able to force their way amongst the bunched ships of the Spanish rear. There was, however, a period, which no doubt seemed long to those concerned, during which the *Captain* was insufficiently supported by the van ships as they came up behind her after tacking. Admiral Thomas Taylor made the interesting point that those ships should not merely have engaged the Spaniards once they were round on the larboard tack, but should have followed Howe's established doctrine (Additional Instructions . . . in Action with the Enemy, Article 6) that when the leading ship came up with the enemy's rearmost ship the second British ship was to pass ahead of her and engage the next enemy ahead, with the remaining ships doing the same in succession.

The result of the pell-mell round the *Captain*, and then ahead and astern of her as well, was that four Spanish ships were captured, two by Nelson in person, though they mounted a total of 382 guns. Cordóva tried to set his fleet a personal example of stout resistance, and the *Santissima Trinidad* was so severely damaged that she actually surrendered, though she eventually escaped.

Cordóva observed that when the improvised line was formed by Nelson and others 'along our line on the starboard side, their ten ships [the rest of the fleet] remained in consequence upon the other side [ie, more to windward], firing at us, in a regular and powerful line, by which manoeuvre they obtained a decision of the combat in their favour'. The *Santissima Trinidad,* he continued, was reduced to a wreck. Four ships were hailed and signalled to give her support, and two of these were amongst those eventually taken. 'In this situation – that is, against four times their force, counting besides numbers the superiority of their fire over ours – I being doubled on and cut off from the greater part of our said ships . . .' Cordóva felt that he ought to have been relieved by the disengaged van or centre ships. 'Whoever considers the state of things, and moreover the rapidity and accuracy with which the English handle their guns, can imagine what must have been our condition by four o'clock.' Admissions such as these show why the victory went to the British fleet, despite the fact that they were outmatched by more than two to one in guns; and by much more in weight of metal.

'A System of Naval Tactics', 1797

Published anonymously and subtitled 'Combining the established theory with general practice, and particularly with the present practice of the British Navy',[95] *A System of Naval Tactics* was an extremely useful and well set out text book, very handy in form, with coloured figures distinguishing the different columns or squadrons in every evolution in red, white and blue.

Part I purported to be a translation of Morogues, though it

was only partly so, as it left out the battle tactics, and rearranged and paraphrased the rest. The sailing diagrams were very well shown. There were tactical and battle sections at the end, which, though still purporting to be from Morogues, were actually drawn from Hoste and Bourdé de Villehuet. These sections are only relevant in so far as they reveal the interest shown at the time in those authors.

Part II gave summaries and extracts from Bourdé de Villehuet, with special reference to his so-called 'order of convoy'.

Part III was described as the 'Tactics of the British Navy', and purported to be 'compiled from official documents with the assistance of several experienced officers'. In the space of twenty pages the anonymous author covered the order of sailing in three columns, six columns and two grand divisions, changes of course and changing from various orders of sailing into the line of battle. For these he provided a summary of current practice as regards distances between ships and squadrons in the various orders, and rules to be observed when changing course or formation. This part was, in fact, a summary of Howe, re-written so as to make it more easily understood.[96] Nothing whatever was said in this part about fighting, no doubt for security reasons.

The Battle of Camperdown

Tactically, Camperdown (11 October 1797) was a Howe battle in the sense that on the British side it was fought with Howe's signal book and accompanying instructions. On the Dutch side it was fought with a numerical signal book containing 795 signals, and of unrivalled simplicity in arrangement – unrivalled even by Howe's.[97] It was based on ten flags, numbered 1–10, with pendants for the hundreds, but the book was thumb-indexed so that the relevant pendant, together with the ten flags, could be seen on each page, with the signal opposite the flags, the flags for the tens being shown along the top.

The battle signals provide for every imaginable contingency though, unlike the French, they are directed entirely towards the actual defeat of the enemy.

Dutch Battle Signals of 1797

No 715	Attack the enemy's van to windward with a superior force of the heaviest ships in the line of battle; light ships to support the attack in lozenge formation on perpendicular of the wind [the lozenge suggests Grenier's influence].
No 717	Attack the enemy's *corps de battaile* to windward with a superior force of the heaviest ships in the line; light ships to support their heavy ships at the head of the line and the rest at the rear.
No 726	Van to crowd sail to attack to windward and engage enemy's van while the centre and rear engage to leeward so as to double.
No 728	Van to cut through the enemy's line and the rest to turn [parallel] with the enemy and engage.
No 729	All ships to cut through the enemy's line individually.
No 730	Enemy's fleet being at anchor in a port or road, to double their van and engage.
No 735	Fleet to retreat fighting.
No 739	Squadron or ships signalled to cut through the enemy's line.
No 741	Admiral sees that he has the advantage; ships, therefore, which are able, to cut through the enemy's line and put them between two lines of fire.

95 *NNM*, Corbett Collection, TUN/96 (formerly T/P/R/1), bookplate of Philip Carteret

96 Sir Julian Corbett, in his *Signals & Instructions* pp75–8, quotes at length from Part III, noting the absence of any mention of breaking the enemy line. The two diagrams which he reproduces, however, are merely copies of those introduced by McArthur on behalf of Hood and later adapted for the Admiralty Signal Book of 1799.

97 *NMM*, DUN/31 (formerly S(D) 31)

No 744   Ships which have doubled enemy to rejoin the
          fleet.
No 748–9 Ships at the head of the line to prevent enemy
          doubling to windward: ditto at rear.
No 750–1 Ditto, to leeward, ditto.
No 752   Enemy is trying to cut our line; ships to close
          up and stop them.
No 754   Close up and stop enemy ships which have cut
          our line from rejoining their main body.
No 755   Enemy has cut our line; ships to form lozenge
          formation to leeward with the wind.

It is astonishing to discover that, although all the ordinary

words and standard terms are in Dutch, the particular
evolutions in the battle signals are in French. This would
seem to show a desire to graft tactical virtuosity, seen as a
French accomplishment, onto the traditional phraseology of
Dutch seamanship. There is no sign whatever of French
wording in the Dutch signals at the start of the American
War.[98] Either they were dazzled with the new French tactics
and signalling (as Kempenfelt was, though in a different
way) or else they had come under French influence as a result
of the Revolution. The signals themselves are in the true
Dutch spirit and seem to be almost in contradiction with those
produced by Morogues and Pavillon. Nor do they read as
mere notional flourishes – paper signals never to be used. On
the contrary, their number and consistency seem to breathe a
truly offensive spirit. In the lozenge formation employed in
signal number 715 we see the influence of Grenier, whose
book, first published in 1787, was no doubt read with great

*The Battle of Camperdown, 11
October 1797, by Philippe-
Jacques de Loutherbourg (dated
1801). NMM*

98 *NMM*, Tunstall Collection, TUN/13 (formerly S/MS/D/1) signal
book of Rear-Admiral Andreis Hartsinck, Mediterranean, 1775–6, old
system with flags in particular positions.

interest in Holland as showing a real breakaway from convention.[99]

Except for the Dogger Bank in 1781 (itself a very second-rate affair), the Dutch had not fought a fleet action since Malaga in 1704, so the signals of 1797 were really intended for the kind of fleet the Dutch no longer possessed.

Admiral Jan Willem de Winter put to sea from the Texel with fifteen ships-of-the-line, very weak in firepower. He made no attempt to manoeuvre when approached by the British fleet but simply formed in line ahead on the larboard tack on a north-easterly course with the wind NW by N, within sight of the Dutch coast. Three heavy frigates were used to support his line. Admiral Adam Duncan had sixteen ships-of-the-line. Neither admiral had any three-deckers. Nevertheless, Duncan's fleet, although badly strained and worn by continuous North Sea cruising and only just recovered from the disruption of the Nore mutinies, was in good fighting form.

Being signalled by his advanced ships that the enemy were in sight to leeward, Admiral Duncan

immediately gave chase, and soon got sight of them, forming line on the larboard tack to receive us, the wind at NW. As we approached them I made the signal for the squadron [fleet] to shorten sail, in order to connect them; soon after I saw land between Camperdown and Egmont, about nine miles to leeward of the enemy, and finding there was no time to be lost in making the attack, I made the signal to bear up, break the enemy's line, and engage them to leeward, each ship her opponent, by which I got between them and the land, whither they were fast approaching. My signals were obeyed with great promptitude.[100]

In his letter to Lord Spencer he wrote: 'For the particulars of our victory [I] shall leave your Lordship to my public letter, and shall just say I was obliged to lay all regularity and tactics aside, we were so near the land, or we should have done nothing.'

Recent writers have been at pains to contradict the idea that Duncan's method of attack anticipated Nelson's at Trafalgar. They need not have worried, for Sir Julian Corbett settled the point more than half a century ago.[101] Duncan had his 'squadron', as he calls it, in two divisions, eight ships in his own division and the other eight under Vice-Admiral Richard Onslow. He saw at once what ought to be done and what in practice could be done. He must get between the Dutch and the land by steering for them as quickly as possible. His ships were in two loosely formed groups; his own division, less the *Adamant* which had previously been detached to observe and was now with Onslow, to the north, and Onslow's division, plus the *Adamant*, to the south. Yet speed must not produce confusion. To break through the Dutch line successfully, each ship must steer for her opposite number in such a way as not to block the fire or movement of flanking ships. To describe Duncan's approach, therefore, as headlong is inaccurate, especially as his speed could in any

Admiral Adam Duncan (1731–1804), first Viscount Duncan. NMM

case scarcely have exceeded three knots in the fresh or moderate breezes recorded in the ships' logs.

In the space of three hours – that is from the beginning of his tactical approach to the start of the battle – Duncan made some thirty signals.[102] He was making fresh signals at roughly six-minute intervals; Keppel made nineteen signals in the eight and a half hours of approach and battle of Ushant. Yet no one would suggest that Duncan was being unduly fussy or trying to exercise an unnecessary degree of control over his subordinates. On the contrary, he was trying to speed up the battle right from the start, his signals being mainly by way of making adjustments of distance and bearing so as to avoid the need for bringing to and dressing his line during the approach. His main signals, apart from a

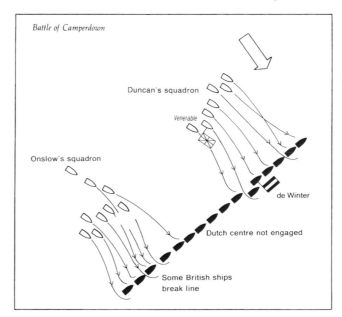

Battle of Camperdown

Duncan's squadron

Venerable

Onslow's squadron

de Winter

Dutch centre not engaged

Some British ships break line

99 A Dutch translation of Grenier by J J Melvill was published at Leyden in 1799 under the title of *Zee-Tactique of Konst des Oodogs te Zee*, but no doubt his work was well known in Holland before that.

100 Julian S Corbett, (and Herbert W Richmond, editor of vols iii and iv), *The Spencer Papers*, London, 1913, four vols, (hereafter cited *The Spencer Papers*), II, 197n. See also *Logs of the Great Sea Fights*, I, editorial comment by Admiral Sturges Jackson.

101 *NMM*, DUN/19 (formerly S(D) 19) is an manuscript notebook of Duncan's signals dated 1795. It contains nothing new.

102 The number is uncertain, owing to discrepancies between the log of the *Venerable*, Duncan's flagship, and the logs of other ships. See also the list of signals made, as recorded by the *Triumph*, NMM, SIG/B/75 (formerly Sm/32).

The Battle of the Nile, 1 August 1798, by Nicholas Pocock (dated 1808). NMM worded very much in the spirit of Howe, they were largely new, and aimed at attaining a still higher efficiency in station-keeping and in avoiding obstructions and collisions. They were drawn in very general terms and without special reference either to cruising or to battle. The seventeen Instructions Respecting the Orders of Sailing which follow were also modelled on Howe's, but were better worded and were illustrated by McArthur in eight plates, with three new ones. Provision was made for the order of sailing in two, three or six columns; the position of the admiral; opening and closing distances; tacking, wearing and bearing up in succession. The four 'Instructions Relating to the Line of Battle' were similar to those in Howe's Additionals. The eleven 'Instructions for Forming the Line of Battle from the Orders of Sailing', again illustrated with folding plates, were quite new. They provided for forming the order of battle in as many ways as practicable from either two, three or six columns, with signals.

The thirty-one 'Instructions for the Conduct of the Fleet, Preparatory to their Engaging, and when Engaged with an Enemy' (a more Howe-like wording even than Howe's), were conservative in tone. A new definition, however, was given to the purpose of a line of battle. Howe, in his Additional Instructions, said only that the line enabled ships to be 'as little as possible exposed to the fire of more than the particular ships corresponding in station with them in the Enemy's Line, or be subject to injury from each other'.

The new instruction (number 2) adopted a far more realistic attitude: 'that the ships may be able to assist and support each other in action; that they may not be exposed to the fire of the enemy's ships greater in number than themselves; and that every ship may be able to fire on the enemy without risk of firing into the ships of her own fleet'. A short table was given of the gradations of making and shortening sail for ships, 'suiting their rate of the sailing to that of the Admiral' (number 9).

When coming onto a parallel course with the enemy to engage, each ship was instructed to be careful to give her next astern 'room to haul up without running on board her' (number 14). The usual instruction was included that small numbers of the enemy beaten out of the line were not to be

pursued 'unless the commander-in-chief, or some other flag officer be among them'. The British ships concerned were instead to assist ships 'much pressed', 'and to continue their attack until the main body of the enemy be broken or disabled', unless specially signalled or instructed (number 15). Instruction number 18 dramatically stated that 'if there should be found a captain, so lost to all sense of honour, and the great duty he owes his country, as not to exert himself to the utmost to get into action with the enemy', he was to be suspended and replaced. Ships were to try to make a simultaneous attack by making or shortening sail, and be careful to avoid collisions (number 26). 'The leading ships must be very cautious not to suffer themselves to be drawn away so far from the main body of the fleet as to risk being surrounded and cut off' (number 27). There were twenty fog signals and nine simple fog instructions, similar to Howe's.

The 'Night Signals', a volume of 41 pages, was a far more elaborate and ambitious work than anything attempted before. There was a detailed index of signals and of 'Private Ships' Signals to the Admiral'. The method adopted was for one to four lights to represent the digits 1 to 4, for false fires or rockets (regardless of numbers) to represent the digit 5 and for guns to represent tens. Thus signal number 48 was represented by four guns, one or more false fires or rockets and three lights. There were seventy signals in all. Included were sailing instructions, battle instructions and 'Instructions for the Conduct of Ships Appointed to Watch the Motions of an Enemy's Fleet in the Night', all three of which were a revision of Howe's, many of the articles being word for word the same.[115]

It is worth asking just what it was that the new books were intended to supersede. Admiral Cuthbert Collingwood, at sea on 17 June 1799 and about to proceed from off Brest to reinforce Keith's fleet in the Mediterranean, issued his flag captain with day and fog signals with 'New Arrangement' printed on the cover and '1795' printed on the title page.[116] Clearly, there were many versions of Howe's, Hood's and St Vincent's signals and instructions in use and the Admiralty books were quite as necessary for securing standardisation as they were for their actual content.

With over 700 ships in the Royal Navy at this time and with continual changes of flag-officers and captains, the Signal Book had to be reprinted frequently during the next three years, even though the date, 1799, remained on the cover label and the short title. The differences were almost entirely typographical and extremely difficult to perceive, especially in the case of the night signals.[117] The Board of Admiralty, moreover, frequently omitted to fill in the date, 'Given under our hands the . . . of . . .', above the space left for the signatures. Second and later printings had twelve diagrammatic plates instead of eleven, and these were interspersed amongst the sailing instructions instead of being grouped together at the end. Handwritten insertions for the period 1799 to 1803 show no evidence of any important change in tactics.

In 1803 a British schooner was captured off Toulon. The lieutenant in command was not entitled to possess the Signal Book, but had evidently made a handwritten copy, which the French captured. The ship, flying the British signal to anchor, was afterwards used by the French in an attempted decoy operation. The Admiralty, therefore, on 4 November, issued a circular letter to all commanders-in-chief, ordering a change in the numbers assigned to the numerical flags and a change in the substitute flag.[118] They enclosed coloured printed slips for pasting over the flag and number columns in the Signal Book, as well as blank uncoloured slips.[119] Private copies of the signals were forbidden for the future and existing copies ordered to be handed in. The new issue of

115 Despite the standardisation now established, occasional anomalies appeared, as for instance *MOD*, NM/45 (formerly *RUSI* 45) which consists of Howe's Instructions of 1793 'given on board the *Lancaster* at Spithead 1st September 1799, to Hon Capt Stopford, HMS *Excellent*, [signed Vice-Admiral Sir] Roger Curtis.' The signature has been verified.

116 *MOD*, Ec/113, issued to Thomas Seccombe, *Triumph*, at sea, 17 June 1793, but without Collingwood's signature.

117 *NMM*, SIG/B/73 (formerly Sp/50), has typographical variants: SIG/B/76 (formerly Sp/102) bears the printer's name, William Winchester & Son, Strand (the flags are those of 1803 and it was issued after 2 May 1805 (Board signatures)); *NMM*, TUN/60 (formerly S/P/R/7), also William Winchester, issued 5 October 1803; *MOD*, Ec/54 issued 1804 (Board signatures). Of the Night Signals, *NMM*, SIG/B/80 (formerly Sp/96) is undated and was issued 1804–05 (Board signatures). All three have typographical differences. DUC/49 (formerly Sp/34), undated, printed by Winchester & Son, was issued 1801–4 (Board signatures). *NMM*, Tunstall Collection, TUN/63 (formerly S/P/R/10) was issued 1805–6 (Board signatures), and *MOD*, Ec/60(a) was issued 29 October 1804.

118 *MOD*, NM/47 (formerly *RUSI* 47), a copy of the 1799 Book issued 5 November 1803, with notes by Sir Julian Corbett inserted which in fact refer to *MOD*, NM/58 (formerly *RUSI* 58).

119 The original number 5 flag, quarterly red and white, now became number 4. It is on the question of whether or not one or more of the coloured slips had the new number 4 painted quarterly white and red by mistake that the dispute about Nelson's Trafalgar Signal depends. See H P Mead, *Trafalgar Signals;* also Captain Holland, 'Number 4 Flag at Trafalgar', *The Mariner's Mirror*, vol 20 (October 1934) no 4, and D B Smith, review article on Mead's book, *The Mariner's Mirror*, vol 22 (1936) no 4

flags took effect in the Mediterranean Fleet on 16 January 1804.

By 1804, if not earlier, the Day Signal Book required expansion, which the Admiralty achieved by inserting an appendix between the day signals and the instructions, containing eighty-two new signals, in two sections.[120] The first section, entitled 'Signals from Senior to Junior Officers', gave printed form to the existing handwritten insertions in the original blanks of the Admiral's signals (ending at 310), and added a further group, numbered 500–506. In the second section, entitled 'Signals to be Added to Those Made From Private Ships, or from Junior to Senior Officers', existing handwritten insertions were again given printed form, and a new group added, numbered 391–417. Although certain tactical refinements were introduced, there was nothing of real importance. Rendezvous signals were added later. The appendix was issued as a separate publication; it was through this that Nelson made known his ideas for the attack at Trafalgar.

Another crisis arose in 1807 when the schooner *St Lucia* was captured in the West Indies and the French obtained 'the whole of the signals'.[121] Sir Alexander Cochrane, the commander-in-chief in the West Indies, changed the numbers and the substitute flag. This, however, does not seem to have gone beyond the West Indies station, as the flags remained the same even when the Admiralty published a completely new edition of the Signal Book in the following year. This edition of 1808 was substantially the same as the original except that the admiral's signals were extended from 310 to 360. This was to give room for expansion, since all signals from numbers 318–60 were left blank. The private ships had their quota increased by being given 175 signals (nos 361–535) instead of the previous eighty. Even so, the last sixty-eight were left blank.[122] There was no appendix, although the rendezvous signals were inserted separately.

It might have been expected that the impact of Nelson's tactics on the Navy as a whole would have resulted in some relevant additions to the Signal Book of 1808, as regards both the signals themselves and the battle instructions. No additions, however, in any way traceable to Nelson can be found. In a tactical sense, the war was as good as won and fleet actions were becoming less and less likely. The Admiralty was, no doubt, content to rely on traditional wisdom rather than promulgating new ideas – excellent when propounded and executed by Nelson but not to be encouraged as standardised procedure.

In 1811 the Admiralty issued a new printing of the day signals with a further alteration of the numerical flags, although no new flags were introduced. Nor was the book itself changed in any important respect. The night signals were not apparently reprinted.[123] The final consolidation by which the book itself and Popham's Telegraph became associated volumes did not take place until after the war, in 1816.

The Battle of Copenhagen

The Danish defence plan at Copenhagen in 1801 was entirely passive and aimed simply at protecting the city and harbour from sea bombardment and from assault by landing parties. At the northern end of the defence area, the difficult harbour approaches were covered by a small squadron of block ships and more effectively by the Three Crowns battery built out on piles above the mud flats. In front of the city itself and stretching well to the south of it was a long strip of mud flats over half a mile wide. Outside these mud flats was the navigable inner channel, about half a mile wide but very narrow at its southern entrance and made difficult at its

Battle of Copenhagen – the British Approach

northern and broader end by a small shoal, right in the middle, near the Three Crowns battery. Further out, beyond the navigable channel and severely limiting its usage, was the Middle Ground Shoal, with shallow-water extensions to the north of it. Instead of arranging movable block ships at both ends of the channel, the Danes provided for covering the harbour approaches at the northern end, and strung out their defences to the south in a single line of warships and floating batteries ranged along the outer edge of the mud flats facing the city. They thus invited an attack by any force able to enter the channel at its narrow southern entrance and then steer north until opposed by the first floating battery or block ship, drawn up, not at right angles and across the bows of the advancing enemy, but merely parallel. In seeking to do no more than protect the very heart of their country, the capital and its port, the Danes adopted a system which denied every principle of true defence.

When first observed at a distance by the British fleet, the Danes were seen to have seven unrigged and dismasted ships-of-the-line and eleven floating batteries of various types ranged in line along the outer edge of the mud flats facing the city. There were two line ships at each end of the line, the other three being spread evenly between them, with the floating batteries filling the gaps. Nelson decided to make his attack from the southern end of the channel. For this, he needed a northerly breeze to carry his force through the narrow channel on the far and eastward side of the Middle Ground Shoal. He then needed a southerly breeze to enable him to round the southern tip of the Middle Ground Shoal, and sail straight up to engage the Danish defences, and if

120 *MOD*, NM/58 (formerly *RUSI* 58), originally issued not later than February 1801 (Board signatures) but may have had the Appendix bound in later. The flags of 1811 are pasted over the top. NM/76 (formerly *RUSI* 76) includes two copies of the Appendix as a separate publication, one issued by Admiral Rainier, Madras, East Indies, 1805 and the other by Cornwallis, probably in 1805.

121 General Memorandum to all captains, *Northumberland*, at sea, 20 April 1807. Typescript copy once in the Mead Collection, documentary source unknown.

122 *NMM*, SIG/B/88 and 89 (formerly 94, two copies) issued 13 July 1808; also SIG/B/91 (formerly Sp/108), issued 16 March 1812.

123 *MOD*, Ec/60 is a copy of the Night Signals printed by Winchester and dated 1806, issued 4 January 1811.

necessary to bombard the city and land troops. By nightfall on 1 April 1801, with the aid of a northerly breeze, he had achieved his first object, and was able to anchor at the southern end of the Middle Ground Shoal, scarcely a mile and a quarter from the nearest and most southerly Danish ship. He was, therefore, able to send Captain Hardy in a boat to take soundings ahead. His force was awkwardly placed in a narrow anchorage, and if, had the Danes been able even on the evening of 1 April to swing three of their ships or

batteries broadside on across the channel, they could have crossed the T of Nelson's advancing ships with a crushing broadside fire. Their attitude, however, was wholly passive, even neglecting night reconnaissance.

For Nelson, everything now depended on a favourable wind from the south, even though this would prevent Sir Hyde Parker with the rest of the fleet supporting him effectively by closing in on the harbour entrance from the north. Left unmolested by the Danes, although he was within

The Battle of Copenhagen, 2 April 1801 (detail), by Nicholas Pocock. NMM gunshot, Nelson gave a dinner party for Rear-Admiral Thomas Graves[124] and several captains, showing himself 'in the highest spirits'. Then, at about 9.30pm he withdrew to the after cabin to draft his orders for the next day's attack,

that is if the wind was propitious. He asked two officers to assist him. One was Captain Thomas Foley (of the *Goliath* at the Nile) in whose ship, the *Elephant*, he now had his flag, his own flagship for the campaign, the *St George,* having too deep a draught for the shallows. The other was Captain Edward Riou whom Nelson had met, apparently, for the first time, on the previous day. Captain Hardy returned at 11pm, having actually sounded round the nearest Danish ship, the Danes apparently having failed to post sentries, let alone send

124 This is the same Thomas Graves who in the American War made the collection of signal books and fighting instructions once in the Royal United Service Institution and now divided between the National Maritime Museum and the Ministry of Defence Library, Whitehall.

out patrol boats after dark. For all his astonishing verve and ardour both in battle and in council, Nelson was an arch-planner when planning was required. It took him until 1am on 2 April before he was finished, having been persuaded to go on dictating from his cot, placed on the deck. Six clerks then began to copy out the orders and finished their work only by 6am. At 7.37 Nelson signalled for all captains so that, by 8 o'clock, each had received a copy of his own orders. The new Admiralty Day Signal Book of 1799 was now about to be used for the first time in a full-dress engagement. These facts are important, as they help to dispel the idea that Nelson acted solely on intuition. With the Danish defence both known and static, every British ship could be given a detailed role in the battle. Nelson's haste was due to the wind having changed during the night, just as required; it was now blowing fresh from SSE.

There was an agonizing moment when the pilots jibbed, their experience being mainly with merchant ships in the Baltic trade, smaller than those now in their care. But, at last, everything was ready. Nelson signalled to weigh, and the squadron began to advance in line ahead. The Danes were unable to do anything until the leading British ship drew level with the southernmost in their own line. Nelson's plan was, with certain special exceptions, for the ships to anchor by the stern in succession in line ahead to engage the Danes, each ship passing on the disengaged side of her next ahead, thus inverting the line according to the standard form. The exceptions were due to the special need for ensuring the early defeat of the two southernmost of the Danish ships-of-the-line which had first to be encountered. A frigate and some smaller ships were detailed to support the attack by going in close and taking their bows. All ships were ordered to have springs on their cables, and the end of the sheet cable taken in at the stern port. Under cover of Nelson's twelve of the line, the smaller vessels were to move still further ahead on the disengaged side. Seven bomb vessels were to anchor about half way along the line and as close to the Middle Ground Shoal as their shallow draught would allow. From there they could fire over the ships-of-the line at the Danish ships and also at the city and harbour. Flat-bottom boats were ready to embark troops for assaulting the Three Crowns battery, as a preliminary to a general bombardment of the city, dockyard and arsenal by the bomb vessels. The last two ships-of-the-line were to push on farthest north and attack the Three Crowns battery. Captain Riou, placed in command of a division of frigates and sloops, had been ordered to act as circumstances might require; he found himself committed to joining in the attack on the battery.

Several things went wrong, but Nelson had reckoned on this, and was prepared to take all risks. The *Agamemnon* grounded on the south-east corner of the Middle Ground Shoal and never entered the inner channel at all. The *Bellona* and *Russell*, going too far to starboard, also grounded, and, though able to fire on the Danes through gaps in the British line, at a range of only half a mile, they were out of position and completely immobilised. The average distance between the two lines during the battle was only a little over a quarter of a mile. Riou's attack on the Three Crowns battery and attendant block ships met with very strong resistance, especially as neither of the two ships-of-the-line ordered to open the attack reached her position, the *Russell* being already aground. Owing to an adverse current, the gun-brigs were unable to push forward to join in the bombardment. Riou

Battle of Copenhagen – Nelson's Attack

Three Crowns battery

Battery

COPENHAGEN HARBOUR

Battery

Middle Ground

Bellona aground

Russell aground

Floating batteries ■

was killed while his ships were withdrawing in response to Sir Hyde Parker's signal 'To discontinue the engagement'.

Nelson continued his bombardment in spite of these reverses and in spite of Sir Hyde Parker's signal.[125] His great advantage lay in the fact that the Danes were stationary, so that some of his ships could shift the weight and direction of their attack. He was able, moreover, to make adjustments in his line by signal.

By about 2pm the battle was largely over. Many of the Danish ships and batteries were silenced. Desultory firing continued from the Three Crowns battery and their supporting ships, both against what were now the three leading British ships and Sir Hyde Parker's advanced ships beating to windward from the north. Some of the Danish ships which had ceased fire, and had their flags shot away, opened irregular fire again when approached by boats sent to take possession of them. The Three Crowns battery was meanwhile exchanging fire with the British across and through some of the Danish ships. It was this irregularity of conduct arising from ignorance of the customs of war (many of the Danes being volunteer landsmen) that caused Nelson to send his famous letter to the Crown Prince. When he eventually withdrew to the north, his own ship and another grounded within a mile of the Three Crowns, which showed how dangerous it would have been to have obeyed Sir Hyde Parker's signal and to have attempted to withdraw while still under fire.

British casualties were 1000 killed and wounded. The Danish total was more than twice that number; in addition, 3500 Danes were taken prisoner.

---

125 Number 39 in the Admiralty Signal Book

CHAPTER NINE

Tactical Development During the Napoleonic Wars

IN 1802 Audibert Ramatuelle published the most stimulating work in the French language on tactics under sail, *Cours Elémentaire de Tactique Navale, Dedié a Bonaparte*.[1] The author, an 'ancien officier de la Marine Militaire', was intent upon dealing with tactics in a practical manner. He regarded Morogues's obtuse angle formation for chasing and retreat as 'impracticable et désavantageuse dans tous les cas', and was critical of recent authors who had included a great many evolutions in their books 'la plupart inutiles, dont le résultat est la complication et la combinaison d'une grande quantité de mouvements dont on ne peut faire aucun usage devant l'ennemi'. Meanwhile, the tactical developments of

---

1 *NMM*, Tunstall Collection, TUN/88 (formerly T/P/F/7), includes a volume of plates

A plate from Audibert Ramatuelle's Cours Elementaire de Tactique Navale, *Paris, 1802, showing his hopelessly impractical designs for signal flags. NMM*

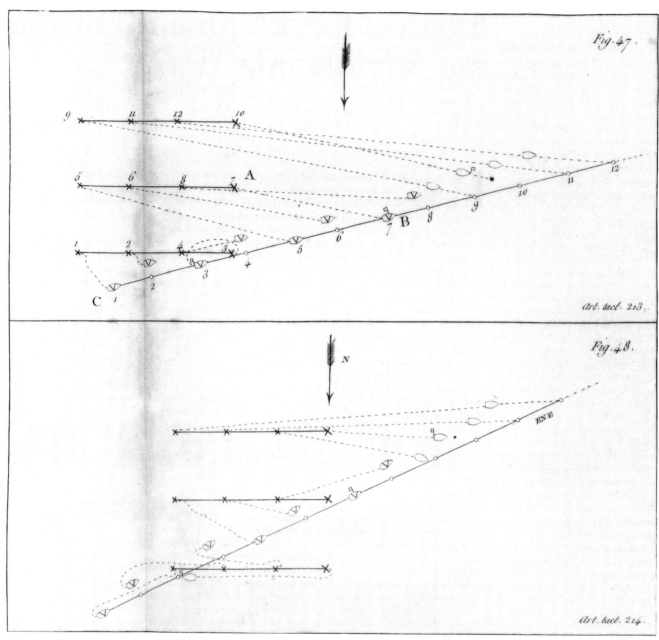

d'Orvilliers, de Guichen and Suffren had passed unnoticed. Few of Ramatuelle's own special ideas for attack may have been really feasible. Though lacking the simplicity and logic of John Clerk's work, however, no officer of the time could have read Ramatuelle's book without advantage.

In discussing battle tactics, Ramatuelle said that the fleet attacking from windward should come to the wind together when still at long range and dress their line carefully, ships taking station from the centre. He cited the example of Byng in describing the unwisdom of attacking from abeam, which obliged the attacking ships to run so much off the wind that they could not control their speed of advance by laying any of their sails aback. Unless a shift of wind was expected, Ramatuelle declared, it was better to attack from abaft the enemy's beam, with the ships lasking down in line of bearing, and thus well able to keep their station by laying sails aback, and able to use their broadsides all the time. In many respects the most interesting and original part of the book is an exposition of how to re-establish the order of sailing or battle after an unfavourable change of wind, and of how to take advantage of a favourable change of wind.[2]

He discussed in great detail the problem of doubling the enemy's van or rear from windward, and proposed various

defences against being doubled, including stationing the frigates in echelon ahead of the van to make it more difficult for the enemy's van to work round to windward. Doubling, he wrote, could also be executed by concentrating the heaviest ships in the centre and using the fastest sailers and frigates to double the enemy's rear. As an alternative, the fleet to windward might concentrate their fire against the weaker ships in the enemy's line, while leaving gaps in their own line arranged so that the enemy's heavy ships were left unopposed.

Another possibility suggested was that the windward fleet might refuse action and tack in succession. The fleets would then pass on opposite courses, but the windward fleet might defile in succession against the enemy's rear and then come to the wind again well astern so as not to lose the weather gage. Should the enemy wear together as a countermove, the van of the windward fleet, having already succeeded in their opening move, mighy tack in succession while the centre and rear formed astern of them and raked the enemy's van.

Further refinements included various ways of doubling the

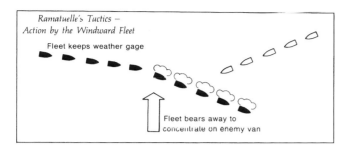

Ramatuelle's Tactics –
Action by the Windward Fleet
Fleet keeps weather gage
Fleet bears away to concentrate on enemy van

enemy's new-formed van once they had worn. If the enemy wore, as previously, the windward fleet might divide, the van and centre passing to leeward and the rear immediately wearing so that they passed astern of the enemy now in reverse order. They might then rejoin their centre and van and concentrate against the enemy's rear. D'Orvilliers's plan at the Battle of Ushant for doubling Keppel's rear by ordering the van to wear failed only, Ramatuelle asserted, because his signal was not seen and acted on quickly enough.

Ramatuelle wrote very clearly of the kind of manoeuvre used by Admiral Pierre Martin against Hotham in 1795. The windward fleet, closing the enemy on opposite course, held just close enough to the wind to retain the weather gage. The fleets would converge practically head on. When fairly close to each other the windward fleet should defile to windward in succession, engaging the enemy's van with every ship of their own while passing out of range of the enemy's centre and rear. Should the enemy counter this by turning away together and running large, the windward fleet could do the same, first in line of bearing and then pursuing in line abreast.

Ramatuelle had some ingenious schemes for the fleet to leeward. If the two vans became extended ahead and the leeward fleet was more numerous, the centre might be able to cut through the enemy's centre. If the gap was wide enough, this could be done together rather than in succession. Similarly, if the leeward's rear could overlap the windward's rear they might be able to double them to windward. A far more ambitious manoeuvre was for the van of the leeward fleet to wear together and then come to the wind in succession and cut through the enemy's line. When through, they would defile along the windward side of the enemy's rear on the opposite course, having gained the windward position. Meanwhile, the centre and rear of the leeward fleet should bear away to leeward, and then take station in the wake of their van. An equally ambitious though simpler move was for the leeward fleet, acting in two divisions, to wear together, the leading division then attacking the enemy's rear while the stern division passed through the gap in the enemy's line. All these schemes would have required superior numbers, a considerable gap in the enemy's line and very quick changes of sail on the part of the leeward fleet. They also required a degree of complacency on the part of the enemy, without which all such plans are difficult to realise in practice.

Ramatuelle's variant of the traditional French method of meeting an attack from windward was so simple that it might

well have produced still better results against the traditional form of British attack. The fleets having begun to pass on opposite courses, the windward fleet was to bear up in order to attack. Beginning to form in reverse order, the new van was to form into line ahead parallel with the leeward fleet while the centre and rear ran large until they came within range. The leeward fleet could now haul to the wind together, coming into line of bearing, and then alter course again so as to resume their order in line of battle ahead. They thus would bring themselves more to windward, nearer the enemy, and thereby throw the deployment of the latter into confusion. The leeward fleet might then defile along their van, and rake their centre and rear as they approached. There might also be a chance for the van of the leeward fleet to cut through the enemy's van.

The experience of de Guichen and Rodney showed that, if both fleets were close-hauled on the same tack, and the centre of the windward fleet broke through a gap in the leeward fleet, there was no immediate or satisfactory countermove. If, however, both fleets had the wind abeam, the centre of the leeward fleet might be able to isolate the ships which broke through by themselves hauling to the wind and then turning away again, thus producing a small encircling movement.

Returning to the case of fleets passing each other on opposite courses, Ramatuelle showed how the leeward fleet, by crowding sail, might double the enemy's rear. Should the windward fleet wear together, reversing direction, the main body of the leeward would be able to haul to the wind and defile across their bows. Meanwhile, those ships of the leeward fleet that had doubled should also wear so that the new van of the windward fleet would still be kept between two fires. Supposing, however, the windward fleet held their course when the doubling movement began, the leeward fleet would then have a fine chance to cripple them. Those ships that had doubled would tack in succession as they rounded the enemy's rear, and get on the same course with them. The remainder, having passed the rear of the windward fleet on the leeward side, would wear together and crowd sail, to draw up level with them. The enemy's rear would thus be placed between two fires. Supposing the windward fleet tacked in succession, thus threatening to double on the doublers, the main body of the leeward fleet should wear together and disengage itself on the opposite course. The ships that had doubled, being now in some danger, should shorten sail, wear together, and form in the wake of their main body, which had just disengaged itself.

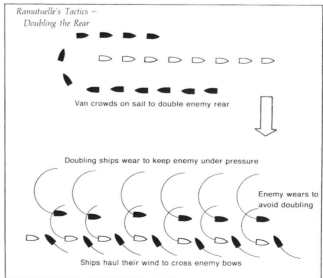

Ramatuelle's Tactics –
Doubling the Rear
Van crowds on sail to double enemy rear
Doubling ships wear to keep enemy under pressure
Enemy wears to avoid doubling
Ships haul their wind to cross enemy bows

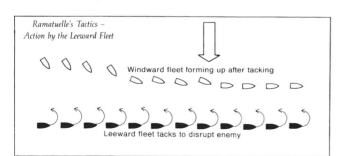

Ramatuelle's Tactics –
Action by the Leeward Fleet
Windward fleet forming up after tacking
Leeward fleet tacks to disrupt enemy

Admiral Sir Robert Calder's action off Cape Finisterre, 23 July 1805, by William Anderson. (see pp246–7)
NMM

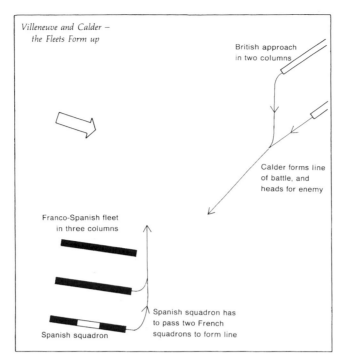

*Villeneuve and Calder –
the Fleets Form up*

British approach
in two columns

Calder forms line
of battle, and
heads for enemy

Franco-Spanish fleet
in three columns

Spanish squadron has
to pass two French
squadrons to form line

Spanish squadron

Fleet sailing large on a course east by south in three columns abreast with the Spanish squadron to starboard. His light squadron formed a fourth column away to larboard. On sighting the British to the north-east, Villeneuve deployed his fleet into a close-hauled line of battle ahead, steering north, with Gravina's starboard squadron leading. In this manner, Villeneuve risked losing the weather gage, since his leading ships, the Spaniards, had to parade right past the rest of the fleet (which had first to turn away to starboard to get into Gravina's water), then follow round astern of him to leeward. In other words, the deployment, though towards the British, was made from the column furthest removed from them. The light squadron, moreover, although nearest the British, had to take station astern of the whole fleet, which required it to run a good distance south before turning up again. Villeneuve presumably did this so as not to risk a shift of wind to the north, which would have enabled the British to steer straight for him.

On first report of the enemy being sighted to the south-west, Calder had turned his fleet towards them, signalling for two columns abreast in close order. Instead of trying to weather them, his intention was to interpose his squadron between Villeneuve and Ferrol and to engage him to leeward. He had two line-of-battle ships with his two frigates about six miles ahead and another 74 chasing a strange sail to leeward. The weather had been thick all day, but now it became worse. Although the two fleets were about sixteen miles apart when mutually sighted by their respective frigates, they were now approaching each other on opposite courses without any visibility at all. When the fog lifted, they were less than ten miles apart and Calder could see that he was outnumbered.

Sir Julian Corbett postulates that what followed shows that Calder shared 'Jervis's mistrust of formal tactics' and that 'He was not of the Howe–Kempenfelt School'.[16] This seems extremely hypothetical in view of what we now know of Jervis's theoretical, though not necessarily formalistic, approach to tactics. Conditions of weather and of relative strength encourage situational empiricism, and it seems unwise to postulate 'schools' when considering situations

which in other ways are not comparable. Calder did not signal his intention to the fleet, presumably because he wished to leave himself with a free hand until the last moment. Unlike Rodney, moreover, he trusted his captains to act to the best advantage when the time came. At a distance of eight miles from the enemy, he recalled his advanced ships and signalled for the line of battle ahead to the SSW, the starboard column leading, and a few minutes later for turning another point to starboard. The fog now closed down again for a quarter of an hour, and when it lifted the two fleets were beginning to pass each other on opposite courses but still out of gunshot.

Calder evidently intended to attack the enemy's rear, or centre and rear, by doubling or otherwise, eschewing all idea of an attack on the van, with or without an attempt to seize the weather gage. Villeneuve divined this easily enough. He, therefore, made a preparative signal for the Combined Fleet to wear in succession as soon as the van heard firing in rear. Gravina, in the leading ship, spotted the signal being repeated by a French frigate and wore at once, followed by the rest of the fleet in good order. The effect of this was to turn the head of the Combined Fleet south, much closer to the British and roughly on a parallel course. As Gravina came down in line ahead, he interposed himself between the British and the rearmost ships of the Combined Fleet, which were still shuffling into station to form line ahead running north. Thus Calder's plan was frustrated in two senses. He could no longer bring off a remunerative attack on the enemy's rear nor obtain the extra benefit of catching them in confusion before they were properly formed. As each new ship appeared astern of Gravina, heading south, the rearmost ships of the Combined Fleet were drawing further away to the north, well covered.

For the moment the fog completely masked Gravina's action. Calder, seeing his leading ship roughly level with the French centre still heading north, he himself being eighth in the British line, signalled the British fleet to tack in succession. The effect of this should have been to bring his van within close range of the enemy, on the same course and about level with the tail ships of their centre. As soon as the leading ship began to tack, Calder signalled to engage the enemy's centre, and soon after for close order. His plan seemed to be going smoothly. However, just as the leading

16 *Campaign of Trafalgar*, pp197–8

*Villeneuve and Calder –
the Engagement*

Villeneuve wears
in succession to
protect rear

Hero tacks, then
engages leading
Spanish ships

French rear not
fully organised

British tack in succession

British ship, the *Hero,* was coming within range, Gravina's flagship appeared out of the fog bearing down on the opposite course and at much closer range. Captain Alan Hyde Gardner, unable either to see or be seen by Calder, and, knowing the prejudice against fleets passing each other in action on opposite courses, tacked again to put himself on the same course as Gravina. As Gravina was leading the line to the south he had been hailed by Villeneuve in passing and ordered to steer WSW, so as to cover the rear of the Combined Fleet. Having fired a poorly aimed broadside at the *Hero,* still in stays, he bore away followed by the ship astern of him which, nevertheless, drove the *Hero*'s next astern out of the line in passing. The ships ahead of Calder all followed the lead given by the *Hero,* Calder himself signalling to tack in succession when his own turn came. Gravina was now in action with the fourth British ship, the *Barfleur,* and as the fog closed down again desultory firing developed all along Villeneuve's line as they headed south. The British rear ships, unable to judge where the enemy were or where the British van had gone, had no means of following the movement marked out by the two tacks in succession. They thus found themselves in action or otherwise on no real tactical basis. All, however, except the ship originally detached in chase, were engaged at various times, some very heavily.

In the Combined Fleet, equal if not greater confusion prevailed. The Spanish van squadron was heavily engaged; it alone met the British ships in any obvious and connected order, and its casualties amounted to nearly twice those of the whole British fleet. The two rearmost ships, fifth and sixth in the whole line, fell to leeward and were captured. The rear of the main Combined Fleet, including those ships which originally constituted the light squadron, was hardly in motion at all.

The Battle of Trafalgar

To understand what happened in the last and greatest battle between two fleets, each moving under sail, it is useful to look at the general intentions of the rival commanders-in-chief. In the Spanish naval base of Cadiz in early October 1805 lay the combined Franco-Spanish fleet under the supreme command of Vice-Admiral Villeneuve, under orders from the Emperor to carry a force of troops into the Mediterranean and land them in Neapolitan territory. The troops were in Cadiz ready for immediate embarkation. This was Villeneuve's first and paramount duty. He could leave Cadiz only if the wind was favourable, that is between northeast and south-east, and he knew that he would have to fight Nelson as soon as he got to sea. Nothing was said in Villeneuve's instructions about the absolute necessity for engaging and defeating the British fleet, so the threatened battle was relegated to an obstacle in the way of obeying unqualified orders to proceed from one place to another.

Villeneuve was a brave man, anxious to do his duty, but he had very natural misgivings about the quality of his ships and crews and their general lack of training. The crews of the Spanish ships were particularly poor. On balance, he knew that his ships were much less efficient and seaworthy than Nelson's. His misgivings were shared by the flag-officers and captains of the Combined Fleet, as was shown at his council of war held on 8 October. On the other hand, his relations with Admiral Gravina, and his Spanish allies generally, were excellent. There were no accusations and no intrigues.

Three considerations led Villeneuve to decide on what

Colonel Desbriere described succinctly as 'un acte de désespoir inutile'. The first was that Nelson had sent Rear-Admiral Sir Thomas Louis to Gibraltar with six of the line for water and victuals. Villeneuve was informed of this by his Gibraltar intelligence; contrary to what Nelson imagined, his main force, assumed to be out of sight, could often be counted from the tower of San Sebastian at Cadiz. Like all commanders of blockaded fleets, Villeneuve could choose his moment to emerge. Commanders of blockading fleets had to be constantly ready, and Nelson was now minus six of the line. Contrary to observed facts, Villeneuve represented himself to be numerically inferior to Nelson, but he realised that with Louis's absence he would never have a better chance of victory.

The second consideration was that the emperor had given secret orders for Villeneuve to be superseded by Vice-Admiral Francois Rosily, and Villeneuve knew or guessed this. The third was that the wind was favourable.

In Nelson's fleet the feeling was utterly different. Although his instruction from the Admiralty included responsibility for the reconstituted Mediterranean command, extending as far west as Cape St Vincent, every detail of their execution depended directly or indirectly on the defeat of the Combined Fleet which he was blockading. Nelson himself saw things in even more absolute terms. He aimed not only at victory but at the complete destruction of the enemy's fighting power. 'I should not doubt of spoiling any voyage they may attempt,' he wrote to Barham, 'yet I hope for the arrival of the ships from England, that as an enemy's fleet they may be annihilated.' [17] To Nelson the coming battle was to be not merely a fulfilment of all the efforts made since 1803 to defeat the French Toulon Fleet, now reinforced by the Spaniards, but a chance 'to take, sink and burn' every single ship in that fleet. 'Let the battle be when it may, it will never have been surpassed.' [18] This extraordinary boast and prophecy illustrates the level at which Nelson's spirit and mind were working. With the possible exception of Suffren, it is inconceivable that any other naval officer of the period

Amiral Pierre Charles Jean-Baptiste Silvestre, Comte de Villeneuve (1763–1806). MM

Admiral Horatio, Lord Nelson (1758–1805), first Viscount Nelson, by Lemuel Francis Abbott. NMM

17 Source unknown
18 Source unknown

should have used such words. Unlike Suffren, however, Nelson possessed that irresistible charm of manner and persuasive power which made even the stupid and the stubborn obey him. To Nelson the tactics necessary to ensure absolute victory must have been a constant preoccupation. The dream battle would be unlike anything that had happened in his own lifetime; unlike anything since Barfleur, and yet far more decisive. And so it proved to be.

During the seventeenth and eighteenth centuries, flag officers aged more quickly than their contemporaries ashore, amongst all of whom the expectation of life was considerably less than it is today. Rodney, Howe, St Vincent and Collingwood were all, in their various ways, men possessed of exceptional powers of endurance, but they all collapsed physically under the strain either of battle or prolonged command. Nelson, aged only just forty-seven when he died, was the youngest British commander-in-chief since Russell to win a major sea battle. It was this combination of supreme ardour and sufficient youth that enabled him to reject with scorn the accepted notion of a 'handsome victory'.

Nelson's fleet was a disparate lot brought together at different times. Only five out of his twenty-seven ships-of-the-line had been with the *Victory* in his original Mediterranean fleet. Only eight of his captains had served with him before. Only his flag captain, Hardy, and three other captains had been with him in his chase across the Atlantic and back again in pursuit of Villeneuve. Only five captains were commanding the same ships they had commanded in 1803. Only five had previously commanded a ship-of-the-line in battle. Of course they were able and experienced men and were assumed to be capable of efficient service in any fleet or squadron to which they might be suddenly appointed. Nevertheless, it was one thing to fit in with the well-known tactical requirements of a well organised fleet, but quite

another to help execute a new system of attack different from anything provided in any British fleet up to that time.

Tactical plans compared

Our understanding of Nelson's tactical plan comes from four sources.[19] The first was a plan of attack, probably written in 1803, and well worth quoting:[20]

The business of a commander-in-chief being first to bring an enemy's fleet to battle on the most advantageous terms to himself (I mean that of laying his ships close on board the enemy, as expeditiously as possible, and secondly, to continue them there without separating until the business is decided), I am sensible beyond this object it is not necessary that I should say a word, being fully assured that the admirals and captains of the fleet I have the honour to command will, knowing my precise object, that of a close and decisive battle, supply any deficiency in my not making signals, which may, if extended beyond those objects, either be misunderstood, or if waited for very probably from various causes be impossible for the commander-in-chief to make. Therefore it will only be requisite for me to state in as few words as possible the various modes in which it may be necessary for me to obtain my object; on which depends not only the honour and glory of our country, but possibly its safety, and with it that of all Europe, from French tyranny and oppression.

If the two fleets are both willing to fight, but little manoeuvring is necessary, the less the better. A day is soon lost in that business. Therefore I will only suppose that the enemy's fleet being to leeward standing close upon a wind, and that I am nearly ahead of them standing on the larboard tack. Of course I should weather them. The weather must be supposed to be moderate; for if it be a gale of wind the manoeuvring of both fleets is but of little avail, and probably no decisive action would take place with the whole fleet.

Two modes present themselves; one to stand on just out of gun-shot, until the van ship of my line would be about the centre ship of the enemy; then make the signal to wear together; then bear up [and] engage with all our force the six or five van ships of the enemy, passing, certainly if opportunity offered, through their line. This would prevent their bearing up, and the action, from the known bravery and conduct of the admirals and captains, would certainly be decisive. The second or third rear ships of the enemy would act as they please, and our ships would give a good account of them, should they persist in mixing with our ships.

---

19 In his manuscript, Tunstall listed a fifth source of information, the famous and mysterious 'Nelson Touch'. Oliver Warner has revealed this for what it really was, the 'Nelson Trunk', to which quite naturally there was a key. See Oliver Warner, editor, *Nelson's Last Diary*, Kent, Ohio, 1971, p20.

20 Corbett, *Fighting Instructions*, pp313-6, with punctuation by Corbett from Clerke & McArthur, II, p427, and Nicolas, VI, p 443.

Nelson's Tactics

Nelson gains weather gage

Fleet wears together and concentrates on enemy van

Nelson bears away to head directly for enemy

Nelson declines weather gage

British ships wear, rearmost first

Breaks enemy line

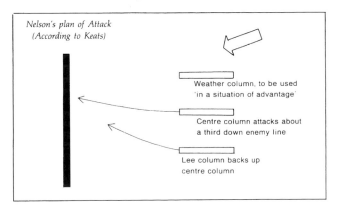

Nelson's plan of Attack (According to Keats)

Weather column, to be used 'in a situation of advantage'

Centre column attacks about a third down enemy line

Lee column backs up centre column

The other mode would be to stand under an easy but commanding sail directly for their headmost ship, so as to prevent the enemy from knowing whether I should pass to leeward or to windward of him. In that situation I would make the signal to engage the enemy to leeward, and cut through their fleet about the sixth ship from the van, passing very close. They being on a wind and you going large could cut their line when you please. The van ships of the enemy would, by the time our rear came abreast of the van ship, be severely cut up, and our van could not expect to escape damage. I would then have our *rear* ship and every ship in succession wear [and] continue the action with either the van ship or the second as it might appear most eligible from her crippled state; and this mode pursued I see nothing to prevent the capture of the five or six ships of the enemy's van. The two or three ships of the enemy's rear must either bear up or wear; and in either case, although they would be in a better plight probably than our two van ships (now the rear), yet they would be separated and at a distance to leeward, so as to give our ships time to refit. And by that time I believe the battle would, from the judgment of the admiral and captains, be over with the rest of them. Signals from these moments are useless when every man is disposed to do his duty. The great object is for us to support each other, and to keep close to the enemy and to leeward of him.

If the enemy are running away, then the only signals necessary will be to engage the enemy on arriving up with them; and the other ships to pass on for the second, third, &c., giving if possible a close fire into the enemy on passing, taking care to give our ships engaged notice of your intention.

The second source of information about the development of Nelson's plan of attack is Sir Richard Keats's account of a conversation with Nelson in late August or early September, 1805:

One morning, walking with Lord Nelson in the grounds of Merton, talking on naval matters, he said to me, 'No day can be long enough to arrange a couple of fleets and fight a decisive battle according to the old system. When *we* meet them,' (I was to have been with him), 'for meet them we shall, I'll tell you how I shall fight them. I shall form the Fleet into three Divisions in three Lines. One Division shall be composed of twelve or fourteen of the fastest two-decked Ships, which I shall keep always to windward or in a situation of advantage, and I shall put them under an Officer who, I am sure, will employ them in the manner I wish, if possible. I consider it will always be in my power to throw them into Battle in any

part I may choose; but if circumstances prevent their being carried against the Enemy where I desire, I shall feel certain he will employ them effectually, and, perhaps, in a more advantageous manner than if he could have followed my orders.' (He never mentioned, or gave any hint by which I could understand who it was he intended for this distinguished service.)[21] He continued – 'With the remaining part of the Fleet formed in two Lines, I shall go at them at once, if I can, about one third of their Line from their leading Ship.' He then said, 'What do you think of it?' Such a question I felt required consideration. I paused. Seeing it he said, 'But I will tell you what *I* think of it. I think it will surprise and confound the Enemy. They won't know what I am about. It will bring forward a pell-mell battle, and that is what I want.' [22]

These two quotations are mainly valuable as evidence of preliminary thinking. Of more immediate importance was Nelson's memorandum written inboard *Victory*, off Cadiz, 9 October 1805. Again it is worth quoting at length:

Secret

Thinking it almost impossible to bring a Fleet of forty Sail of the Line into a Line of Battle in variable winds, thick weather, and other circumstances which must occur, without such a loss of time that the opportunity would probably be lost of bringing the Enemy to Battle in such a manner as to make the business decisive; I have therefore made up my mind to keep the Fleet in that position of sailing (with the exception of the First and Second in Command) that the Order of Sailing is to be the Order of Battle; placing the fleet in two Lines of sixteen Ships each, with an Advanced Squadron of eight of the fastest sailing Two-decked Ships *which* will always make, if wanted, a Line of twenty-four Sail on whichever Line the Commander-in-Chief may direct.

The Second in Command will, after my intentions are made known to him, have the entire direction of his Line to make the attack upon the Enemy, and to follow up the blow until they are Captured or destroy'd.

If the Enemy's fleet should be seen to Windward in Line of Battle and [in such a position] that the two Lines and the Advanced Squadron can fetch them, they will probably be so extended that their Van could not succour their Rear.

I should therefore probably make the 2nd in Command's signal to Lead through, about their Twelfth Ship from the Rear (or wherever he could fetch, if not able to get as far advanced); my Line would lead through about their Centre, and the Advanced Squadron to cut two, three or four Ships Ahead of their Centre, so far as to ensure getting at their Commander-in-Chief, on whom every effort must be made to capture.

---

21 Corbett's note: it was certainly not Keats himself, though afterwards Nelson meant to offer him command of the squadron he intended to detach into the Mediterranean. In the expected battle Keats, had he arrived in time, was to have been Nelson's second in the line; Nelson to Sir Alexander Hall, October 15 1805.

22 Nicolas, VII, p241

The Cadiz Memorandum –
Attack from Leeward

Enemy fleet of 42 ships

Advanced squadron
○○○○○○○○

○○○○○○○○○○○○○○○○○
Nelson's squadron

○○○○○○○○○○○○○○○
Second-in-command's squadron

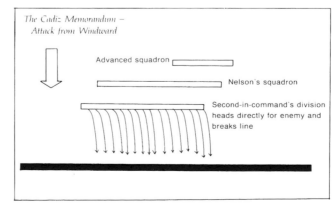

The Cadiz Memorandum –
Attack from Windward

Advanced squadron

Nelson's squadron

Second-in-command's division heads directly for enemy and breaks line

The whole impression of the British Fleet must be to overpower from two to three Ships ahead of their Commander-in-Chief, supposed to be in the Centre, to the Rear of their fleet. *I will suppose twenty Sail of the Enemy's Line to be untouched*; it must be some time before they could perform a manoeuvre to bring their force compact to attack any part of the British fleet engaged, or to succour their own Ships; which indeed would be impossible, without mixing with the ships engaged.[23]

Something must be left to chance; nothing is sure in a Sea Fight beyond all others. Shots will carry away the masts and yards of friends as well as foes; but I look with confidence to a Victory before the Van of the Enemy could succour their Rear and then that the British Fleet would most of them be ready to receive their twenty Sail of the Line, or to pursue them, should they endeavour to make off.

If the Van of the Enemy tacks, the Captured Ships must run to Leeward of the British fleet; if the Enemy wears, the British must place themselves between the Enemy and the Captured and disabled British Ships; and should the Enemy close, I have no fears as to the result.

The Second in Command will in all possible things direct the Movements of his Line, by keeping them as compact as the nature of the circumstances will admit. Captains are to look to their particular Line as their rallying point. But, in case Signals can neither be seen or perfectly understood, no Captain can do very wrong if he places his Ship alongside that of an Enemy.

The Divisions of the British fleet will be brought nearly within gun shot of the Enemy's Centre. The signal will most probably then be made for the Lee Line to bear up together, to set all their sails, even steering sails, in order to get as quickly as possible to the Enemy's Line and to Cut through, beginning from the twelfth Ship from the Enemy's Rear. Some Ships may not get through their exact place, but they will always be at hand to assist their friends; and if any are thrown round the Rear of the Enemy, they will effectually complete the business of Twelve Sail of the Enemy.

Should the Enemy wear together, or bear up and sail large, still the twelve Ships composing, in the first position, the Enemy's Rear, are to be *the* object of Commander-in-Chief; which is scarcely to be expected, as the entire management of the Lee Line, after the intentions of the Commander-in-Chief is [sic] signified, is intended to be left to the judgment of the Admiral Commanding that Line.

The Remainder of the Enemy's fleet, 34 Sail, are to be left to the Management of the Commander-in-Chief, who will endeavour to take care that the movements of the Second in Command are as little interrupted as possible.

The Memorandum introduces three major ideas: the separate role and responsibility of the second-in-command; the concentration of the attack on the enemy's rear, with their centre contained and thus prevented from interfering, and with their van unable to interfere until too late; concealment of the direction of attack until the very last moment. None of these ideas was new. A fourth and new idea is specifically mentioned: crowding sail, something entirely contrary both to contemporary means and traditional practice. Here a calculation is made of the risk of collision involved, and of the heavy sailers, not merely of dropping astern, but being out of alignment altogether. Against this is the dual advantage of presenting a quickly moving target to the enemy and of covering more quickly the space in which enemy fire must be endured without reply. Added to this is the association of speed with concealment of the point of attack. Not only will the enemy be kept guessing, but they will have less time in which to guess.

Yet a further idea, thought by Sir Julian Corbett to be implied in the Memorandum, when put alongside Nelson's latest known order of sailing, dated 10 October, is that he intended his three-deckers to lead the attack in the manner earlier suggested by Morogues. This, Sir Julian points out, was still valid even if Nelson and Collingwood, after leading their respective lines in the order of sailing, dropped back to second and third, respectively, when changing over to the order of battle. Nelson's conjectural order of battle shows:

Van or Weather squadron

First division
1 *Téméraire* 98
2 *Victory* 100
3 *Neptune* 98
4
5

Second division (six ships)
7 *Britannia* 100
8
9
10
11
12

Rear or Lee squadron

First division (eight ships)
1 *Prince* 98
2 *Mars* 74
3 *Royal Sovereign* 100
4 *Tonnant* 80
5 *Belleisle* 74
6
7
8

Second division (seven ships)
9 *Dreadnought* 98
10
11
12
13
14
15

In the battle itself, the order of engaging the enemy is still undetermined. We know for certain that in the weather line it was the *Victory*, *Téméraire*, *Neptune*, with the *Britannia* possibly sixth. For the lee line it was certainly the *Royal Sovereign* with the *Mars* and *Belleisle* either second and third or third and second. The *Dreadnought* was probably tenth and the *Prince* about fourteenth.

In Nelson's Memorandum of 9 October there is a note in the upper margin after the words 'The signal will most probably *then* [interlinear addition] be made for the Lee Line to bear up together and set all their sails even steering sails . . . '. It reads '*Vide* instructions for Signal Yellow with Blue fly, page 17, eighth Flag Signal Book, with reference to appendix'.

Sir Julian Corbett, when writing *The Campaign of Trafalgar*, suggested that since the signal did not appear in the

---

23 The words in italics were added by Mr Scott, Lord Nelson's Secretary. In the upper margin of the paper Lord Nelson wrote and Mr Scott added to it a reference, as marked in the text – 'the enemy's fleet is supposed to consist of 46 Sail of the Line, British fleet of 40 – if either is less, only a proportionate number of Enemy's ships are to be cut off; B to be ¼ superior to the E cut off.' Nicolas, VII, p89.

printed appendix it must have been added by Nelson himself.[24] His supposition was correct; the full text of the signal is in a copy of the Appendix in the Tunstall collection.[25] It reads:

Signals from Senior to Junior Officers by Lord Nelson.
Single Flag Yellow with Blue Fly. Cut through the Enemy's line and engage them on the other side.
NB: This Signal is to be repeated by all the Ships. The Number of the Enemy's ship under the Stern of which the Van Ship is to pass and engage will be pointed out by Signal counting from the Enemy's Rear. The Ships being prepared are to make all possible sail (keeping their relative Bearings and close order) so that the whole may pass thro' the Enemy's line as quick as possible and at the same time. It is recommended to cut away the Studding sails if set, to prevent confusion and fire. Each ship will of course pass under the Stern of the one she is to engage if circumstances permit, otherwise to refer to the Instructions Page 160 Article 31 [for breaking the enemy's line]. The Admiral will probably advance his Fleet to the van of theirs before he makes the Signal in order to deceive the Enemy by inducing them to suppose it is his intention to attack their Van.

The National Maritime Museum possesses a copy of the Signal Book of 1799, issued between 2 May 1805 and 10 February 1806,[26] marked 'Exd TMH' [Examined TM Hardy]. It was most probably used in Nelson's fleet at Trafalgar though attempts to identify it with the *Victory* appear optimistic rather than scholarly. It is the only one of many examined which has a yellow and blue flag for single flag signalling on page 17, though actually in the ninth and not the eighth position. In the blank signal space opposite, is the pencilled insertion: 'Cut through the Enemy's line and engage close on the other side . . . NB: This signal to be repeated by all the ships (*vide* Appendix)'. It would seem, therefore, that the signal inserted in the appendix must be accepted as having a very close relationship to Nelson's Memorandum, the reference in the signal quoted immediately above being taken as collateral evidence. In that case it was certainly valid twelve days before the battle, and must be accepted as the fourth source of information about Nelson's tactical plans.[27] One reason for the uncertainty which still persists about Nelson's actual tactics in the battle and the greater uncertainty still as to the actual text of the instructions under which it was fought, is that Nelson had no captain of the fleet. His personal staff consisted of a civilian and a divine: John Scott, his public or admiral's secretary, who was killed in the battle, and the Revd Alexander John Scott, the ship's chaplain, who acted as his private secretary, interpreter and intelligence officer. An efficient first-captain would have taken care to preserve all copies of battle instructions and orders of battle and sailing, leaving the chaplain to attend, as indeed he did most faithfully, to Nelson's personal affairs. After the battle, Hardy of course was much too busy superintending the repair of the ship and keeping off a lee shore to collect and title the tactical papers. His own captain's clerk, equivalent of a secretary, was also killed in the battle.

On the day of the battle, Captain Henry Blackwood of the *Euryalus* acted as an *ad hoc* chief of staff. He was a captain of ten years' seniority with a great fighting record. Nelson had given him command of the whole of the inshore forces, Advance squadron and frigates included. He had even refused the offer of a ship-of-the-line in place of his frigates. On the morning of the battle he spent three hours with Nelson in the *Victory*, leaving only when the ship was already under fire. Nelson, he wrote, 'not only gave me the command of all the frigates [in the battle itself], for the purpose of assisting disabled ships, but he also gave me a latitude seldom or ever again, that of making any use I

pleased of his name in ordering any of the sternmost line of battle ships to do what struck me as best.' In using his frigates 'to complete the destruction of the enemy whether at anchor or not', Blackwood was ordered not to think about saving ships or men, destruction being the chief object, not prizes or prize money. This last command reveals the full measure of Nelson's demand for absolute victory.

Villeneuve's final instructions, dated 21 October, the actual morning of the battle, are practically a repetition of those he issued before leaving Toulon at the end of March. They are in very general form until the paragraph about probable British tactics is reached, in which we read:

The enemy will not confine himself to forming on a line of battle with our own and engaging us in an artillery duel, in which success is frequently with the more skilful but always with the more fortunate; he will endeavour to envelop our rear, to break through our line and to direct his ships in groups upon such of ours as he shall have cut off, so as to surround them and defeat them.

This is so startlingly accurate that one is at first inclined to think that the French Government must have been operating a very efficient intelligence service. It seems more likely, however, that this was no more than an appreciation made by Villeneuve himself, using his personal experience at the Battle of the Nile, the pocket manuscript signal book found in the captured schooner *Redbridge* during the previous year, and perhaps the 1804 edition of Clerk's *Essay on Naval Tactics*. A tactical study of the Battle of Camperdown would also have been useful. What seems equally surprising is that, having told his flag officers and captains what to expect, he vouchsafed nothing whatever about how to combat it. A final reference to fleet regulations and to the Specific Instructions for the Repetition of Signals, printed at the head of the Signal Book for the Use of the Fleet, sounds completely inadequate, especially when followed by the remark, 'They contain excellent maxims with which you must be fully conversant and dispositions which I am keeping in force to their full extent.'

In fact Villeneuve had nothing better to offer than he had had in the previous year. French tactics, so far as actual fighting was concerned, were completely barren. Spanish tactics, vigorous and resourceful on paper, were equally barren at sea.

Villeneuve's only contribution to executive tactics, his Squadron of Observation of twelve ships organised in two divisions and commanded by Gravina in the 112-gun *Principe de Asturias*, should have proved a decisive element once the

---

24 p449n

25 *NMM*, Tunstall Collection, TUN/61 (formerly S/P/R/8). This copy of the Appendix also contains a list of 'Additional Rendezvous by Lord Nelson' marked 'not in force' and two pages of 'Lord Nelson's Additional Rendezvous 26 September 1805' evidently written before joining the fleet off Cadiz on 29 September. This, however, does not mean that this copy of the Appendix was used in the *Victory*, though equally it does not preclude this.

26 *NMM*, SIG/B/76 (formerly Sp/102); Board signatures indicate date limits, otherwise undated. This is a later printing by William Winchester & Son, Strand.

27 Article 31 on p160 of this copy of the Signal Book has two separate pencilled additions: 'Signal 27 with a White Pendant is for the ships to break through the enemy wherever they can, filled in by Mr [?]'; and 'Signal 27 with a Dutch Pendant, according to the masthead it is hoisted'. Article 31 of the Fighting Instructions is for the execution of signal no 27, breaking the line. Curiously enough, this signal, which was printed in full in the original issue of 13 May 1799, is left blank in the later printing, the words being inserted in ink; this may have been a security measure. It is also curious that the variations of the signal noted above should appear as additions to the instruction and not to the signal itself.

The Battle of Trafalgar, 21 October 1805; beginning of the action, by Nicholas Pocock. NMM

Battle of Trafalgar –
Nelson's Squadron Attacks

Santissima Trinidad
Bucentaure
Neptune
Victory *Redoutable*

Santa Ana *Royal Sovereign*

A little before 8am Nelson signalled for his four frigate captains to come on board the *Victory*. Meanwhile, he held on his course followed by his division, steering for what seemed to be roughly the centre of the whole Combined fleet, Gravina included. This was also the point of greatest concavity in the enemy line and where there was still a gap between the *Santa Ana* and the four ships ahead of her, all of which had fallen considerably to leeward. Ahead of these were the *Redoutable* (74), the *Bucentaure* (80), flying Villeneuve's flag, and the *Santissima Trinidad* (120), flying the flag of Rear-Admiral Don Báltasar Cisneros. The Combined Fleet had steerage way of about half a knot and if Nelson intended to cut through between the thirteenth and fourteenth ship, the place was between the *Redoutable* and her next astern, but actually he seemed to be steering for the *Santissima Trinidad*, the four-decker being the obvious big ship to attack in that part of the enemy's line. Apart from signals to individual ships to make more sail and get into station, even including one for the *Mars* to reach her station at the head of Collingwood's line (without, of course, hope of response), Nelson gave no further orders. At 10.50am he 'telegraphed' to Collingwood, 'I intend to pass or go through the enemy's line to prevent them getting into Cadiz.' This was purely informative but shows the way his mind was working.

As the *Victory* came within random gunshot, Nelson dismissed his frigate captains, though Blackwood remained a little longer. They were to hail the ships astern, telling them it was Nelson's intention 'to cut through the enemy's line about the thirteenth or fourteenth ship, and then to make sail on the larboard tack for their van'.

Nelson appears to have had three courses of action in mind. He could do as he told his frigate captains. This, however, would not necessarily enable him to attack and overwhelm the enemy's commander-in-chief. With as yet no flags showing, where could he be, now that Gravina's junction made the true centre of the enemy's line in doubt? Nelson could, on the other hand, attack the *Santissima Trinidad*, and so at least make certain of a flag officer. Still a little uncertain, he had the *Victory* turned slightly to larboard as if to attack the enemy's van. There seems no doubt about this turn. Was he looking for Villeneuve, or a good place to cut through, or was it a feint? Although the newly discovered signal with the yellow and blue flag was never made, we cannot today, in the light of this new evidence, entirely dismiss the possibility that some notion of a feint entered into his calculations. At this point the nearest ships of the Combined Fleet opened fire with full broadsides, all flags were hoisted, and the commander-in-chief was at last revealed in the *Bucentaure*, twelfth

31 Cd. 7120, 1913

in the line. Nelson at once had the *Victory* turned to starboard and ran back along the enemy's line under heavy fire. The wind had almost dropped and he could hardly have been doing two knots. Angling to pass astern of *Santissima Trinidad* and Villeneuve, he was foiled by the *Bucentaure* being brought up to the wind on the *Santissima Trinidad*'s larboard quarter. At the same time, Captain Lucas brought the *Redoubtable* up astern of the *Bucentaure*. The *Victory* bashed through between them, but the immediate attempt of the following British ships to push through in succession was defeated. Ships became locked together and drifted to leeward, so that the original gap ahead of the *Santa Ana* opened wider. The remaining ships of Nelson's division were brought into action with great skill wherever there seemed a disengaged enemy to attack or a British ship to be relieved. Codrington was not far wrong in saying that they 'scrambled' into action as best they could. Nelson had succeeded in bringing the leading ships of both divisions to a very close action on the lee side of the enemy without having a single one disabled during the approach, the speed of which – even in case of the fastest sailer – never exceeded three knots. The pell-mell was now joined and just as Nelson guessed, planned and hoped, the enemy's van failed completely to help their centre and rear.

The prolonged controversy at the beginning of the present century about the tactics actually pursued by Nelson from dawn on 21 October until about 2.30pm, when most of the British ships were engaged, tended to centre on a question that had already been answered. The question was in what order did the two columns attack and how did this fit in with Nelson's Memorandum of 9 October? The answer had already been given by Colonel Edouard Desbrière in his monumental and exhaustive work, *La Campagne Maritime de 1805: Trafalgar*, published under the direction of the historical section of the French General Staff in 1907. Unfortunately, it was not read in England, though Sir Julian Corbett used it as his main authority for the French and Spanish side when writing *The Campaign of Trafalgar*, published in 1910. By 1912 the controversy was still so much alive that the Admiralty appointed a committee 'for the purpose of thoroughly examining and considering the whole of the evidence relative to the tactics employed by Nelson at the Battle of Trafalgar'. It took this committee fifteen months to produce their report, which provided the British counterpart of what Desbriere had already discovered and it in no way modified his main conclusions.[31] Their findings, moreover (though there was no such admission) confirmed the work of Sir Julian Corbett. The secretary to the committee, William Gordon Perrin, then Admiralty librarian, supplied a schedule of all the British signals made, just as he had done for Corbett's book three years previously.

The British fleet never formed properly in the technical sense of the word. Forming meant not only getting the ships into the right order but also at the standard distance from each other. To do this, it was necessary for ships already in their stations not to set so much sail that the ships out of station had no chance to gain it. Once formed, it was also necessary to avoid crowding sail so much that the heavy sailers were unable to keep up by crowding sail themselves. None of these considerations seemed to matter much at Trafalgar. All ships crowded sail. Ships at a distance had no chance to get into station, and of those the heavy sailers found themselves dropping further and further astern. As Nelson saw it, the most important thing was for all ships to reach the enemy as quickly as possible.

The British fleet triumphed because, once the pell-mell became general, the gunnery of the individual ships was better than that of the individual ships of the Combined Fleet.

In the pell-mell, moreover, the British captains were better at handling their ships so as to give mutual support, and in making such movements as were possible in order to bring superior fire to bear at angles from which their opponents could not so effectively reply. As far as bravery was concerned, there was little to choose between the two fleets. Apart from ships which sank with their crews still on board, the Combined Fleet suffered enormous casualties. The *Redoubtable* had 490 men killed, the *Bucentaure* 197, the *Santissima Trinidad* 216. Twenty-six flag-officers and captains were either killed or wounded. These are terrible figures. Many ships refused to surrender until either there were no men left capable of firing guns or no guns left capable of being fired. The *Victory* had fifty-seven men killed, the highest figure for any British ship.

The following propositions seem worth considering. If Nelson had realised that when Villeneuve wore the Combined Fleet towards Cadiz he was not intending to return to the port and take refuge, there would have been more time for the British to form. Furthermore, had both divisions of the British fleet been better formed, the impact on the Combined Fleet would have been greater and the victory more complete. It can be argued, however, that by launching his attack at once, slow though his advance was bound to be, Nelson forced the Combined Fleet out of their stride and paralysed Villeneuve's capacity for initiative. More particularly, Nelson's immediate advance prevented Villeneuve from taking preliminary steps to start an enveloping movement, either by tacking his van or by using Gravina's Squadron of Observation against Collingwood's lee division. It becomes a question, therefore, of how much was lost and how much was gained by Nelson's precipitate action. Opinions about this will always tend to reflect the personal temperament of those who give them.

Strachan and Dumanoir

In a brief action on 4 November 1805 Sir Richard Strachan, with an 80, three 74s and four frigates, captured Rear-Admiral Dumanoir's squadron of four French ships surviving from Trafalgar. The French squadron also, on paper, comprised an 80 and three 74s, but with no frigates. Dumanoir had been obliged to jettison twenty guns to keep the *Formidable* afloat. He was being slowly overhauled, though still fairly well ahead, when he was forced to haul to the wind and form in line of battle. This, surprising as it may seem, was because the two leading British frigates, ahead of the rest of the squadron, had gained enough distance to begin firing broadsides into his sternmost ship.

Strachan's fourth ship, the *Namur,* was still a good way astern, so he hailed the captains of the two ships already up with him to say that he intended to attack the enemy's rear. Strachan himself took the lead in the *Caesar* (80), and, being to windward, brought to opposite Dumanoir, who was second in the French line. Thus the main action began with these three British ships-of-the-line engaging the three rear French ships, the leading French ship being so far left alone. Meanwhile, all four British frigates, ignoring custom and tradition, came up on both quarters of the rearmost French ship and joined in the battle. Dumanoir signalled his squadron to tack in succession, so as to relieve the pressure on his rear and try to cut off the *Namur,* now approaching. It was a very bold manoeuvre and splendidly executed in the face of broadsides at close range. Strachan also tried to tack but had to wear, and this produced temporary confusion. With his three ships round, and with the *Namur* now into place as second in his reversed line, he soon began to get the upper hand. By this time, the frigates were engaging the

French on the leeward side of their line, thus putting them between two fires. After nearly four hours of fighting all four French ships surrendered, Dumanoir's *Formidable* being in danger of sinking. Only two masts out of a total of twelve remained standing in the French ships at the end. British casualties were twenty-four killed and 116 wounded.

Duckworth and Leissègues

After a remarkable series of strategic movements which took him from the post-Trafalgar blockade of Cadiz to the West Indies, Sir John Duckworth in the *Superb* (74), under Captain Richard Keats, found himself on the morning of 6 February 1806 in sight of a French squadron at San Domingo. He had besides an 80, four more 74s, a 64 and three frigates. His squadron included Rear-Admiral the Hon Alexander Cochrane and Rear-Admiral Thomas Louis. The French squadron, which was ready for sea, was commanded by Vice-Admiral Corentin Leissègues in the 120-gun *Impérial*; he had with him an 80, three 74s and three frigates. On sighting an enemy force to windward, Leissegues made sail to the west and formed in line ahead with the *Impérial* second and with the frigates in line ahead inshore of him. Duckworth immediately chased and with the wind almost directly astern of him, gained rapidly. Dividing his force as he went, he took four ships with him, including Cochrane's, against the French admiral and ordered Louis to attack the French rear with the remaining three ships.

Although the sailing powers of the British ships varied considerably, they were all in action in the end, despite the surprising statement that the French were running at 8 knots. As the *Superb* began to engage (with the band playing God Save the King on the poop), Keats hung a portrait of Nelson on the mizzen stay, where it remained untouched throughout the battle, though splattered with the blood and brains of a seaman killed close beside it. Smoke soon covered the scene, yet despite confusion and collisions with each other, the

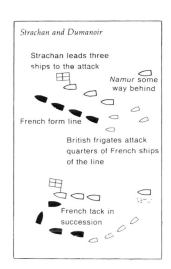

Strachan and Dumanoir

Strachan leads three ships to the attack

Namur some way behind

French form line

British frigates attack quarters of French ships of the line

French tack in succession

Admiral Sir John Thomas Duckworth (1748–1817), artist unknown, in the style of Sir William Beechey. NMM

Duckworth's action off San Domingo, 6 February 1806, by Nicholas Pocock (dated 1806). NMM

British ships fought with great energy and skill. Only the French frigates escaped; three of the line were taken and the *Impérial* and another wrecked.[32]

British tactical memoranda

Even before Trafalgar, the character of the war at sea had begun to change. Between the recommencement of the war in 1803 and its conclusion in 1814, Trafalgar was the only great naval battle. Apart from Strachan's defeat of Dumanoir, almost a part of Trafalgar itself, and Duckworth's overwhelming victory at San Domingo, tactical encounters by battle fleets were at an end. Shore bombardments, combined operations, attacks on ships at anchor and the constant watch on blockaded forces, absorbed the activities of substantial British fleets. However, the great efforts Napoleon made to acquire a new fleet with which to challenge the grip Britain had gained on the sea ensured that efficiency in battle tactics remained vital. There were at least three important British tactical documents issued for operations which in fact never took place.

Admiral Gambier, when proceeding to the Baltic in 1807, issued instructions for a fleet of thirty of the line. The orders of sailing and of battle were the same, with the fleet in three squadrons of ten ships each, and each squadron in two divisions of five ships each. Gambier place himself in the centre of the fleet, but the commanders of each subdivision of the van and rear were stationed in such a manner as to lead his own subdivision when his own squadron led the fleet. The effect was to strengthen the extremities, rather in the manner of the seventeenth century.

The commodores were to command their own subdivisions 'subject to the general direction' of their squadron commanders. The distance between ships was to be two cables, 'increasing that distance according to the state of the weather', no doubt a precaution for North Sea conditions. When sailing in two columns, which were referred to as grand divisions, the leading ship of the starboard (windward) division was to keep the admiral two points on her weather bow. The leading ship of the lee division was to keep the leader of the weather column two points before her beam, 'when sailing large, abreast of her'. These tactics were never really tested, because the Danes, as in 1801, declined a fleet action.

Collingwood's General Order of 23 March 1808 was much more detailed and, superficially at least, more reminiscent of the ideas of Nelson.

Collingwood's Tactics – Attack from Leeward

Weather column breaks line one quarter from the van

Lee column crowds on sail, then breaks through the centre

Should the enemy be found formed in order of battle with his whole force, I shall notwithstanding probably not make the signal to form the line of battle; but, keeping in the closest order, with the van squadron attack the van of the enemy, while the commander of the lee division takes the proper measures, and makes to the ships of his division the necessary signals for commencing the action with enemy's rear, as nearly as possible at the same time that the van begins. Of his signals, therefore, the captains of that division will be particularly careful.

If the squadron has to run to leeward to close with [the] enemy, the signal will be made to alter the course together, the van division keeping a point or two more away than the lee, the latter carrying less sail; and when the fleet draws near the enemy both columns are to preserve a line as nearly parallel to the hostile fleet as they can.

In standing up to the enemy from the leeward upon a contrary tack the lee line is to press sail, so that the leading ship of that line may be two or three points before the beam of the leading ship of the weather line, which will bring them to action nearly at the same period.

The leading ship of the weather column will endeavour to pass through the enemy's line, should the weather be such as to make that practicable, at one fourth from the van, whatever number of ships their line may be composed of. The lee division will pass through at a ship or two astern of their centre, and whenever a ship has weathered the enemy it will be found necessary to shorten sail as much as possible for her second astern to close her, and to keep away, steering in a line parallel to the enemy's and engaging them on their weather side.

A movement of this kind may be necessary, but, considering the difficulty of altering the position of the fleet during the time of combat, every endeavour will be made to commence battle with the enemy on the same tack they are; and I have only to recommend and direct that they be fought with at the nearest distance possible, in which getting on board of them may be avoided, which is always disadvantageous to us except when they are flying.

As Corbett pointed out, Gambier and Collingwood appear to have over-reacted to Nelson's bold tactics, which had counted upon the tactical incapacity of the Franco-Spanish fleet. The 'Nelson Touch' had come to be seen as a formula calling for swift attack, dual organisation and independent divisional control. No direction was given about concentration of force, or about breaking the line in one place or all places. There remained no vestige of the use of an advance squadron to contain part of the enemy force. The focus of attack on the enemy van contradicted centuries of experience that tactical necessity obliged the attacking fleet to seek to isolate the enemy rear. In the case of the attack from leeward, 'one cannot help wondering how far the leading ships after passing the line would have been able to lead down it [the enemy's line] before they were disabled'. However, 'the addition is interesting as the first known direction as to what was to be done after breaking the line in line ahead after Rodney's method.'[34]

A more constructive, if over-elaborate, use of the Nelson approach was made by the Hon Sir Alexander Cochrane who, while commander-in-chief on the Leeward Islands station, issued a Memorandum on 'Modes of Attack from the

32 A very detailed tactical plan of the action showing the track of all the British ships appears in the ship's log of the *Donegal*, which was sold at Hodgsons on 14 December 1962 (Lot 595).

33 G L Newnham, *A Selection from the Public and Private Correspondence of Vice-Admiral Lord Collingwood*, Collingwood, 1828, pp313-4

34 Corbett, *Fighting Instructions*, pp324-5

Windward, etc'.[35] Corbett saw in it the influence of Rodney, Clerk of Eldin and Morogues, although 'his excellent device for inverting the line after passing through the enemy's fleet' was Cochrane's own idea.

When an attack is intended to be made upon the enemy's rear, so as to endeavour to cut off a certain number of ships from that part of their fleet, the same will be made known by signal no 27,[36] and the numeral signal which accompanies it will point out the headmost of the enemy's ships that is to be attacked, counting always from the van, as stated in page 160, Article 31 (Instructions).[37] The signal will afterwards be made for the division intended to make the attack, or the same will be signalled by the ship's pennants, and the pennants of the ship in that division which is to begin the attack, with the number of the ship to be first attacked in the enemy's line. Should it be intended that the leading ship in the division is to attack the rear ship of the enemy, she must bear up, so as to get upon the weather quarter of that ship; the ships following her in the line will pass in succession on her weather quarter, giving their fire to the ship she is engaged with; and so on in succession until they have closed with the headmost ship intended to be attacked.

The ships in reserve, who have no opponents, will break through the enemy's line ahead of this ship, so as to cut off the ships engaged from the rest of the enemy's fleet.

When it is intended that the rear ship of the division shall attack the rear ship of the enemy's line, that ship's pennants will be shown; the rest of the ships in the division will invert their order, shortening sail until they can in succession follow the rear ship, giving their fire to the enemy's ships in like manner as above stated; and the reserve ships will cut through the enemy's line as already mentioned.

When this mode of attack is intended to be put in force, the other divisions of the fleet, whether in order of sailing or battle, will keep to windward just out of gunshot, so as to be ready to support the rear, and prevent the van and centre of the enemy from doubling upon them. This manoeuvre, if properly executed, may force the enemy to abandon the ships on his rear, or submit to be brought to action in equal terms, which is difficult to be obtained when the attack is made from windward.

When the fleet is to leeward, and the commanding officer intends to cut through the enemy's line, the number of the ship in their line where the attempt is to be made will be shown as already stated.

If the ships after passing the enemy's line are to tack, and double upon the enemy's ships ahead, the same will be made known by a blue pennant over the Signal 27;[38] if not, they are to bear up and run to the enemy's line to windward, engaging the ship they first meet with; each succeeding ship giving her fire, and passing on to the next in the rear. The ships destined to attack the enemy's rear will be pointed out by the number of the last ship in the line that is to make this movement, or the pennants of that ship will be shown; but, should no signal be made, it is to be understood that the number of ships to bear up is equal in number to the enemy's ships that have been cut off; the succeeding ships will attack and pursue the van of the enemy, or form, should it be necessary to prevent the enemy's van from passing round the rear of the fleet to relieve or join their cut-off ships.

If it is intended that the ships following those destined to engage the enemy's rear to windward shall bear up, and prevent the part of their rear which has been cut off from escaping to leeward, the same will be made known by a red pennant being hoisted over the Signal 21,[39] and the number of

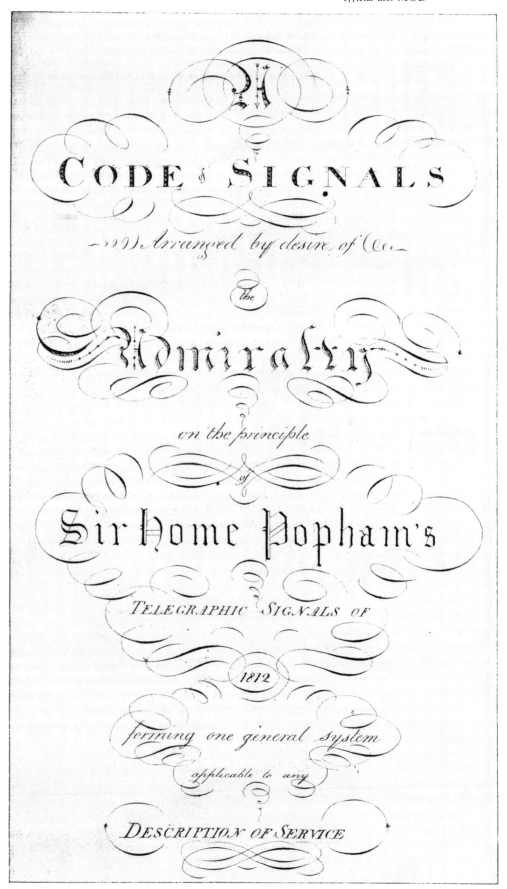

Below and overleaf:
The title page and heavily corrected copy of Admiral Sir Home Popham's telegraphic signals, used when adapting his system for official use. MOD

CODE of SIGNALS

Arranged by desire of the

Admiralty

on the principle

of

Sir Home Popham's

TELEGRAPHIC SIGNALS OF

1812

forming one general system

applicable to any

DESCRIPTION OF SERVICE

---

35 Ekins, *Naval Battles*, pp394–7, and reprinted in Corbett, *Fighting Instructions*, pp330–4

36 'Break through the Enemy's line, in all parts, where it is practicable and engage on the other side . . .'

37 Variations of the attack with explanations.

38 Collingwood's own variation of the signal.

39 'Attack the enemy on the bearing indicated; or any number of their ships of war separated from the main Body of their fleet.'

quotes with approval the observation of an elderly French naval officer writing shortly after Trafalgar that Nelson's attack could have been defeated by a really firm double line 'with reserve squadrons on the wings stationed in such a manner as to bear down most easily upon the points too vigorously attacked'. He also approves of a Spanish tactician's observation that 'where two fleets are equally well trained, that which attacks in this [Nelson's] manner must be defeated'.[47] Given the incapacity of Napoleon's navies, the Admiralty were absolved from the need to work out counter-countertactics. Rightly, no doubt, they came to the conclusion that Nelsonian tactics were for Nelson only, and had no place in an official signal book.

The signal which particularly inspired Corbett's cricitism was class I signal 1.7, to 'cut through the enemy's line in the order of sailing in two columns'. From the detailed instruction and the accompanying diagram, it is easy to see that this was an attempt to regularise Nelson's attack at Trafalgar, and at the same time make it less risky. 'The Admiral will make known what number of ships [counting] from the rear at which the lee division is to break through their line.' The leading ship of each division, having broken through, would, unless otherwise directed, engage the ship astern of which she had just passed from leeward. The second ship in each division, and the rest in succession, would cut through one ahead of where their next ahead cut through, and engage from leeward the enemy ship astern of which they had just passed. Each ship was thus to break through one place ahead of its next ahead and so form an inverted line along the leeward side of the enemy. A special advantage of this procedure was that, as each fresh ship passed along the line seeking its place to break through, she could fire on those enemy ships already engaged from leeward by the ships of her own fleet which had already broken through. In this way the enemy were gradually and continuously doubled.

A great deal of attention was given to the preparation of this signal. An extremely long explanatory gloss in the draft had been entirely struck out, together with the diagram, and replaced with a new text and diagram. The working shows quite clearly that the attack was to be made on the enemy's centre or rear. The diagram, however, omitted to show the enemy van. In editing for the Navy Records Society his *Fighting Instructions*, Corbett thought this indicated an attack on the enemy's van and centre leaving their rear 'free to come into immediate action with an overwhelming concentration on the lee division'. He wrote, 'It is the last word of British sailing tactics, and surely nothing in their whole history, not even in the worst days of the old Fighting Instructions, so staggers us with its lack of tactical sense.' [48] Five years later he had seen his mistake and was able to take a more favourable view in his *Campaign of Trafalgar*.[49]

The original draft included a signal to 'cut through the enemy's line in the order of sailing on [sic] three columns', with a short explanation, the whole being erased.[50] Another original signal which did not survive, together with a long explanation and diagram, was to 'attack the enemy in the order of sailing on [sic] two columns to windward and to leeward'. 'This is a new evolution and it may in the first instance mislead the enemy as to its purpose.' The weather

division was to engage the enemy's van and centre by inverting the line along the enemy's windward side while the lee division did the same thing from the rear along the enemy's leeward side. No doubt it was rejected as being too complicated to bring off with any degree of certainty. Instead, the signal for Howe's attack on the enemy's rear by inverting the line was retained. His other device, of attacking the enemy's rear in succession, was also in the original draft, but was erased.

The experience of the Nile and possibly of Copenhagen is reflected in signals to 'anchor and engage the enemy', 'attack the enemy at anchor, passing between them and the land', 'attack the enemy at anchor, passing on the outside of them', 'anchor within the enemy and engage', and 'anchor on the outside of the enemy and engage'. Only the first of these was provided with a detailed explanation, representing about three-quarters of the original draft.

> The ships are to have springs on their bower anchors, and the sheet cable taken through the stern port, to be prepared to anchor without winding, if they should go to the attack with the wind aft; the boats should be hoisted out, and hawsers coiled in the launches with the stream anchor ready to warp them into their stations or to assist other ships that may be in want of assistance.

No doubt this signal might be combined with one of the others.

Howe's signal, 'ships are to take stations for their mutual support and engage the enemy as they get up with them', was provided with a lengthy gloss which survived revision almost intact. It included the warning, 'but the leading ships must be very cautious not to suffer themselves to be drawn so far from the body of the fleet as to risk their being surrounded and cut off'.

A sign of the times was the group of signals dealing with the falling of 'shells'. There were also nineteen very good signals, clearly meant to be made by private ships reporting on various movements of the enemy. The thumb-indexing of the signal numbers included marginal notes on the subject matter. A scheme for distance signals in class 3 made use of a sixteen-sided table; the symbols to be flown included a ball, discovered by means of two further tables. Signals in the three main classes for which it was deemed necessary, had their distance signal numbers in a column next to the class number.

The instructions for the orders of sailing, with diagrams, were almost exactly the same as for 1799. The instructions for forming the line of battle omitted the diagrams and all the explanatory matter derived from the Hood-McArthur books. The battle instructions were reduced from thirty-one to nineteen, with no important differences. Ships' names in the list of the navy were signalled by the Union Jack over two other flags, etc. The fog signals and instructions were much the same as for 1799.

The night signals and instructions were also much the same as for 1799, though the night battle instructions were reduced from eleven to two. The real change was in the actual method of signalling, which was entirely new. There were sixty-three main signals, including three preparatory. The first fifteen were made with lights alone, up to four, hung in different patterns. The next fifteen were the same light patterns over again, but distinguished by 'one or more rockets or false fires'. The third fifteen were distinguished by 'one or more guns', and the last fifteen by 'one or more guns' and 'one or more rockets or false fires'. There were also nineteen signals at anchor. The scheme depended entirely on the visibility of the lights and was put forward in an effort to curtail the use of guns, rockets and false fires. The first fifteen

47 Corbett, *Fighting Instructions*, pp338 9; quotations from Mathieu-Dumas, *Précis des Evénements Militaires: Pièces Justificatives*, XIV, p408, and Fernandez Duro, *Armada Española*, VIII, p353.

48 Corbett, *Fighting Instructions*, p342

49 pp445-6

50 In the original draft, *MOD*, Ec/51, 'on' is also used for the attack with two columns.

signals with lights alone certainly included the most important ones: for altering course, tacking and wearing; forming line of battle and engaging; number 13, 'Engage as best able'; number 14, 'Being disabled in battle'; and number 15, 'Passed enemy without taking possession', which might be used after an enemy had surrendered.

Popham's *Telegraphic Signals for the use of His Majesty's Fleet: 1816* was based entirely on the scheme he first established in 1812. He employed nine flags, numbered 1 to 9; five cornets lettered A to E; five guidons (triangular flags) lettered F to K; and four pendants, lettered L to O. The shapes and colours were the same as those used in 1812. The main body of the signals consisted, as before, of words and sentences, arranged in alphabetical order. Each of them could be signalled with a three-flag hoist. Starting with flags 4 and 5, which represented A in the Telegraph, the remaining flags, guidons and pendants were used to cover all the words and sentences. Thus all words and sentences beginning with the letter P were signalled by a three-flag hoist beginning with the character guidon G followed by two numbers. As before, syllables and parts of words were included, but these were now split up and placed at the end of the appropriate alphabetical sections. They included blanks for the insertion of words having 'local signification'. For these particular signals a four-flag hoist was necessary. As before, the lists of military and geographical terms were quite small. There was also a complete list of the navy with call numbers, exactly as in the Signal Book, and at the end of the book there was placed a system of semaphore signalling.

In order to spell out a word or words letter by letter, the 'orthographical pendant', white with a red cross, was first hoisted. The arrangements were then much as before. Character flag number 1 equalled A, and so on; number 9 equalled I and J, letter I equalled S and Z; L equalled U and V; and

A plate from the Chevalier Delarouvraye's Traité sur l'art des Combats de Mer, *Paris, 1815, illustrating tactics for placing an enemy line of battle between two fires.*

mainly organisational, and for carrying out the usual evolutions in the order of sailing and changing into or out of the order of battle in line ahead. There were only five battle instructions of any interest. Number 233 was for the *corps de réserve* to be ready to attack and double on either the van or rear of the enemy, as directed. Nothing further was indicated, thus showing lack of any interest in the tactical possibilities of an advance squadron as conceived by Howe and Nelson. Number 234 indicated that, when the admiral made the signal for cutting the enemy's line, the ship which cut through first was to be followed by 'les autres', without specifying what group or division, their duty being to prevent the enemy ships cut off from rejoining their fleet. Number 235 instructed that, should the particular ship ordered to cut the enemy line be unable to do so, the first ship which found it possible should do so, and those astern should follow her. Number 236 indicated that the *corps de réserve* must be ready to intervene at the point where the enemy's line was cut. The rest of the fleet, on seeing the *corps de réserve* ready to support them, were to close the enemy and attempt to cut their line. Number 237 further advised that 'traversing' the enemy's line by cutting through the intervals, presumably between each enemy ship, was generally too dangerous. It could best be executed when the enemy were already doubled and between two fires, as for instance when the *corps de réserve* had doubled on them.

It is quite clear from the signals and instructions, and from the further tactical material which followed, that the lessons derived both from the more remote and the more immediate past admitted of only two kinds of attack. The first was doubling, with or without the intervention of the *corps de réserve*, and the second was cutting the enemy's line through a single gap or traversing by several ships together.

The book ended with a long chapter of 'Explications'. Section I discussed the history and principles of signals and established the value of tactics based on Hoste and Morogues. Section II discussed tactics, defined as 'l'art de ranger les vaisseaux de guerre sur une ou plusiers lignes, et de régler tellement leur mouvements, qu'ils puissent, avec promptitude, prendre la situation la plus favourable à l'attaque et à la défense'. Ships, it stated, were weakest at their bow and stern, and strongest on their broadsides, and various consequences were seen to flow from this, including the importance of concentration of fire. Section III expressed approval of Morogues's acceptance of 'l'ancien système des signaux', and rejection of the numerical system, including La Bourdonnais's method, after the vote taken at each of the naval bases.

Admiral Raper's new system of signals

Although by the time Admiral Henry Raper published in 1828 his *New System of Signals, by which colours may be wholly dispensed* with the Napoleonic wars were long over, his book is an interesting place at which to conclude this study because of the information in his preface, and other glimpses he provides of wartime practice. Raper (1767–1847) went to sea at the age of thirteen, was Lord Howe's signal lieutenant at the First of June, and was liaison officer with the Portuguese squadron subsequently attached to the Channel Fleet. He did not offer his system to the Admiralty until 1815 and, despite continued efforts, it was never accepted. Subsequent publication was presumably at the author's expense. A little earlier, possibly in 1827, he presented a somewhat different manuscript copy to the Duke of Clarence, then Lord High Admiral.[53]

Raper started from the proposition that colours had always been difficult to distinguish and hence a source of confusion in signalling. To get over this difficulty, he proposed a completely new set of colour symbols. He employed ten flags measuring 12ft by 10ft, a substitute flag of upright shape flown lengthwise to the mast, ten pendants, a triangular substitute flag, a short substitute pendant and a 27ft substitute pendant. Under ordinary conditions, these were used for ordinary numerical signalling with a full list of the navy and a telegraph vocabulary of words and sentences in addition. The only difference in principle from the Admiralty book was that the signals were placed in twelve classes. When colours could not be distinguished, it was possible to change over immediately to signalling by shapes. These consisted of:

1	Any flag	6	Ball
2	Upright flag	7	Irregular triangular flag
3	Triangular flag	8	ditto inverted
4	Pendant	9	Long pendant
5	Short pendant	0	Weft

He gives the following example of usage:

A ship having important intelligence rejoins the fleet and signals her number, – consisting of three pendants. The nearest ship to her is still too far off to distinguish the colours but can tell from the *shapes* used that the signal falls within Class XI, 'Ships Numbers'. She at once hoists a ball, to request Shapes, on which the stranger signals her number by *Shapes* alone. In the particular case illustrated, the number 932 is made with a long pendant (9), a triangular flag (3) and the upright flag (2).

To make the distant signal 'Pass through the [enemy's] line where the greatest interval appears', which is no 79 in Class III (Engaging and Evolutions), the signal will be an irregular triangular flag (7), over the short substitute pendant (for the class) and the long pendant for 9. The class pendant is thus flown between the two flags indicating the signal number. Telegraphic signals at a distance are made by signalling the page number and the number of the word or sentence on the page.

The feasibility of Raper's scheme depended on the shapes being clearly distinguishable when colours were not. The colours of his pendants were very confusing anyhow, though this was inevitable in view of the number employed. Raper failed to gain acceptance for his scheme mainly because he was too late, Popham having already secured the Admiralty's acceptance of his Telegraph. Besides, the Admiralty Signal Book of 1816 already contained a system for distant signalling which, if clumsier than Raper's, was designed to fit in with the main signalling system. Raper's handicap was that he wanted to reorganise the whole system in order to make it fit in with his distance arrangements, at the same time introducing a number of new and by no means desirable pendants. He did alter his scheme as time went on, in the light, no doubt, of useful criticisms. Nevertheless, it seems odd that after submitting it to the Admiralty for the last time in 1825 and offering it in similar form to the Duke of Clarence, two or three years later, he should have completely recast it before publication. He acknowledged, in detail, the complete change made in the distant signal shape, but said nothing about the changes made in the classes of signals, the signals themselves and the method of subjects and method of presentation.[54] In both versions he supplied diagrams and explanatory notes for his evolutionary signals. These also are changed in the published version. On the whole, the easier version in the form presented to the duke seems better arranged.

53 *NMM*, Tunstall Collection, TUN/29 (S/MS/Rap/1)
54 p92

Index